Ecotechnology and Pollution Control

Ecotechnology and Pollution Control

Edited by **Bruce Horak**

SYRAWOOD
PUBLISHING HOUSE
New York

Published by Syrawood Publishing House,
750 Third Avenue, 9th Floor,
New York, NY 10017, USA
www.syrawoodpublishinghouse.com

Ecotechnology and Pollution Control
Edited by Bruce Horak

International Standard Book Number: 978-1-68286-251-3 (Hardback)

Printed in China.

Contents

Preface

This book was inspired by the evolution of our times; to answer the curiosity of inquisitive minds. Many developments have occurred across the globe in the recent past which has transformed the progress in the field.

Ecotechnology focuses on minimizing any ecological disruption resulting from human activities. Pollution control is a primary area of concern in ecotechnology. This book covers the varied aspects of ecotechnology by presenting the latest methods and techniques in the field. Some of the topics covered in this text are water management, waste management, technology and equipment for the environment, etc. It aims to provide a holistic view of this field to the readers and broaden the scope of the research in this discipline.

This book was developed from a mere concept to drafts to chapters and finally compiled together as a complete text to benefit the readers across all nations. To ensure the quality of the content we instilled two significant steps in our procedure. The first was to appoint an editorial team that would verify the data and statistics provided in the book and also select the most appropriate and valuable contributions from the plentiful contributions we received from authors worldwide. The next step was to appoint an expert of the topic as the Editor-in-Chief, who would head the project and finally make the necessary amendments and modifications to make the text reader-friendly. I was then commissioned to examine all the material to present the topics in the most comprehensible and productive format.

I would like to take this opportunity to thank all the contributing authors who were supportive enough to contribute their time and knowledge to this project. I also wish to convey my regards to my family who have been extremely supportive during the entire project.

Editor

1

AGRICULTURAL POLLUTION AND WATER QUALITY IN SMALL RETENTION RESERVOIR IN KORYCIN

Joannna Szczykowska[1], Anna Siemieniuk[1], Józefa Wiater[1]

[1] Department of Technology in Environmental Engineering and Protection, Faculty of Civil and Environmental Engineering, Bialystok University of Technology, ul. Wiejska 45 b, 15-351 Białystok, Poland, e-mail: j.szczykowska@gmail.com

ABSTRACT

The study aimed at determining the changes in the trophic status of the processes associated with the risk of eutrophication in small retention reservoir in Korycin, characterized by agricultural performance of direct and indirect catchment. The study was conducted using the surface water samples that were collected systematically every month over four hydrological years (2008 to 2014) from three research points. Mean annual concentration of total phosphorus varied from 0.641 mg P/dm^3 during research conducted in 2007/2008 to 0.874 mg P/dm^3 in 2013/2014 showing an increasing trend from year to year. Taking into account the calculated average values, an upward trend can be seen along with particular years of the study from the annual average value of 1.44 mg N/dm^3 determined in the first year to the value of 2.66 mg N/dm^3 recorded in the last year of the study. It was observed during the study that in non-flowing parts of Korycin reservoir, plankton developed more abundantly than in the central fragments, where the flow of water is more intensive. A mild growth of phytoplankton in the waters of Korycin reservoir is provided by relatively low concentrations of chlorophyll "a" found during all research periods. Concentrations of chlorophyll "a" in waters of the reservoir were within the range of 4.08 to 5.21 g/dm^3. At a Schindler coefficient > 2, this value, based on the Decree of the Minister of the Environment of 2011, should not exceed 7 and therefore waters of the Korycin reservoir during the research can be classified as the first class. Values of the general trophic level (Trophic State Index) during the first three years of the study ranged from 67.15 to 68.65, which enabled to count waters of the reservoir in Korycin to eutrophic ones. In the last year of the analyzes, this coefficient increased to a value of 72.43, classifying its waters as hypertrophic. The largest share in such a situation was expressed by TSI (TP), which ranged from 97.34 to 101.82, always classifying the waters to hypertrophic ones. In contrast, the indicator TSI (TN) ranged from 59.71 to 68.57 and it classified the reservoir waters to eutrophic, while TSI (Chl) to mesotrophic.

Keywords: small water retention reservoir, pollutions, biogenic compounds, trophic status.

INTRODUCTION

Water deficit in rural areas is a major problem in Polish agriculture. Agrotechnical procedures significantly alter the ability of soil to retain water and additionally as a result of hasty melioration works, it may lead to lowering the groundwater level. Beneficial impact on water resources by controlling the water outflow in the main stream or controlling using hydro-devices was confirmed by many scientists [1, 2]. Implementing the program of small water retention development in Podlasie region since 1995 leads to a gradual increase in available water resources, particularly for agriculturally used soils. The activities undertaken in this area are one of the elements that meet the demands of the Water Framework Directive, with the very important issue of maintaining a "good ecological status of waters" [3]. Although significant quantities of nutrients are naturally present in the soil, they are too few to achieve high yields of crops. This creates a need for the supply nutrients in the form of various fertilizers to plants cultivated on agricultural lands. Many-year use of organic fertilizers on the same surface causes an increase in the nitrogen content in soil and should be included in the development plans of fertilization. The proper calculation of the components balance in the soil is particularly important, because it aims at evaluating the risk of nutrients loss. A positive balance of nutrients means that their smaller amount was absorbed by plants and taken along with crops than was introduced in a form of fertilizers. The components that remained in the soil after harvest can be washed out by rainfall and infiltration to groundwater and surface waters. On agricultural areas, both surface and subsurface runoff of biogenic substances from agricultural fields is considered the greatest threat to water quality [4, 5]. The amounts of nitrogen and phosphorus compounds removed from a catchment undoubtedly depends on its performance manners, share of cultivated fields, especially arable lands, total area of the catchment, and intensity of agriculture, taking into account the size of fertilization [6].

The study aimed at determining the changes in the trophic status of the processes associated with the risk of eutrophication in small retention reservoir in Korycin, characterized by agricultural performance of direct and indirect catchment.

AREA AND METHODS OF STUDY

The small retention reservoir with capacity of 481 thousand m^3, surface area of 6.8 hectares, and shoreline length of 1000 meters, was put for using in September 2002 in Korycin. The indirect catchment of the reservoir that was located on the river Kumiałka, is dominated by agricultural lands covering 84.9% of the municipality Korycin. Accumulation of water in the reservoir is 195 cm, its average depth 165 cm, width of the reservoir 145 m, and the length 440 m. Referring to soil typology, the largest area of the direct catchment is occupied by brown soils, pseudo-podsolic soils, as well as chernozems and alluvial soils in river valleys. Twenty hectares of adjacent areas were put into use for recreation and sports facilities, tourism, and hospitality. The main purpose of hydraulic engineering structures in Korycin is regulation of water balance and it also plays the role of the storage and flood reservoir, while in summertime it serves for recreation and sports.

The study was conducted using the surface water samples that were collected systematically every month over four hydrological years (2008 to 2014) from three research points. The selection and placement of measurement and control points in a longitudinal section of the reservoir was a result of the need to capture the changes in selected physicochemical indicators of water samples. The first point was located within the inlet, the second in the middle, and the third at the mouth of water outlet from the reservoir. The water samples were subject to following determinations: total nitrogen, total phosphorus, and chlorophyll "a", according to obligatory methodology [7, 8].

At the same time, the trophic state of analyzed reservoir was evaluated according to the concentration criteria and based on the calculated trophy indices TSI (Trophic State Index) by Carlson and Kratzer and Brezonik [9]. Trophy indices were calculated after Carlson trophic TSI (Chl) in mg/dm^3, TSI (TP) in µg/dm^3, according to Kratzer and Brezonika TSI (TN) in mg/dm^3 as well as the overall trophy of the reservoir TSI was determined using the mean value obtained from the three calculated indicators: TSI (Chl), TSI (TP), TSI (TN). Water transparency measured by Secchi disk was skipped, because it is of little importance in dam reservoirs, when assessing the trophic status. Water inflow during heavy rainfall brings a large amount of suspensions, which cause turbidity of retained water. The reduction in transparency due to this fact is not related to the development of phytoplankton [10, 11]. The values of trophic status indicators to assess due to Carlson are as follows: TSI < 40 – oligotrophy, 40–50 mesotrophy, 50–70 eutrophy, > 70 hypertrophy [8]. The assessment of the eutrophication degree is made on the basis of average annual values of biogenic indicators for flowing waters [9].

RESULTS AND DISCUSSION

The presence of nitrogen in surface waters is influenced by pollution inflowing from the point and area sources, development of aquatic vegetation, oxygen and temperature conditions. Nitrogen is a key element in monitoring the functioning of aquatic ecosystems and its contents in surface waters shows fluctuations during the year [12]. During the research, it was observed that the concentrations of total nitrogen were low in summer, during periods of increased vegetation, and under conditions of favorable to biochemical changes. During the growing season, lowering nitrogen concentration was observed, which was affected by developing autotrophic aquatic organisms that are likely to absorb nitrogen compounds and built them into their cells. In winter, when biochemical processes involving microorganisms are inhibited, the concentration of total nitrogen reached higher values. Taking into account the calculated average values, an upward trend can be seen along with particular years of the study from the annual average value of 1.44 mg N/dm^3 determined in the first year to the value of 2.66 mg N/dm^3 recorded in the last year of the study (Tables 1-4). Based on the analysis of total nitrogen concentrations, it can be concluded that during all test series they satisfied the requirements of the quality of the first class water [13], wherein the maximum values were obtained in samples collected in the third measurement point located near the water outflow from the reservoir.

Eutrophication is the most important anthropogenic factor affecting the proper functioning of aquatic ecosystems. The smaller the reservoir, the greater the impact of substances originating from the catchment on water quality in the reservoir. During the studies, a clear seasonal increase in the incidence of total phosphorus in autumn, winter, and spring was recorded, while a significant reduction in the concentration during summer was observed. The achieved results are confirmed by literature data [4, 5, 10, 14].

Both phosphates and polyphosphates are sorbed on the particles of clay minerals and may form complex compounds with metal ions present in water, especially iron, calcium, magnesium, and aluminum with different pH-dependent coefficients of water solubility [12]. The chemical transformations of phosphorus compounds in water often lead to their precipitation and accumulation in bottom sediments, in which mineral forms are predominant. Phosphorus exchange processes between water and sediments depend on the oxygen conditions, temperature, pH, redox potential, and the surface of bottom sediments to a large extent. In the period of elevated temperatures and under anaerobic conditions, much faster return of phosphates to the water column occurs [15]. During the autumn and spring periods of circulation, vertical distribution of phosphorus is not uniform due to the re-suspending the bottom sediments [12]. Mean annual concentration of total phosphorus varied from 0.641 mg P/dm^3 during the research conducted in 2007/2008 to 0.874 mg P/dm^3 in 2013/2014 showing an increasing trend from year to year. Water in the dammed reservoirs does not flow uniformly with the entire surface, the stream can occupy only a small percentage of the reservoir surface. It was observed during the study that in non-flowing parts of Korycin reservoir, plankton developed more abundantly than in the central fragments, where the flow of water is

Table 1. Water testing results for reservoir Korycin in 2007/2008

Korycin 2007/2008		First point	Second point	Third point	Mean value
Tested parameter	Unit	min–max mean	min–max mean	min–max mean	min–max mean
Total nitrogen	mg·N/dm^3	0.11 – 2.22 1.17	0.32 – 2.65 1.49	0.40 – 2.90 1.60	0.28 – 2.59 1.44
Total phosphorus	mg·P/dm^3	0.009 – 0.971 0.486	0.046 – 1.122 0.584	0.109 – 1.588 0.849	0.055 – 1.227 0.641
Chlorophyll „a"	µg/dm^3	2.52 – 5.40 3.96	2.98– 5.68 4.33	3.23 – 4.67 3.95	2.91 – 5.25 4.08

Table 2. Water testing results for reservoir Korycin in 2009/2010

Korycin 2009/2010		First point	Second point	Third point	Mean value
Tested parameter	Unit	min – max mean	min – max mean	min – max mean	min – max mean
Total nitrogen	mg·N/dm^3	0.15 – 2.59 1.37	0.38 – 2.83 1.61	0.49 – 3.24 1.87	0.34 – 2.89 1.62
Total phosphorus	mg·P/dm^3	0.011 – 0.984 0.498	0.053 – 1.242 0.648	0.124 – 1.618 0.871	0.063 – 1.281 0.672
Chlorophyll „a"	µg/dm^3	2.58 – 5.94 4.26	3.02 – 6.07 4.55	3.34 – 6.37 4.86	2.98 – 6.13 4.56

Table 3. Water testing results for reservoir Korycin in 20011/2012

Korycin 2011/2012		First point	Second point	Third point	Mean value
Tested parameter	Unit	min. – max mean	min. – max mean	min. – max mean	min. – max mean
Total nitrogen	mg·N/dm³	0.17 – 2.62 1.40	0.39 – 2.91 1.65	0.51 – 3.43 1.97	0.36 – 2.99 1.68
Total phosphorus	mg·P/dm³	0.016 – 0.989 0.503	0.057 – 1.278 0.668	0.138 – 1.623 0.881	0.070 – 1.297 0.684
Chlorophyll „a"	μg/dm³	2.63 – 6.14 3.96	3.03 – 6.15 4.33	3.42 – 6.69 3.95	3.03 – 6.33 4.68

Table 4. Water testing results for reservoir Korycin in 20013/2014

Korycin 2013/2014		First point	Second point	Third point	Mean value
Tested parameter	Unit	min – max mean	min – max mean	min – max mean	min – max mean
Total nitrogen	mg·N/dm³	0.21 – 3.12 1.67	0.46 – 4.68 2.57	0.78 – 6.69 3.74	0.48 – 4.83 2.66
Total phosphorus	mg·P/dm³	0.023 – 1.454 0.739	0.099 – 1.522 0.811	0.156 – 1.988 1.072	0.093 – 1.655 0.874
Chlorophyll „a"	μg/dm³	3.16 – 6.73 4.95	3.58 – 6.98 5.25	3.96 – 7.22 5.59	3.57 – 6.98 5.26

more intensive. Every small retention reservoir shows different susceptibility to degradation. The larger the share of arable lands in the catchment area with small amounts of forest areas, the greater the supply of nutrients, and thus the higher concentrations determined in water [16, 17].

Accumulation of material brought by the river occurs as a result of slow water flow through the reservoir. Precipitating sediments are a mixture of organic and mineral compounds aggregating in front of a dam. If a flow brings waters rich in nitrogen and phosphorus, a deficit of oxygen conditioning the anaerobic metabolism, can quickly appear in benthic zone. Initially, there are no visible changes in the surface layer of the reservoir until the content of biogenic substances in the dammed water does not rise so that algae and phytoplankton would massively grow, i.e. a "water bloom" will occur. Determination of the chlorophyll "a" concentration is a direct indicator of the trophic status of waters, thus it is widely used [8]. Chlorophylls have the ability to absorb the visible light in the range of blue and red color as well as they reflect green light, which gives the plants a green color. The chlorophyll content is an indicative, that is relatively simple to evaluate, but burdened with some disadvantages. The errors may arise from the fact that the concentration of the pigment may vary, depending on physiological properties of algal cells or their species affiliation. The algae cells under poorer light conditions contain more chlorophyll than those exposed to more intense radiation [19]. These results can hardly coincide with reality also due to disregarding the biomass of macrophytes, although they do not significantly affect the obtained results [19]. A mild growth of phytoplankton in the waters of Korycin reservoir is provided by relatively low concentrations of chlorophyll "a" found during all research periods. Concentrations of chlorophyll "a" in waters of the reservoir were within the range of 4.08 to 5.21 g/dm³. At a Schindler coefficient > 2, this value, based on the Regulation of the Minister of the Environment of 2011, should not exceed 7, and therefore, waters of the Korycin reservoir during the research can be classified as the first class [20].

Values of the general trophic level (Trophic State Index) during the first three years of the study ranged from 67.15 to 68.65, which enabled to count waters of the reservoir in Korycin to eutrophic ones. In the last year of the analyzes, this coefficient increased to a value of 72.43, classifying its waters as hypertrophic. The largest share in such a situation was expressed by TSI (TP), which ranged from 97.34 to 101.82, always classifying the waters to hypertrophic ones. In contrast, the indicator TSI (TN) ranged from 59.71 to 68.57 and it classified the reservoir waters to eutrophic, while TSI (Chl) to mesotrophic.

Contamination of water reservoirs with phosphorus compounds is a frequently occurring problem and quite difficult to remove, since in general it results from the distributed nature of its external sources, especially for agricultural catchments. These seasonal changes indicate the release of

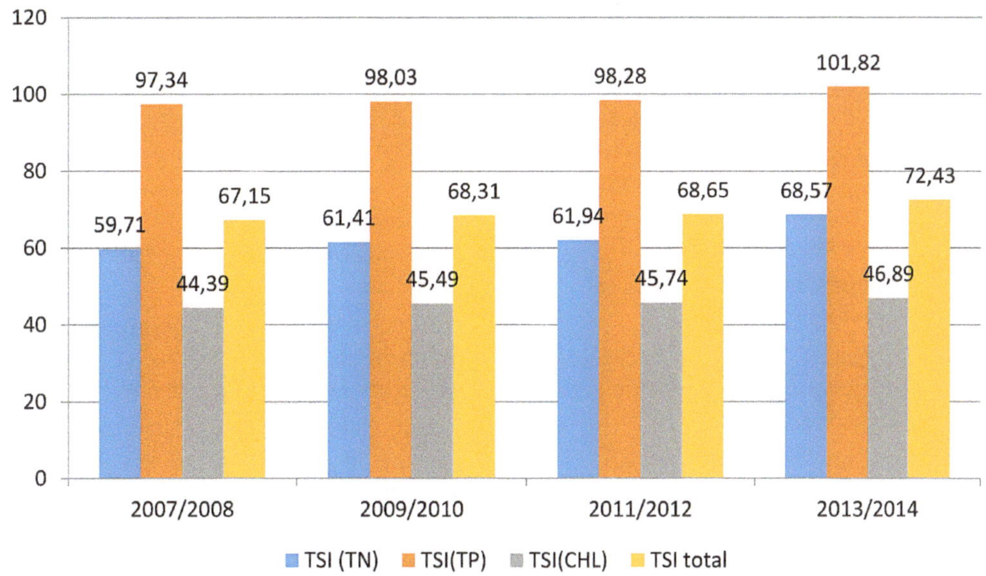

Figure 1. Annual average trophic level of the Korycin reservoir in particular years of research

phosphates from bottom sediments of previously accumulated organic matter. The secondary pollution of the reservoir is also evidenced by the concentration of total phosphorus in the individual sampling points. During the research, the lowest concentration of total phosphorus was present in the waters flowing into the reservoir, while the highest one was found in the samples collected around the outflow. The values of TSI (TN), TSI (TP), TSI (Chl), and total TSI increasing from year to year, clearly indicate a gradual, slow degradation of water quality in the analyzed reservoir.

CONCLUSIONS

1. Considering the calculated mean values of total nitrogen, it can be observed an increasing trend along with subsequent years of the research, from the average annual value of 1.44 mg N/dm^3 determined in 2007/2008 to 2.66 mg N/dm^3 in 2013/2014 season.
2. The mean annual concentration of total phosphorus varied from 0.641 mg P/dm^3 in the initial period of the study to 0.874 mg P/dm^3 in the last year, showing a continuous upward trend.
3. The values of the overall trophy during the first three years of the research ranged from 67.15 to 68.65, which enabled to count waters of the Korycin reservoir to eutrophic, whereas in the last year, this coefficient increased to a value of 72.43, classifying the water to hypertrophic ones.
4. During the research, the lowest concentration of total phosphorus was present in the waters flowing into the reservoir, while the highest one was found in samples collected around the outflow and values of TSI (TN), TSI (TP), TSI (Chl), and total TSI increasing from year to year, clearly indicate a gradual and slow degradation of water quality in the analyzed reservoir.

REFERENCES

1. Bykowski J., Stefański Cz., Fiedler M. 2004. Dynamika zmian uwilgotnienia czarnych ziem w warunkach piętrzenia wody w rowie melioracyjnym. Rocznik AR w Poznaniu, 357, 29–34.
2. Sojka M., Murat-Błażejewska S. 2007. Możliwość poprawy bilansu wodno-gospodarczego zlewni rzeki Małej Wełny. Zesz. Probl. Post. Nauk Rol. PAN, 519, 289–300.
3. Szczykowska J., Siemieniuk A. 2014. The present condition of small water retention and the prospects of its development using the example of the Podlaskie Voivodship. J. Ecol. Eng. 15 (3), 90–96.
4. Kajak Z. 2001. Hydrobiologia – Limnologia. Ekosystemy wód śródlądowych. Wydawnictwo PWN, Warszawa.
5. Szczykowska J., 2009. Occurrence of elements contaminating the low-retention reservoirs on agricultural areas. Polish Journal of Environmental Studies – Series of Monographs, vol. 3, 75–80.
6. Skwierawski A. 2005. Ocena stanu małych zbiorników wodnych na terenach wiejskich. Cz. I. Metoda waloryzacji małych zbiorników. Zeszyty Probl. Post. Nauk Rol., 506, 391–402.

7. Hermanowicz A.L., Dojlido J., Dońska W., Kosiorowski B. 1999. Fizykochemiczne badanie wody i ścieków. Arkady, Warszawa.
8. PN-86-C-05560/02 Woda i ścieki. Badania zawartości chlorofilu w wodach powierzchniowych. Oznaczanie chlorofilu alfa w glonach planktonowych metodą spektrofotometryczną monochromatyczną z poprawką na feopigmenty alfa.
9. Carlson R.E. 1977. A tropic state index for lakes. Limnology and Oceanography, 22.
10. Szczykowska J., Siemieniuk A., Wiater J. 2013. Sezonowe zmiany stanu troficznego zbiorników retencyjnych. Ekonomia i Środowisko, 2, 107–116.
11. SzczykowskaJ ., Siemieniuk A., Wiater J. 2013. Problemy ekologiczne zbiorników małej retencji na Podlasiu. Ekonomia i Środowisko, 4, 234–244.
12. Wetzel R.G. 2001. Limnology lake and river ecosystem. Academic Press, USA, 205–220.
13. Rozporządzenie Ministra Środowiska z dnia 9 listopada 2011 r. w sprawie sposobu klasyfikacji stanu jednolitych części wód powierzchniowych oraz środowiskowych norm jakości dla substancji priorytetowych.
14. Kowal A.L., Świderska-Bróż M. 1997. Oczyszczanie wody. Wydawnictwo Naukowe PWN, Warszawa – Wrocław.
15. Dojlido J. 1995. Chemia wód powierzchniowych. Wyd. Ekonomia i środowisko, Białystok.
16. Wiater J., Siemieniuk A., Szczykowska J. 2012. Rola zbiornika małej retencji w kształtowaniu jakości wód powierzchniowych. Gaz Woda i Technika Sanitarna, 279–282.
17. Jeppsen E., Jansen J.P., Söndergaard M. 2003. Role of sediment and internal loading of phophorus in shallow lakes. Hydrobiologa 506-509(1-3), 135–145.
18. Tolstoy A. Chlorophyll a in relation to phytoplankton volume in some Swedish lakes. Arch. Hydrobiol. 85(2), 133–151.
19. Neverova-Dziopak E. 2010. Podstawy zarządzania procesem eutrofizacji antropogenicznej. Wydawnictwo AGH, Kraków.
20. Rozporządzenie Ministra Środowiska z dnia 23 grudnia 2002 r. w sprawie kryteriów wyznaczania wód wrażliwych na zanieczyszczenia związkami azotu ze źródeł rolniczych. Dz. U. 2002 r., Nr 241, poz. 2093.

2

SEASONAL CHANGES IN PHOSPHORUS LOAD FLOWING OUT OF SMALL AGRICULTURAL CATCHMENTS

Krzysztof Pulikowski[1], Katarzyna Pawęska[1], Aleksandra Bawiec[1]

[1] Institute of Environmental Engineering, Wrocław University of Environmental and Life Sciences, pl. Grunwaldzki 24, 50-363 Wrocław, Poland, e-mail: krzysztof.pulikowski@up.wroc.pl

ABSTRACT

In this article distribution of monthly phosphorus loads flowing out of two agricultural catchments which are located in different physiographic conditions of Lower Silesia was analysed. Loads of phosphorus runoff from the catchment located in the piedmont part of Lower Silesia in each month rarely exceed 0.10 kg P · ha^{-1}. The size of annual load is determined by loads obtained in two months of early spring. Much lower loads obtained for lowland catchment, located near Wroclaw. Values calculated for each month rarely exceed the value of 0.01 kg P · ha^{-1}. Culmination of loads bringing away is a bit more extended in a time compared to the catchment located on Sudety Mts. Foreland. Much higher loads are observed during the period from January to April – this period has a major impact on the size of phosphorus load that flows out from this catchment during whole hydrological year.

The obtained results clearly indicate that the threat of watercourses and water reservoirs supply in phosphorus compounds from agricultural land is periodic and it is particularly high during early spring. Phosphorus load flowing out from the analyzed catchments is very diverse. From facility located on Sudety Foothill in hydrological year, during research period, flowed away average 0.81 kg P · ha^{-1}. Significantly lower values were obtained for second facility and it was average 0.15 kg P · ha^{-1} during a year. The size of load discharged during a year is largely determined by amount of phosphorus load flowing out during winter half of the year (from XI to IV). In case of foothill catchment in this period flowed out average 0.56 kg P · ha^{-1}, which presents 69% of annual load and in lowland catchment this percentage was even slightly higher and was 73%.

Keywords: water, catchment, season, load, phosphorus.

INSTRODUCTION

Phosphorus compounds flowing into surface water pose a significant risk of eutrophication, and thus contribute to the degradation of these resources. This significantly limited the usefulness of this kind of water to be used for various purposes like water supply of population and industry, recreation, etc. Supply of phosphorus to surface water as a limiting element of primary production should be subject to special control and restriction. Currently, much more attention is paid to reducing the flow of nitrogen into surface waters, especially in the nitrate form, which reduces their suitability for the use for purposes connected with supply of the population with drinking water.

One of the sources of phosphorus contaminating surface water is agricultural land [Pulikowski 2004; Sojka et al. 2008; Czaban 2009;]. Phosphorus compounds used as fertilizers are adsorbed on soil particles and hardly dissolve in the soil. Phosphates enter into surface water with suspensions, mainly as a result of erosion. It is estimated that the loss of phosphorus during fields fertilization ranges from 0.1 to 5.0 % and is a significant source of phosphorus in surface water [Dojlido 1995]. In breeding farms plants needs of phosphorus can be fully covered by natural fertilizers. Reduction of mineral fertilization ensures the reduction of excess of this component in the soil, which poses a threat to groundwater, while not significantly affect the yield increase [Pietrzak 2005]. In most cases, a reduction in phosphorus fertilization does not reduce the profits of the farmer, but reduces costs, and it is always beneficial to the environment [Sapek 2008].

The phosphorus content in the water flowing from the small lowland catchments, in contrast to nitrate, has no relationship with the indicator of soils bonitation [Witkowski 1997] and content of phosphorus in the soil [Gardner 2002] but is correlated to the share of crops fields in catchment and inclines [Ekholm et al. 2000] as well as arable lands [Skorbiłowicz 2010]. The concentration of phosphorus also shows a positive (not always significant at $p = 0.05$) correlation with the volume of runoff [Pulikowski 2002], which confirms the high proportion of surface runoff in shaping the size of phosphorus load flowing out from the catchment. Runoff, groundwater and water captured in soil pores from areas located by the river are responsible for moving 98% of orthophosphate outflowing from the catchment [Banaszuk et al. 2009]. The average annual concentration of total phosphorus in lowland rivers may even exceed 2 mg P · dm^{-3} [Dąbrowska, 2008]. In case of catchment with disordered water and wastewater management an important source of phosphorus in surface water may be domestic wastewater discharged in an uncontrolled way [Pulikowski et al. 2011a].

Excess of phosphorus in surface water is a global problem. Phosphorus once introduced into the environment cumulates and can be removed only in the process of remediation [Sapek 2011]. One way to reduce the flow of this component into surface water may be the use of materials absorbing this component in wastewater treatment plants or reservoirs where runs pollutions from small agricultural catchments [Karczmarczyk, Bus 2014]. Another solution is to use on the inlet to the reservoir a small primary reservoir [Dąbrowska et al. 2014].

RESEARCH METHODOLOGY

The study analyzed the results of research carried out in three consecutive hydrological years 1999/2000, 2000/2002 and 2001/2002 at two different facilities located in the lower Silesia. One of them – Stare Bogaczowice – is located on the border of the Bolkow-Walbrzych Foothill and Wałbrzyskie Mountains, in the region of Central Sudetes. The study included ditch catchment with total area of 29 hectares. This catchment is situated 400–500 meters above sea level on the northern and north-eastern slopes. Declines of this area are very large, ranging from 5–8%. The catchment area includes arable fields, partially drained [Pulikowski 2004]. During the study period rainfalls were higher than long-term average value (655 mm). Year 2001/2002 is worth attention because during this period very high annual precipitation, up to 1044 mm, was noted. This was connected to an extremely high rainfall that occurred in August (348 mm) (Table 1). The average annual temperature in each year was higher than the average of the long term and ranged from 8.4 °C to 8.6 °C (Table 2).

The second research facility (Szewce) is located in the Silesian Lowland near Wrocław. The property is situated in the catchment at an altitude of 114–132 m above sea level Declines in the area range from 2.5 to 4.0‰. The study included the catchment area of 100.7 hectares that is an arable land, of which nearly 40% is drained by drainage. On this area occur brown soils with particle size of light and medium loam. There are also heavy clays and small amounts of well decomposed peat [Pulikowski 2004].

During the two years of study rainfalls occurred close to the long-term average of 587 mm.

Table 1. Monthly and annual precipitation totals (mm) at particular experimental object

Meteorological station (object)	Years	Months												Year
		XI	XII	I	II	III	IV	V	VI	VII	VIII	IX	X	
Szczawno-Zdrój (Stare Bogaczowice)	1999/2000	40	26	66	51	65	17	112	56	175	44	69	17	738
	2000/2001	45	20	34	40	70	77	50	80	201	114	122	21	874
	2001/2002	52	57	29	42	35	34	100	64	160	348	65	58	1044
	2001/2002	32	21	21	40	16	27	28	40	63	108	50	48	494
Ligota Piękna (Szewce)	1999/2000	43	31	50	40	94	11	64	25	127	41	28	8	562
	2000/2001	42	33	16	21	79	30	53	85	170	44	107	34	714
	2001/2002	29	32	27	68	37	46	48	48	49	107	36	59	586

Table 2. Mean monthly and annual air temperatures (°C) at particular experimental object

Meteorological station (object)	Years	Months												Year
		XI	XII	I	II	III	IV	V	VI	VII	VIII	IX	X	
Szczawno-Zdrój (Stare Bogaczowice)	1999/2000	1.7	0.2	-2.0	1.8	2.9	10.7	14.0	16.7	15.0	17.0	11.9	11.3	8.4
	2000/2001	6.3	1.4	-0.6	0.3	2.5	6.7	13.4	13.8	17.9	18.1	11.0	11.9	8.6
	2001/2002	1.7	-3.7	-0.8	4.4	4.1	7.5	15.6	16.8	18.3	18.6	11.4	7.0	8.4
Wrocław-Strachowice (Szewce)	1999/2000	3.0	1.9	-0.3	3.8	5.0	12.1	15.7	17.9	16.5	19.0	13.4	12.5	10.0
	2000/2001	6.8	2.5	0.6	1.1	3.5	8.0	14.8	15.1	19.2	19.3	12.5	12.7	9.7
	2001/2002	3.4	-1.7	0.6	4.9	5.3	9.0	17.0	18.2	20.1	20.6	13.6	8.2	9.9

Year 2000/2001 in which there was a high annual precipitation of 714 mm is noteworthy. It was the result of, among others, high rainfall that occurred in July (170 mm) (Table 1). The average annual temperature in each year was higher than the average of the long term and ranged from 9.7 °C to 10.0 °C (Table. 2).

Measurements were taken in measurement cross-sections equipped with triangular Thomson's overflow, staff gauge and limnigraph. Water samples for determination of phosphorus were taken 1–2 times a month.

The calculation of the phosphorus load flowing out from the catchment were based on daily measurements of runoff and periodic analyzes of the composition of the outflowing water. For periods between successive terms of chemical analyzes, phosphorus concentration was determined by linear interpolation:

$$Ł = \sum_{t=1}^{t=T} (\overline{Q}_t \cdot C_t^{\,I})$$

where: Ł – load of contaminants outflowing from catchment during a month, $kg \cdot ha^{-1}$,

$C_t^{\,I}$ – determined the average daily value of the concentration of the examined component, for the days of the period between the execution of two consecutive analyzes determined by linear interpolation, $kg \cdot m^{-3}$,

$\overline{Q}_t$ – average daily volume of unit outflow, $m^3 \cdot ha^{-1}$,

t – next days in the month.

RESULTS AND DISCUSSION

The load is an objective parameter in order to compare the share of different sources, e.g. point and surface in water pollution of the river or reservoir. The most common calculation period is a year, less common - a day, and in the case of area sources, unit load referenced in addition to the catchment area is in use. While in the case of point sources, for example outflow of the treated wastewater from wastewater treatment plant load is not subject to significant fluctuations over the year, whereas in the case of area sources this variability is very important. Accordingly, the phosphorus load variability was analyzed in subsequent months over 3 consecutive hydrological years.

Loads of phosphorus runoff from the catchment located in the piedmont part of Lower Silesia in each month rarely exceed 0.10 kg $P \cdot ha^{-1}$ (Figure 1.). About size of the annual load decide loads obtained in one or two months. In the first two years there are months of February and March, a period in which the early spring thaw begins, time of intense flow of water from the catchment. During this period, the monthly load exceeds even 0.40 kg $P \cdot ha^{-1}$ (Figure 1). A slightly different situation was found in 2001/2002, the maximum monthly charge was obtained for August, which is directly related to the extremely high rainfall that was recorded in this month (Table 1).

Much lower loads were determined for lowland catchment, located near Wroclaw. The values calculated for each month rarely exceeded the value of 0.01 kg $P \cdot ha^{-1}$. The culmination of load flowing out is a bit more spread in time comparing to catchment located on Sudety Foothill. Much higher loads were observed during the period from January to April. This period has a major impact on the size of phosphorus load, which flows under these conditions throughout the hydrological year. Seasonal nature of phosphorus runoff from catchment is closely related to volatility of the water volume discharged by the river which decides about its size, while the maximum concentration of phosphorus was found in the summertime [Skorbiłowicz, Ofman 2014]. A similar seasonal distribution of loads is obtained for example for nitrogen [Pulikowski et al. 2011b].

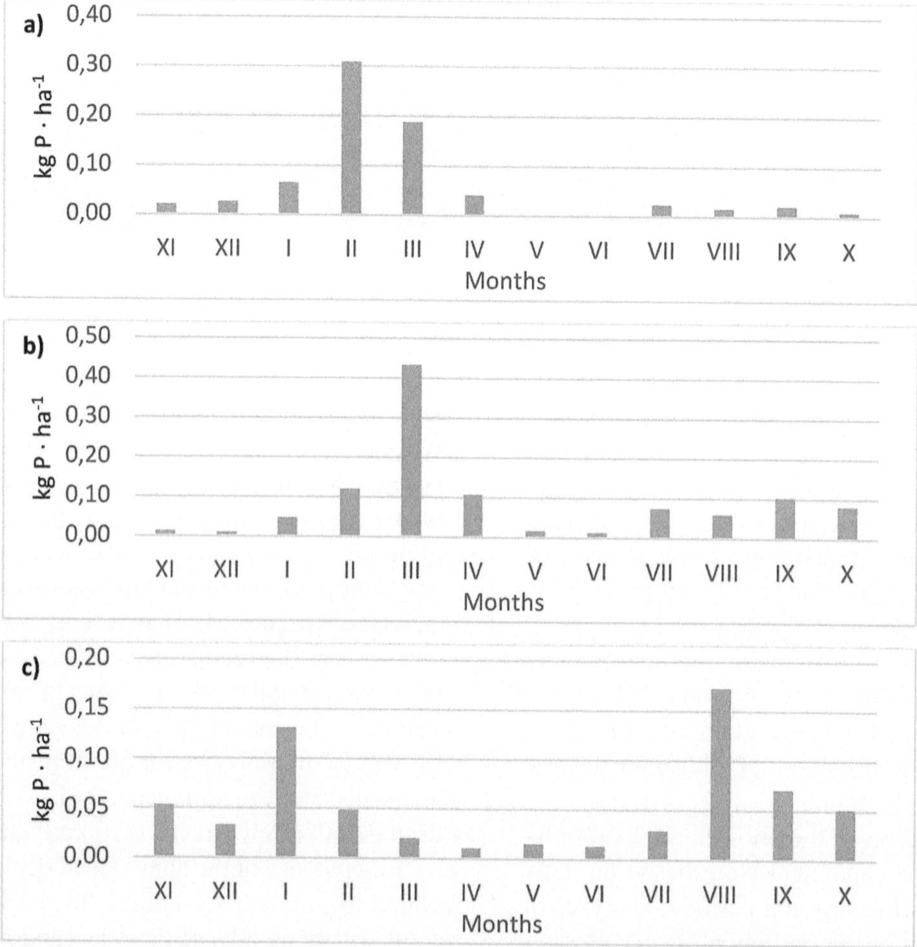

Figure 1. Monthly loads of phosphorus runoff from one hectare of agricultural land in the Sudeten Foreland (Stare Bogaczowice) in the hydrological years: a – 1999/2000, b – 2000/2001, c – 2001–2002

The results clearly indicate that the threat of supply watercourses and water reservoirs in phosphorus compounds from agricultural land, is periodic and is particularly high during early spring.

Phosphorus load flowing out from the analyzed catchment is very diverse (Table 3). In the case of a facility located in the Sudety Foothill in the hydrological year 2000/2001 float away even more than 1 kg P · ha^{-1}, during study period flowed away average 0.81 kg P · ha^{-1}. Significantly lower values were obtained for the second object, averaged over the year flowed 0.15 kg P · ha^{-1}, and the range of variation was small – from 0.12 to 0.18 kg P · ha^{-1}. A similar value (0.13 kg P · ha^{-1}) was obtained by Witkowski [1997b]. Such large differences in loads between these objects is likely due to the varying amount of rainfall, and thus a greater unit outflow in piedmont catchment.

The size of load discharged during the year is largely determined by the amount of phosphorus flowing out in the winter (XI and IV). In case of piedmont catchment at this time flowed away on

Table 3. Periodic and annual loads of phosphorus outflow from the catchment area of agricultural. kg P · ha^{-1}

Object	Years	Winter half-year	Summer half-year	Year
Stare Bogaczowice	1999/2000	0.64	0.07	0.71
	2000/2001	0.73	0.34	1.07
	2001/2002	0.35	0.30	0.65
	Mean	**0.56**	**0.25**	**0.81**
Szewce	1999/2000	0.09	0.03	0.12
	2000/2001	0.13	0.05	0.18
	2001/2002	0.13	0.03	0.16
	Mean	**0.11**	**0.04**	**0.15**

average 0.56 kg P · ha^{-1} (Table 3), which represents 69% of annual load, in the case of lowland catchment, this percentage was slightly higher and amounted to 73%. Studies in Canada have shown that in the same period of melting snow, flows 20% of the annual phosphorus load [Su et al. 2011].

CONCLUSION

Loads of phosphorus runoff from the catchment located in the piedmont part of Lower Silesia in each month rarely exceed 0.10 kg P · ha^{-1}. The size of the annual load is determined by the loads obtained in the two months of early-spring. Much lower loads were determined for lowland catchment, located near Wroclaw. The values calculated for each month rarely exceed the value of 0.01 kg P · ha^{-1}. The culmination of loads outflowing is a bit more spread in time compared with the object located at the foothills of the Sudety. Much higher loads were observed during the period from January to April. This period has a major impact on the size of the phosphorus load, which flows under these conditions throughout hydrological year.

The results clearly indicate that the threat of supply watercourses and water reservoirs in phosphorus compounds from agricultural land, is periodic and is particularly high during the early spring.

Phosphorus load flowing out from the analyzed catchment is very diverse. From a facility located in the foothills of the Sudety during hydrological year, average in the study period flowed 0.81 kg P · ha^{-1}, significantly lower values were obtained for the second object, on average, in the year flowed 0.15 kg P · ha^{-1}. About the size of load discharged during the year largely determines the amount of phosphorus flowing out in the winter (XI and IV). In case of piedmont catchment at this time flowed away on average 0.56 kg P · ha-1 (Table 3), which represents 69% of annual load, in the case of lowland catchment, this percentage was slightly higher and amounted to 73%.

REFERENCES

1. Banaszuk P., Krasowska M., Kamocki A. 2009. Źródła azotu i fosforu oraz drogi ich migracji podczas wezbrania roztopowego w małej zlewni rolniczej. Woda-Środowisko-Obszary Wiejskie, 9(4), 5–26.
2. Ekholm P., Kallio K., Salo S., Pietiläinen O.P., Rekolainen S., Laine Y., Joukola M. (2000) Relationship between catchment characteristic and nutrient concentrations in an agricultural river system. Water Research 34(15), 3709–3716.
3. Dąbrowska J. 2008. Ocena zawartości związków azotu i fosforu w wodach rzeki Trzemy. Infrastruktura i ekologia terenów wiejskich, 7, 57–68.
4. Dąbrowska J., Kempa O., Markowska J., Sobota J. 2014. Mściwojów reservoir – study of a small retention reservoir with an innovative water self-purification system. J. Ecol. Eng. 15(2), 7–16.
5. Czaban S. [ed.]. 2009. Assessment of anthropogenic impacts on water bodies in agricultural catchment. Monografia, seria Współczesne Problemy Inżynierii Środowiska nr 15, Wyd. UP we Wrocławiu, pp. 80.
6. Dojlido J.R. 1995. Chemia wód powierzchniowych. Wyd. Ekonomia i Środowisko, Białystok, pp. 342.
7. Gardner C.M.K., Cooper D.M. Hughes S. 2002. Phosphorus in soils and field drainage water in the Thame catchment, UK. The Science of the Total Environment, 282/283, 253–262.
8. Karczmarczyk A, Bus A. 2014. Testing of reactive materials for phosphorus removal from water and wastewater – comparative study. Ann. Warsaw Univ. of Life Sci. – SGGW, Land Reclam. 46 (1), 57–67.
9. Pietrzak S. 2005. Optymalizacja wykorzystania azotu i fosforu w gospodarstwach prowadzących chów bydła mlecznego na Podlasiu. Woda-Środowisko-Obszary Wiejskie, Rozp. nauk. i mongr., 13, 129.
10. Pulikowski K. 2002. Zależność między wartościami wskaźników zanieczyszczenia a objętością przepływu w małych ciekach. Zesz. Probl. PNR Warszawa, 487, 295–303.
11. Pulikowski K. 2004. Zanieczyszczenia obszarowe w małych zlewniach rolniczych. Zesz. Nauk. AR we Wrocławiu, ser. Rozprawy CCXI, 479, 137.
12. Pulikowski K., Orzepowski W., Pokładek R. 2011. Water quality in agriculturally used catchments in Lower Silesia. Wyd. UP we Wrocławiu, Monografia CXXXVII, ser. Współczesne problemy inżynierii środowiska XVIII, 94.
13. Pulikowski K., Czyżyk F., Pawęska K., Strzelczyk M. 2011. Sezonowe zmiany wielkości ładunku azotu odpływającego z mikrozlewni użytkowanych rolniczo. Infrastruktura i Ekologia Terenów Wiejskich, 10, 161–171.
14. Sapek A. 2008. Nawożenie fosforem a jego skutki w środowisku, Woda-Środowisko-Obszary Wiejskie, 8, 2b(24), 127–137.
15. Sapek A. 2009. Fosfor w łańcuchu pokarmowym człowieka a środowisko w Polsce. Inżynieria Ekologiczna, 21, 62–72.
16. Skorbiłowicz M. Czynniki i procesy kształtujące obieg składników mineralnych w wodach rzecznych zlewni górnej Narwi. Rozprawy Naukowe nr 197. Wyd. Politechniki Białostockiej, 2010.

17. Skorbiłowicz M., Ofman P. 2014. Seasonal changes of nitrogen and phosphorus concentration in Supraśl river. Journal of Ecological Engineering 15(1), 26–31.

18. Sojka M., Murat-Błażejewska S., Kanclerz J. 2008. Ocena oddziaływań antropogenicznych o charakterze obszarowym na jakość wód Strugi Dormowskiej. Acta Scientiarum Polonorum, Formatio Circumiectus 7(2), 15–28.

19. Su J., J. van Bochove E., Thériault G., Novotna B., Khaldoune J., Denault J.T., Zhou J., Nolin M., C., Hu C., X., Bernier M., Benoy G., Xing Z.S., Chow L. 2011. Effects of snowmelt on phosphorus and sediment losses from agricultural watersheds in Eastern Canada. Agricultural Water Management 98, 867–876.

20. Witkowski D. 1997a. Wybrane czynniki kształtujące jakość wód powierzchniowych w małych zlewniach nizinnych. Roczn. Glebzn. t. XLVIII, 3/4, 5–21.

21. Witkowski D. 1997b. Składniki nawozowe w dwóch małych zlewniach o zróżnicowanych właściwościach glebowo-rolniczych. Roczn. Glebzn. t. XLVIII 3/4, 23–36.

3

CLOPYRALID DISSIPATION IN THE SOIL CONTAMINATED WITH HEAVY METALS

Mariusz Kucharski[1], Olga Kalitowska[1]

[1] Institute of Soil Science and Plant Cultivation - State Research Institute, Department of Weed Science and Tillage Systems, Orzechowa 61, 50-540 Wrocław, Poland, e-mail: m.kucharski@iung.wroclaw.pl; o.kalitowska@iung.wroclaw.pl

ABSTRACT

The aim of the studies was to determine the influence of copper and zinc contamination on clopyralid dissipation in soil. The experiment was carried out in laboratory conditions (plant growth chamber). Clopyralid was applied to three different soils [similar textures, pH, organic carbon content and contrasting copper and zinc content: soil natural contaminated with Cu and Zn (S1), soil with natural low Cu and Zn concentration (S2) and soil S21 prepared in the laboratory (S2 soil additionally contaminated with Cu and Zn salts in the amounts equivalent to contamination level of S1 soil)]. Soil samples were taken for analyses for 1 hour (initial concentration) and 2, 4, 8, 16, 32, 64 and 96 days after treatment. Clopyralid residue was analysed using GC/ECD (gas chromatography with electron capture detector). Good linearity was found between logarithmic concentration of clopyralid residues and time. The differences in Cu and Zn content influenced the clopyralid decay in soil. The values of DT_{50} obtained in the experiment ranged from 21 to 27 days. A high concentration of Cu and Zn in soil slowed down clopyralid degradation (the DT_{50} value was higher – 25–27 days).

Keywords: heavy metals, Cu, Zn, herbicide, soil, dissipation rate, clopyralid.

INTRODUCTION

"Głogów II" Copper Smelter and Refinery, a branch of KGHM Polska Miedź S.A., is the source of soils contamination with heavy metals, especially with copper and zinc. In this region, high content of these metals in the soils also results from the type of bedrock [Jaworska, Dąbkowska-Naskręt 2012]. Direct neighborhood of the copper smelter and arable land effects on the structure of their use. A common way of the development of this land is cultivation of industrial plants [Kabała et al. 2010]. High content of copper and zinc influences a number of soil biological and chemical properties, which can modify the behavior of organisms and substances which are already present in the soil or the ones introduced to it [Valery, Eugene 1998, Nwuche, Ugoji 2008]. The latter possibility can be exemplified by herbicides, which, as one of the basic components of modern technology of plant cultivation, are an alien element introduced to the environment. After fulfilling their weed control function in crops, herbicides should undergo degradation to simple substances, not toxic to the environment. Decomposition of herbicide active substances in the soil dependent on many factors, such as temperature and soil moisture, the type of soil, the content of organic carbon and clay fraction, as the dominant ones [Sedeghi et al. 2000, Li et al. 2003, Rodriguez-Liebana et al. 2013]. Metals like copper or zinc can effect the process of herbicide degradation, which can take place through catalyzed degradation reaction, forming complexes with active substances, which can modify the processes of sorption, transport in the soil and susceptibility to decomposition [Guo et al. 2009]. The process of affecting herbicides by metals can be of an indirect character, through the influence on soil microorganisms, e.g. on their population and activity, and therefore, affect the rate of herbicide decay [Wang et al. 2007, Xu et al.2008].

Clopyralid, 3,6-dichloro-2-pyridinecarboxylic acid, selected to the tests, is a phenoxy acid herbicide, which was first developed and produced by U.S. Dow AgroSciences Ltd, and was registered for many annual and perennial broadleaf weeds control in different crops. It is absorbed by the leaves and roots of weed and moves rapidly through the plant, and affects plant cell respiration and growth. This herbicide is generally active in the soil and maybe persistent in soils under anaerobic conditions and with low microorganism [Corredor et al. 2006, Zhao et al. 2011].

The aim of the work was the assessment of the effect of copper and zinc contamination on clopyralid degradation in the soil.

MATERIALS AND METHODS

Laboratory studies

The research involved controlled experiments. The soil samples were collected from cultivated fields, from 0–15 cm arable layer. Two types of soil, featuring naturally diversified content of copper and zinc, were used in the investigation. One of the soils (S1), of increased copper and zinc content, came from the field situated in the vicinity of "Głogów II" Copper Smelter, while the second sample (S2), characterized by low content of Cu and Zn, was collected from a field near Wrocław. In order to eliminate the effect of the soil properties on herbicide degradation, the soil S2 was selected so that its basic parameters (C_{org}, granulometric composition and pH) were similar to the properties of S1 soil. The studies also included additional soil S21 prepared in the laboratory (S2 soil additionally contaminated with Cu and Zn salts in the amounts equivalent to contamination level of S1 soil). The analysis of basic physical and chemical parameters of the soils, as well as the content of heavy metals was done in accredited chemical and agricultural laboratory. The basic parameters of the soils were shown in Table 1.

After sieving the soil through a 2-mm sieve, it was stored in covered trays in a greenhouse for 10 days and regularly mixed. Five days before the experiment was established, S21 soil was prepared by adding Cu and Zn salts to S2 soil in the amounts corresponding to the level of S1 soil contamination. Soil moisture was measured before the start of the trials by heating to dryness for 24 h at 105 °C and determining the difference in weight. Soil moisture was set at 60% of field capacity, checked at regular intervals and adjusted with distilled water to the initial level.

Soil samples were transferred into 90 mm diameter and 85 mm height pots that were placed in growth chambers; each variant had four replicates. Day/night temperature regimes were 20/10 °C (average temperatures recorded at the end of April and May – typical term of clopyralid application on the field in Poland) and light intensity was 320 ±10 $\mu mol \cdot m^{-2} \cdot s^{-1}$ photosynthetic photon flux, with 15 h day length. Two days after placing the pots into the growth chambers, the commercial formulation of clopyralid (herbicide Lontrel 300 SL; Dow AgroSciences®, USA) at a dose 90g of active substance per hectare was applied. Application of herbicide was done using a stationary chamber sprayer equipped with a mobile nozzle TeeJet XR 11003-VS. The nozzle was operated at pressure of 200 kPa and speed 3.6 $km \cdot h^{-1}$ producing a spray volume of 250 $l \cdot ha^{-1}$. Herbicide doses and spraying conditions were the same as for field conditions. Soil samples (one pot containing ca. 250 g of soil = one sample and one replica) were taken for analysis one hour after herbicide application and 2, 4, 8, 16, 32, 64 and 96 days after treatment. The samples taken from the experiment were well mixed and stored in polyethylene bags at minus 20 °C until sample extraction.

Clopyralid determination – analytical method

The analytical procedure consisted of three crucial steps:

1) Extraction of the analyzed substance from the matrix.

   Soil samples were homogenized. Soil portions (3×10g) were mixed with 3×2.5g of Diatomaceous Earth (Dionex® ASE® Prep DE) and

Table 1. Soil characteristics

Soil	pH (1n KCl)	C_{org} [%]	Sand [%]	Silt [%]	Clay [%]	Cu [$mg \cdot kg^{-1}$]	Zn [$mg \cdot kg^{-1}$]
S1	6.4	2.21	15	25	60	278.2	93.6
S2	6.2	2.09	19	24	57	6.8	3.6
S21	6.5	2.09	19	24	57	275.9	94.2

transferred into stainless steel cells. Extraction was done using accelerated solvent extraction on fully automated Dionex ASE 350 extractor (Dionex®, CA, USA) [extraction solvent – acetone (33 mL per cell for two cycles of extraction), temp. 40 °C, extraction time – 20 min. and pressure – 0.2 MPa)]. The combined extracts (from three cells) were then slowly evaporated under a nitrogen stream until dry.

2) Extract cleaning.
Dry residues were dissolved in 10 mL of water (acidified to pH = 2) and subjected to SPE (Solid Phase Extraction). 3 mL-capacity cartridges and sorbent bed – 0.5 g of octadecyl, 40 μm particle size (Bakerbond®, J.T.Baker®, Philipsburg, NJ, USA) were used. The sorbent was preconditioned with water (3 mL) and then methanol (3 mL). 10 mL of extract solution was loaded with small portions onto the cartridge and the eluate was discarded. Analytes were eluted with two portions of methanol (2×1 mL). To eluate was added 20 mL of water, moved to separator and extracted with two portions of dichloromethane (2×10 mL). Collected extract was evaporated to dryness. Next, dry residue was dissolved in methanol (2 mL) and esterified with diazomethane (injection solvent).

3) Final determination.
A gas chromatograph Varian CP 3800 equipped with software Varian GCMS 2000 and the electron capture detector (Varian®, CA, USA) was used to carry out the final determination. Throughout the entire experiment, a VF-5ms capillary column (30 m ×0.25 mm ×0.25 μm film thickness) (Varian®, CA, USA) was used. Nitrogen was used as carrier gas at a flow rate of 1.5 mL/min. Chromatographic separation was performed at the column oven where the initial temperature was held at 120 °C for 7 min. Then, the temperature ramped at a rate of 10 °C/min. to 200 °C and was held for 2 minutes. Finally, the second ramp was continued at 5 °C/min. to 230 °C, which was held for 20 min. Injector and detector temperatures were set at 230 and 300 °C, respectively. Aliquots of 1 μl of the samples were injected.

The recovery of clopyralid from soil was determined by analyzing fortified samples. The analyses were carried out at four concentration levels in three replicates. The average recovery for all concentration were 97.2%. The quantification limit of the method was 0.0010 mg kg^{-1}.

Each soil sample was analyzed three times. Repeatability of the analytical results was satisfactory, with relative standard deviation (RSD) not exceeding 8.9% of the mean values. All experimental data was calculated using the statistical program Statgraphics Centurion, version XV and Excel (MS Office 2010).

RESULTS AND DISCUSSION

The detailed course of clopyralid degradation in the soils was show in Figure 1. The initial clopyralid concentration (analysed 1 hour after application) amounted for all samples to 0.283 ± 0.0203 $mg\cdot kg^{-1}$. The process differed according to the content of copper and zinc in the soil. The slowest rate of clopyralid dissipation was observed in S21 soil (additionally contaminated by the two examined metals). Clopyralid proved to decompose only a bit faster in S1 soil (of high natural Cu and Zn content). Significantly faster degradation was recorded in the case of the soil characterized by low Cu and Zn content (S2). The differences in clopyralid degradation were maintained in the course of research. Only 96 days after introduction of the herbicide, clopyralid the concentration in all the soils was similar and it ranged, respectively: 3.8% of initial concentration – S1 soil, 6.2% – S21 soil and 2.6% – S2 soil.

The correlation observed in the research, i.e. slowing of herbicide active substance degradation in the soil contaminated with heavy metals, was confirmed in the research by other authors [Said, Lewis 1991, Pieuchot et al. 1996, Suhadolc et al. 2004, Dictor et al. 2008, Wróbel, Sadowski 2012].

As it was already mentioned in the introduction, clopyralid dissipation takes place mainly in a biological way (soil microorganisms). Numerous investigations prove that higher concentration of metals in the soil does significantly reduce the activity of soil microorganisms, which can slow herbicide degradation [Giller et al. 1998]. In similar research by the authors (not yet published), it was possible to state that the presence of copper and zinc in the soil accelerates metazachlor decomposition, which follows mainly the chemical way.

Diversified chemical structure and metabolic processes of herbicides do not allow to assess the effect of heavy metals on decomposition of these compounds. Similar research should be conducted for particular active substances.

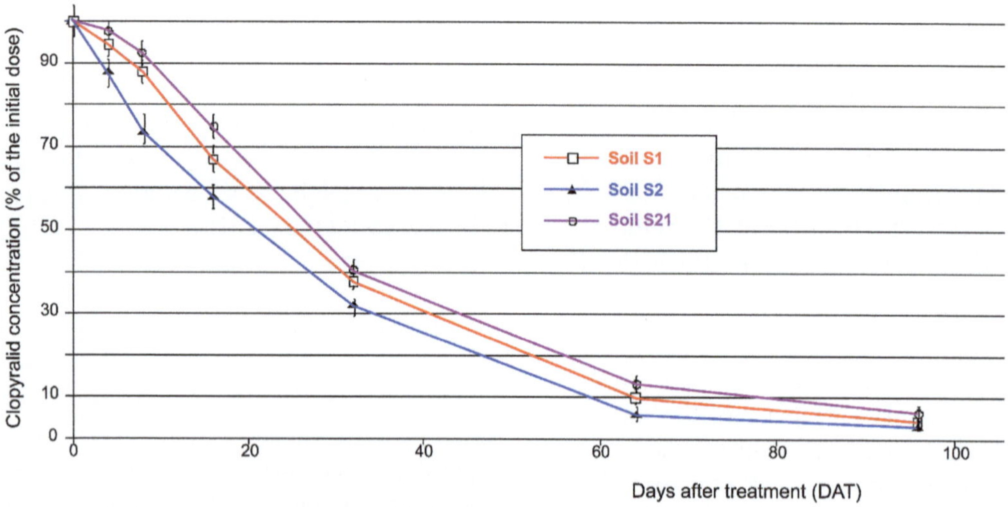

Figure 1. Degradation of clopyralid in soils. Vertical bars represent ± standard errors of means (n = 3)

The degradation data was plotted. Good linearity was found between logarithmic concentration of clopyralid residues and time, indicating first-order rates of degradation with correlation coefficients (R^2) about 0.99 for soils. The DT_{50} values (graphically derived by interpolating the values between successive residue measurements) varied from 21.1 days for S2 soil, to 27.6 days for S21 soil. The DT_{50} values obtained after fitting the curves to first-order kinetics were similar (19.5 and 25.3 days, respectively) (Table 2). The same model of kinetics equation was noticed for clopyralid and other herbicides (Allen, Walker 1987, Ravelli et al. 1997, Ma et al. 2004, Cuevas et al. 2007, Hu et al. 2011, Kucharski, Sadowski 2011, Zhao et al. 2011).

Significant differences in degradation rate of the herbicide in soil in the first period after treatment influenced the DT_{50} indicator. The DT_{50} values for soils in this experiment are consistent with the data presented by Smith and Aubin [1998], who measured the dissipation of clopyralid in three prairie soils (clay, clay loam and sandy loam) at different temperatures. They found that dissipation increased as increased temperature (from 10 to 30 °C), and showed first-order kinetics. Half-lives ranged from 10 to 47 days, and were similar in the clay and sandy loam soils, and greatest in the clay loam soil at higher temperatures.

Different half–life period of clopyralid is significant, but the obtained results do not exceed the range of values published by European Food Safety Authority (EFSA). On the basis of the research results it is possible to state that clopyralid belongs to substances of medium stability in the soil. Its half–life, obtained due to laboratory tests, ranged from 13 to 65 days, while in field experiments it was shorter and amounted to 2–24 days [EFSA Sci. Rep. 2005].

CONCLUSIONS

1. Clopyralid belongs to the group of moderate persistent compounds, its half-life (DT_{50}) in the soil, in this research, ranged from 21 to 27 days.
2. Graphical representation of clopyralid degradation in the soil course takes the form of exponential curve (rate equation of first-order kinetics).
3. The increase in copper and zinc content in the soil influenced clopyralid degradation and DT_{50} value. Slower degradation, and therefore higher DT_{50} value was observed in the soil with higher Cu and Zn content (DT_{50} = 25–27 days).

Table 2. Equations for clopyralid degradation curves and values of DT_{50}

Object	Equation of degradation curve (for average values)	DT_{50} [days]	
		Graphically determined	Calculated from equation
Soil S1	$C_t = 0.299 e^{-0.0292t}$, $R^2 = 0.9959$	25.32 ± 2.12	23.73
Soil S2	$C_t = 0.291 e^{-0.0356t}$, $R^2 = 0.9901$	21.13 ± 1.76	19.47
Soil S21	$C_t = 0.293 e^{-0.0274t}$, $R^2 = 0.9918$	27.64 ± 2.31	25.29

Explanation: C_t – acetochlor concentration at time „t", R^2 – correlation coefficient.

Acknowledgements

The work has been prepared as a part of 2.6 activity in the Institute of Soil Science and Plant Cultivation – State Research Institute Long-term Program (funded by the Polish Ministry of Agriculture and Rural Development).

REFERENCES

1. Allen R., Walker A. 1987. Influence of soil properties on the degradation rates of metamitron, metazachlor and metribuzin in soil. Pestic. Sci., 18, 95–111.
2. Corredor M.C., Rodriguez-Mellado J.M., Ruiz-Montoya M. 2006. EC(EE) process in the reduction of the herbicide clopyralid on mercury electrodes. Electrochim Acta, 51, 4302–4308.
3. Cuevas M.V., Cox L., Calderon M.J., Hermosin M.C., Fernandez J.E. 2007. Chloridazon and lenacil dissipation in a clayey soil of the Guadalquivir river marshes (southwest Spain). Agric. Ecos. Environ., 124(3–4), 245–251.
4. Dictor M., Baran N., Gautier A., Mouvet C. 2008. Acetochlor mineralization and fate of its two major metabolites in two soils under laboratory conditions. Chemosphere, 71, 663–670.
5. EFSA Scientific Report. 2005. Conclusion on the peer review of clopyralid, 50, 1–65.
6. Giller K.E., Witter E., McGrath S.P. 1998. Toxicity of heavy metals to microorganism and microbial processes in agricultural soils: A review. Soil Biol. Biochem., 30, 1389–2359.
7. Guo B.Y., Wei W.L., Lin J.M. 2009. The study of the interactions between heavy metals with sulfonylurea herbicides using ACE. J. Chromatographic Sci., 47, 116–120.
8. Hu J.Y., Zhen Z.H., Deng Z.B. 2011. Simultaneous determination of acetochlor and propisochlor residues in corn and soil by solid phase extraction and gas chromatography with electron capture detection. Bull. Environ. Contam, Toxicol., 86, 95–100.
9. Jaworska H., Dąbkowska-Naskręt D. 2012. Influence of Głogów Copper Works on the content of mobile forms of copper and zinc in arable soils. J. Elem., 17, 57–66.
10. Kabała C., Karczewska A., Kozak M. 2010. Przydatność roślin energetycznych do zagospodarowania gleb zdegradowanych. Zesz. Nauk. UP Wroc., Rol., XCVI, 576, 97–118 [In Polish].
11. Kucharski M., Sadowski J. 2011. Behaviour of metazachlor applied with additives in soil – laboratory and field studies. J. Food Agric. Environ., 9(3&4), 723–726.
12. Li H., Sheng G., Teppen B.J., Johnston C.T., Boyd S.A. 2003. Sorption and desorption of pesticides by clay minerals humic acid-clay complexes. Soil. Sci. Soc. Am. J., 67, 122–131.
13. Ma Q.L., Rahman A., James T.K., Holland P.T., McNaughton D.E., Rojas K.W., Ahuja L.R. 2004. Modeling the fate of acetochlor and terbuthylazine in the field using the Root Zone Water Quality Model. Soil Sci. Soc. Am. J., 68, 1491–1500.
14. Nwuche C.O., Ugoji E.O. 2008. Effects of heavy metal pollution on the soil microbial activity. Int. J. Environ. Sci. Tech., 5, 409–414.
15. Pieuchot M., Perrin-Ganier C., Portal J.M., Schiavon M. 1996. Study on the mineralization and degradation of isoproturon in three soils. Chemosphere, 33, 467–478.
16. Ravelli A., Pantani O., Calamai L., Fust P. 1997. Rates of chlorsulfuron degradation in three Brazilian oxisoils. Weed Res., 37, 51–59.
17. Rodriguez- Liebana J.A., Mingorance M.D., Pena A. 2013. Pesticide sorption on two contrasting mining soils by addition of organic wastes: Effect of organic matter composition and soil solution properties. Colloids and Surfaces A: Physicochem. Eng. Aspects, 435, 71–77.
18. Said W.A., Lewis D.L. 1991. Quantitative assessment of the effects of metals on microbial degradation of organic chemicals. Appl. Environ. Microbiol., 57, 1498–1503.
19. Sadeghi A.M., Isensee A.R., Shirmohammadi A. 2000. Influence of soil texture and tillage on herbicide transport. Chemosphere, 41, 1327–1332.
20. Smith A.E., Aubin A.J. 1989. Persistence studies with the herbicide clopyralid in prairie soils at different temperatures. Bull. Environ. Contam. Toxicol., 42, 670–675.
21. Suhadolc M., Schroll R., Gattinger A., Schloter M., Munch J.C., Lestan D. 2004. Effects of modified Pb-, Zn-, and Cd- availability on the microbial communities and on the degradation of isoproturon in a heavy metal contaminated soil. Soil Biol. Biochem., 36, 1943–1954.
22. Valery B., Eugene K. 1998. Soil surface geochemical anomaly around the copper-nickel metallurgical smelter. Water Air Soil Pollut., 103, 197–218.
23. Wang Y.P., Shi J.Y., Wang H., Lin Q., Chen X.C., Chen Y.X. 2007. The influence of soil heavy metals pollution on soil microbial biomass, enzyme activity, and community composition near a copper smelter. Ecotoxicol. Environ. Saf., 67, 75–81.
24. Wróbel S., Sadowski J. 2012. Interaction of copper and herbicide in contaminated soil under remediation. Fres. Environ. Bull., 21, 599–603.
25. Xu J., Yang M., Dai J., Cao H., Pan C., Qiu X., Xu M. 2008. Degradation of acetochlor by four microbial communities. Bioresource Technol., 99, 7797–7802.
26. Zhao P., Wang L., Chen L., Pan C. 2011. Residue dynamics of clopyralid and picloram in rape plant rapeseed and field soil. Bull. Environ. Contam. Toxicol., 86, 78–82.

4

EVALUATION OF LANDFILL LEACHATE POLLUTION: FINDINGS FROM A MONITORING STUDY AT MUNICIPAL WASTE LANDFILL

Magdalena Daria Vaverková[1], Dana Adamcová[1]

[1] Faculty of Agronomy, Mendel University in Brno, Zemědělská 1665/1, 613 00 Brno, Czech Republic, e-mail: magda.vaverkova@uake.cz; dana.adamcova@mendelu.cz

ABSTRACT

The monitoring that was carried out at the landfill specialized in leachate, groundwater and surface water. There were 6 sampling sites. The observed parameters were pH, BOD_5, COD_{Cr}, conductivity. Leachate reached the high values in all observed parameters. Groundwater samples were collected at two monitoring wells and the sampling site (A, B, C). Surface water was collected from two sampling sites (D, E). The pH showed slightly acid values at all sampling points. The pH of surface water was slightly acid to neutral. Both BOD_5 and COD_{Cr} values remained stable over the reporting period. The average conductivity value at sampling points D, E remained constant. In line with the Czech National Standard ČSN 75 7221 "Classification of Surface Water Quality" sampling point D belongs to II Water Quality Class – slightly polluted water and sampling site E to Class I water quality – unpolluted water. The authors believe that the fluctuations occurring with regard to the values of certain samples were not caused by the operation of the landfill itself, but were a result of the intense agricultural activity nearby the landfill.

Keywords: groundwater monitoring, leachate monitoring, municipal solid waste landfill, surface water, water monitoring.

INTRODUCTION

Landfills represent a widespread and significant threat to groundwater quality, human health and even some of the ecosystems due to their nature of operation and abundance. In communal language landfill means waste disposal on land. However, technically the International Solid Wastes Association [Bagchi 1994] defines landfill as "the engineered deposit of waste onto or into land in such a way that pollution or harm to the environment is prevented, and through restoration of land provided which may be used for other purpose". However in works by Kerndorff et al. [1992], Lee and Jones-Lee [1993], Massing [1994], Godson and Moore [1995], Mato [1999], Heron et al. [1998], Mikac et al. [1998], Riediker et al. [2000] the environmental impact of the landfill leakage, particularly on groundwater quality, has been noticed several times regardless of an ideal site selection and a monitoring network design. Therefore, evaluation of potential risks associated with groundwater contamination due to landfills is of great importance in the design of such facilities. Designs of landfill liner systems, detection and assessment of the extent of contaminants in groundwater and risk assessment for human health and environment are the three main relevant issues. Groundwater quality monitoring systems are the main link among them since they help to determine the likelihood, and severity of contamination problems. Therefore, a reliable and efficient monitoring network design is of great importance in the overall design of a landfill. However, because of the numerous and significant uncertainties involved, more often it is difficult to ensure that a specific network will detect all of the contaminants released from the landfill. Uncertainties that have great influence on reliability of the monitoring network are size and location of the possible contaminant leak and spatial variability of the hydrogeological characteristics, which make groundwater flow and contaminant paths hard to predict. Locations, depth and number of monitoring wells, chemical characteristics of contaminants, and sampling are also significant parameters that affect the reliability of a monitoring network [Buket et al. 2005].

In practice, monitoring network design is controlled and structured by institutional regulations. European Community and US Environmental Protection Agency (USEPA) regulations are widely recognized and applied in many countries. These regulations require installation of a sufficient number of the detection monitoring wells to detect a contaminant leak before it crosses the compliance boundary. Minimum requirements are three down gradient wells and one up gradient well. The post closure monitoring time mentioned is 30 years while the position, number (more than the minimum requirement) and depth of the monitoring wells are proposed by the landfill owners or operators and by local authorities [Buket et al. 2005].

Landfill leachate is a potentially polluting liquid, which unless returned to the environment in a carefully controlled manner may cause harmful effects on the groundwater and surface water surrounding a landfill site. The reasons for monitoring are to provide assurance that the landfill operation does not cause harm to human health or the environment. The leachate formation occurs when soluble components are dissolved (leached) out of a solid material by percolating water. Leachate may also carry insoluble liquids (such as oils) and small particles in the form of suspended solids. Depending on the waste types, further contaminants may be introduced as a result of biodegradation of wastes. Almost any material will produce leachate if water is allowed to percolate through it. The quality of leachate is determined primarily by the composition and solubility of the waste constituents. If waste is changing in composition, for example due to weathering or biodegradation, then leachate quality will change with time. This is particularly the case in landfills containing municipal waste [Salem et al. 2008].

Our research team has been involved in the investigation of environmental problems of pollutants produced and released from landfill facilities. As part of such research efforts, a preliminary study was conducted to measure the emission concentrations and leachate from municipal solid waste (MSW) landfill S-OO3 Štěpánovice. Measurements were carried out in the years 1997–2010. The results of this field measurement study are used to diagnose the landfill influence on the nearest surrounding.

From the literature view it can be concluded that papers usually contain only random leachate characterization [Wu et al. 2004, Bila et al. 2005, Kurniawan et al. 2006, Laitinen et al. 2006]. Only some authors have tried to characterize leachate composition during the landfill exploitation [Lo 1996, El-Fadel et al. 2002, Statom et al. 2004, Kulikowska and Klimiuk 2008]. Therefore, the purpose of this study was to search for the effects of a landfill on the characterization of major components in leachate from a Czech MSW landfill, namely the landfill site in Štěpánovice near Klatovy. Data related to nine years of monitoring (2002–2010) are shown in this study.

MATERIAL AND METHODS

Field investigation - site description

Leachate used in this study was collected from MSW landfill located in Czech Republic (Figure 1, Figure 2), 1 km north of Štěpánovice commune and 1 km south of Dehtín commune (GPS coordinates of the landfill – 49°26'15.934"N, 13°16'55.352"E). The MSW landfill has been operating since summer 1996. It is situated in the north part of widely opened valley directed towards W-E. The bottom part of this area is restricted with a nameless stream being the right tributary of Úhlava River. The upper part of the area is covered with woodland vegetation predominated by *Pinus sylvestris*. The south slope is used for agriculture (Figure 1).

Figure 1. MSW Štěpánovice [www.mapy.cz]

Figure 2. Location of the research [www.maps.com, modified by Vaverková and Adamcová 2013]

The landfill is located at the north slope from the valley axis. In the past, the landfill area was used as the meadow. In terms of maintenance, the landfill is classified in the S-category - other waste. The landfill has a total authorized volume of about 569 000 m^3; at the moment, it is being used to dispose mixed municipal waste. The landfill (Figure 3) is formed by three sub-landfills: landfill A (closed in 2003, area 8 750 m^2); landfill B (operating from 2003, area 26 000 m^2); landfill C (that will operate after closing part B). Total volume of both (A,B) parts of the landfill is 289 000 m^3. Planned service life of the facility is up to year 2018.

At the moment, the municipal waste treated at the MSW landfill Štěpánovice comes from Klatovy city with satellite towns and villages for a total of about 37 725 inhabitants. The total amount of waste deposited in the landfill in 2004–2011 was in detail described in the article "Can vegetation indicate a municipal solid waste landfill's impact on the environment?" [Vaverková and Adamcová 2014].

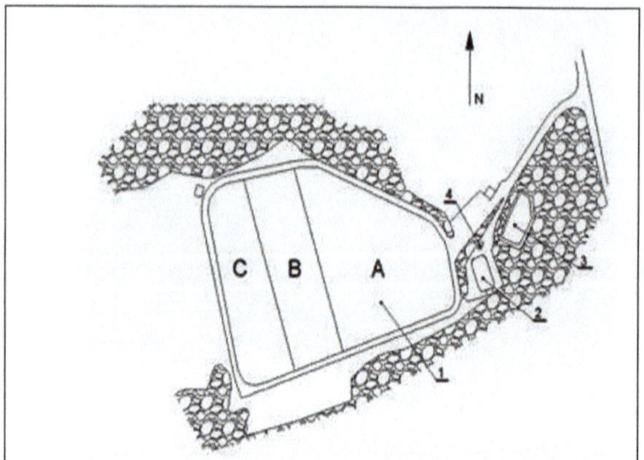

1 – landfill. 2 – leachate pond, 3 - rainwater reservoir, 4 - entrance gate

Figure 3. Map of three sub-landfills [Vaverková and Adamcová 2014]

The activities carried out in this landfill are: landfilling of municipal waste, biogas measurement and monitoring, leachate management. Apart from municipal waste the following types of waste are deposited at the landfill: fluid waste, faecal matter, hazardous substances, radioactive and toxic waste. In 1996–2011 approximately 25% of disposed waste was of organic origin. The rest was glass, paper, plastics, metals, textiles and inorganic remains like construction waste, etc.

MONITORING STUDY ON LEACHATE

Landfill monitoring is a set of activities, thereby the effects of a landfill on neighboring environment are monitored and behaviour of individual parts of a landfill is observed. MSW landfill Štěpánovice is equipped with monitoring system. In line with the Czech National Standard ČSN 83 8036 the landfill Štěpánovice is monitored throughout the whole duration of its operation. The control programme and monitoring of the landfill as set in the Czech National Standard ČSN 83 8036 involve monitoring of quality and quantity of leachate and monitoring of groundwater and surface water in the vicinity of the landfill.

Methods of monitoring of leachate, groundwater and surface water are set in the Czech National Standard ČSN 83 8036. The Standard stipulates that quality and quantity of leachate are examined on the outflow from the internal drainage system into the pond. Sampling frequency is determined with regard to the riskiness of the landfill. In terms of quality of surface water and groundwater (water in the vicinity of the landfill) the measurement concerns the following: (i) level and quality of groundwater in the vicinity of the landfill, in particular with respect to potential contamination of substances contained in the extracts from stored waste, (ii) quality of surface water, into which flows the drainage system of the landfill and eventual external drainages, at the places above and below the drainage outflow.

The research specializing in monitoring of quality of leachate, groundwater and surface water at the MSW landfill Štěpánovice at set time intervals took place in the years 2002–2010. Monitoring of leachate was realized via samples taken from leachate pond (Figure 4). The samples were collected via sampler from the surface of the leachate pond. Analyses of the leachate samples were carried out twice a year (spring, autumn). The assessed quality parameters of the leachate in the research were as follows: pH, electrical conductivity, chemical oxygen demand, dichromate (COD_{Cr}), biological oxygen demand, 5 days (BOD_5). An accredited laboratory carried out measurement and subsequent analysis of the samples. The pH parameter and electrical conductivity value were determined directly at the sampling place.

In order to determine the parameters of quality of groundwater and surface water, monitoring wells and sampling locations were used (Figure 4). The samples were collected from the wells in a dynamic state, after short-time drainage (approx. triple exchange of water in the well, or to stable state). Sampling methods were in accordance with the Czech National Standard ČSN ISO 5667, part 11 (groundwater). Surface water samples were collected from the surface of nameless streams – left tributary of Točnický Creek (Točnický potok). All samples were collected into airtight glass or plastic containers based on the analyzed component and immediately transported to accredited laboratory for analysis.

Sampling points and monitoring wells for leachate, groundwater and surface water were defined in the area of the landfill. Their distribution and frequency were established by accredited laboratory. The allocation of sampling points and monitoring wells for the assessment of leachate, groundwater and surface water at the landfill MSW Štěpánovice is shown in Figure 4.

Leachate samples (sampling point F) were removed from the surface of the leachate pond. Groundwater samples were collected from sites that were identified in the following manner: A – old well (under the landfill), B – new well (above the landfill) and C – bedrock. Surface water samples were collected from sites: D – stream below

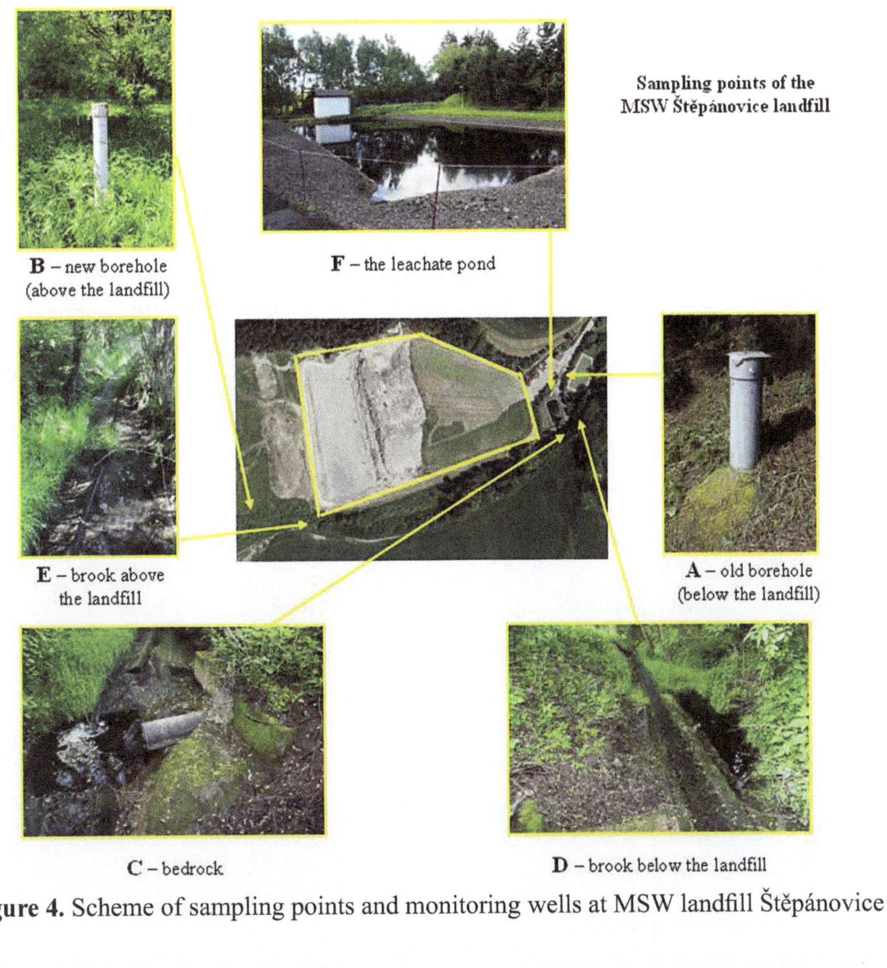

Figure 4. Scheme of sampling points and monitoring wells at MSW landfill Štěpánovice

the landfill (less than 2 m from sampling point C – the bedrock) and E – stream above the landfill.

Samples of leachate, groundwater and surface water were taken within time period of 9 years (April 2002 - August 2010). The analyses of the samples were carried out by the company SOM, s.r.o., Mníšek pod Brdy and Health Institute seated in Pilsen, the Center of laboratories Klatovy (accreditation No. 150/2007).

RESULTS AND DISCUSSION

The research was divided into three parts. The specifications of particular parts of the research are described in Table 1.

The obtained values were assessed pursuant to the criteria set forth in the Methodological Guide of the Ministry of Environment of the Czech Republic – „Soil and Groundwater Contamination Criteria" (1996), according to the Czech National Standard ČSN 75 7221 „Classification of Surface Water Quality)" and ČSN 75 7111 "Drinking water".

Monitoring – pH

The course of pH values measured for leachate (sample from the surface of the leachate pond) for the reporting period is shown in the chart (Figure 5). The pH figure ranged from 7.64 to 8.83 pH. The average pH figure was 8.15 – the leachate showed alkaline values during the reporting period.

The courses of measured values of pH for groundwater (monitoring wells A, B and sampling point C) for the reporting period are shown in the graph (Figure 6).

The pH ranged from 6.07 to 7.76 in groundwater samples at monitoring well A. The average pH value was 7.08 – groundwater showed slightly alkaline values. With regard to sample A, it was not possible to determine the pH value on 1 September 2008, since the sample could not be collected due to drought.

On 27 December 2002 the pH value could not be determined, because it was impossible to collect the sample as the monitoring well was frozen. The pH of the sample from monitoring well B ranged from 6.4 to 7.36. The average pH value was 6.66 - groundwater showed slightly acid values.

The pH of sample C (sampling point) ranged from 6.51 to 7.54. The average pH figure was 6.92 – groundwater at the sampling point showed slightly acid values.

The measured pH values for surface water (sampling points D, E) during the reference period are shown in the graph (Figure 7).

Table 1. Specification of particular parts of the research

Research	Specialization	Parameters	Period	Sampling point
I	Leachate water	pH, BOD_5 [mg/l], COD_{Cr} [mg/l]	4/2002 – 8/2010	F
		Conductivity [mS/m]	5/2005 – 8/2010	F
II	Groundwater	pH, BOD_5 [mg/l], COD_{Cr} [mg/l]	4/2002 – 8/2010	A, B, C
		Conductivity [mS/m]	5/2005 – 8/2010	A, B, C
III	Surface water	pH, BOD_5 [mg/l], COD_{Cr} [mg/l]	4/2002 – 8/2010	D, E
		Conductivity [mS/m]	5/2005 – 8/2010	D, E

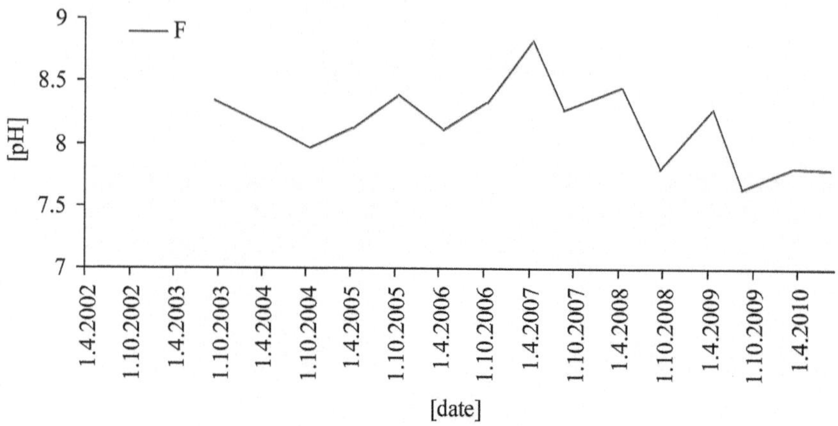

Figure 5. pH of leachate – F

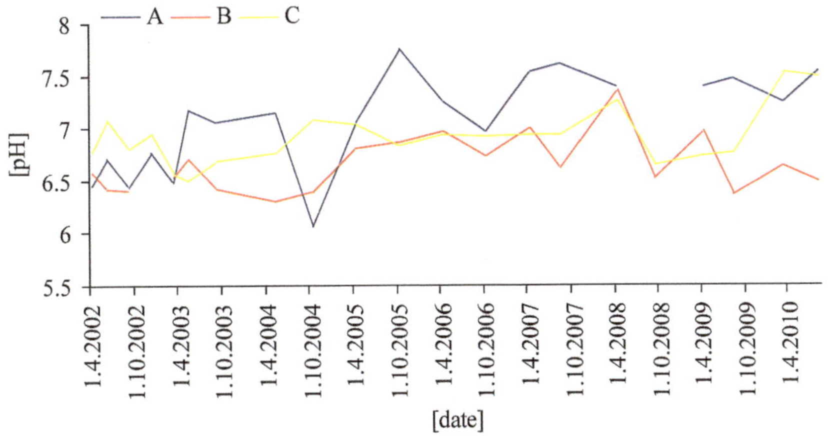

Figure 6. pH of groundwater– A, B, C

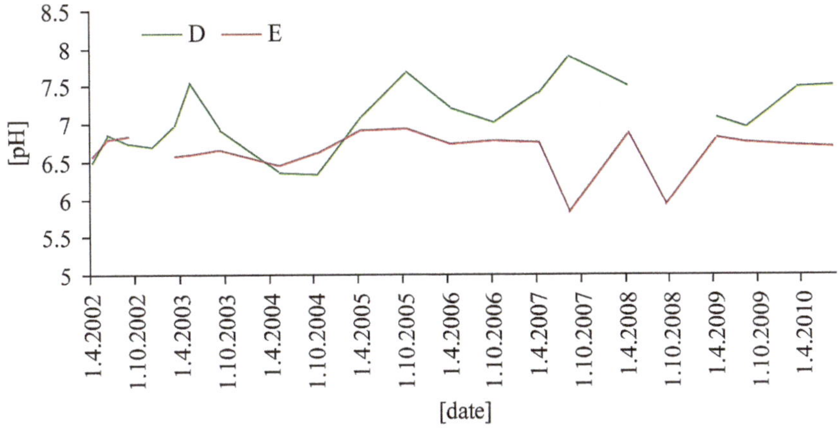

Figure 7. pH of surface water – D, E

The pH of sample D ranged from 6.33 to 7.9. The average pH value was 7.09 in groundwater – water showed slightly alkaline values. With regard to sample D, it was not possible to determine the pH value on 1 September 2008, since the sample could not be collected due to drought. The pH of sample E ranged from 5.82 to 6.94. The average value was 6.65 – the water showed slightly acid values. On 27 December 2002 the pH of sample E was not determined, because it was impossible to collect the sample as the monitoring well was frozen.

The measured pH values in the period 2002–2010 at monitoring wells and sampling points A, B, C, D, E, according to the Czech National Standard ČSN 75 7111 „Drinking Water", approved on 5 January 1989, did not exceed the indicator for the reaction of water – pH 6–8; only samples from sampling point F exceeded these values. Since this is leachate water (water that passed through the BOD_5 of the landfill), such a result is expectable.

Monitoring – BOD_5

The development of BOD_5 values in leachate (sampling from the surface of the leachate pond) within the reporting period is presented in the following graph (Figure 8).

BOD_5 values ranged from 5.6 – 370 mg/l in leachate collected from the surface of the leachate pond (F). The mean value for BOD_5 at sampling point F was 103.24 mg/l. Sampling point F showed highest BOD_5 figures of all sampling sites and monitoring wells.

The course of the BOD_5 values in groundwater (monitoring wells A, B and sampling point C) is displayed in the following chart (Figure 9).

The values of BOD_5 at monitoring well A ranged in the reporting period from 2.0 to 4.2 mg/l. The average BOD_5 value was 2.12 mg/l. With regard to sample A, it was not possible to determine the BOD_5 value on 1 September 2008, since the sample could not be collected due to drought. On 27 December 2002 the BOD_5 value for sample A

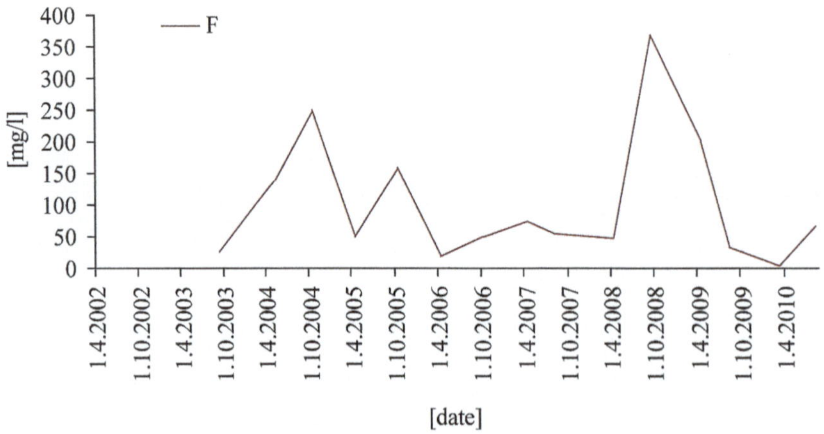

Figure 8. BOD_5 leachate – F

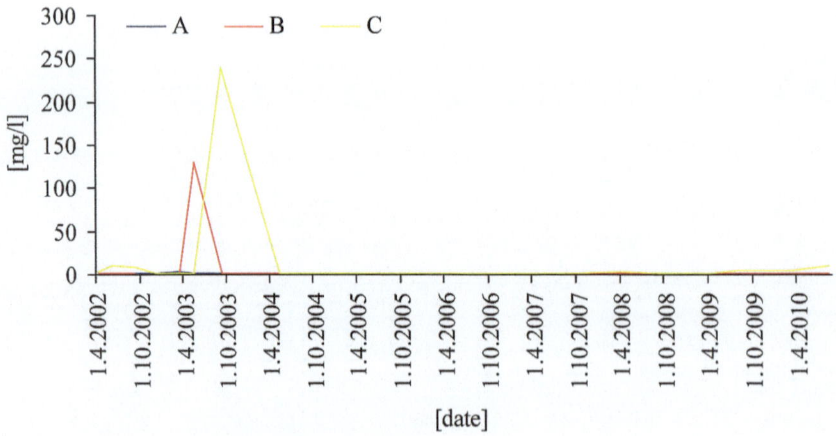

Figure 9. BOD_5 groundwater – A, B, C

was not determined, because it was impossible to collect the sample as the monitoring well was frozen. At monitoring well B the values of BOD_5 ranged from 2.2 to 130 mg/l and the mean BOD_5 value reached 8.41 mg/l. At sampling point C the BOD_5 values ranged from 2.0 to 240 mg/l, while the average figure was 14.87 mg/l.

The Methodological Guide of the Ministry of Environment of the Czech Republic – "Soil and Groundwater Contamination Criteria" (1996) does not stipulate parameters for BOD_5 in groundwater.

The development of BOD_5 values in surface water at sampling points D and E is presented in the following graph (Figure 10).

BOD_5 values at sampling point D fluctuated in the reporting period from 2.0 to 4.4 mg/l. The mean BOD_5 value reached 2.31 mg/l. With regard to sample D, it was not possible to determine the BOD_5 value on 1 September 2008, since the sample could not be collected due to drought. At sampling point E the values of BOD_5 ranged from 2.0 to 150 mg/l and the average figure was 9.62 mg/l. On 27 December 2002 the BOD_5 value for sample E was not determined, because it was impossible to collect the sample as the monitoring well was frozen.

When comparing the measured parameters for sampling points D and E (surface water) with the parameters set forth in the Czech National Standard ČSN 75 7111 "Classification of Surface Water Quality", it may be stated that sampling point D – relying on average values - is classified in II Water Quality Class (<4 mg/l). Thus, the water is slightly polluted – surface water condition that has been affected by human activities; however, water quality criteria attain values that enable the existence of a rich, balanced and sustainable ecosystem.

Sampling point E is classified according to the measured mean values in IV Water Quality Class (<15 mg/l). That means heavily polluted water – surface water condition that has been affected by human activities to such an extent that

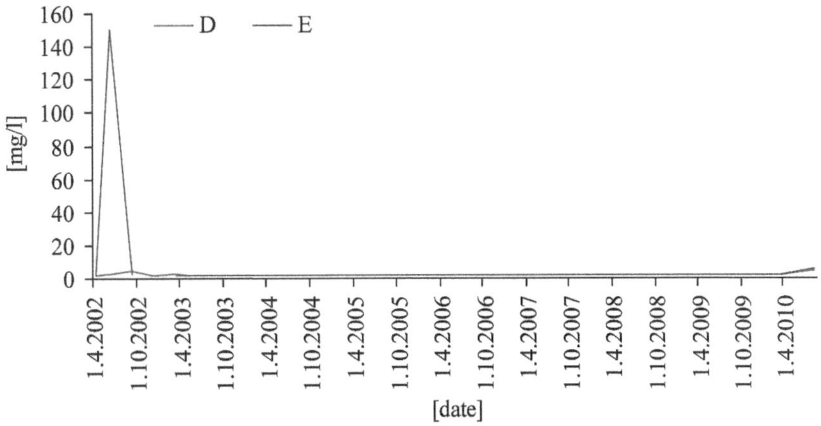

Figure 10. BOD_5 surface water – D, E

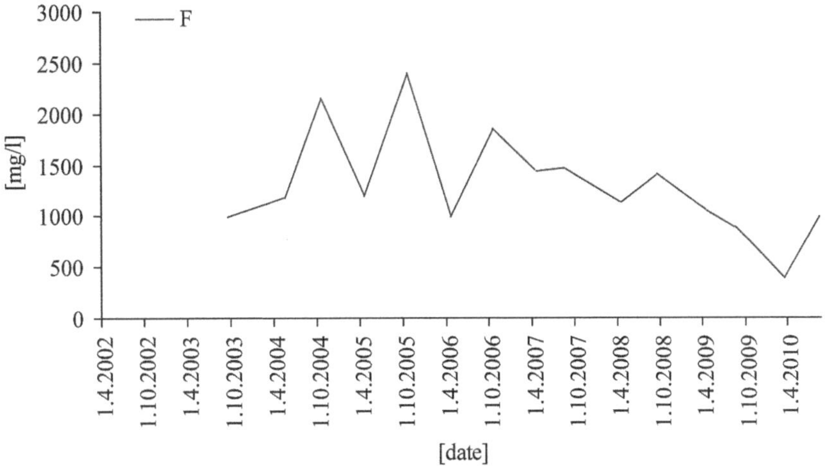

Figure 11. COD_{Cr} leachate – F

water quality criteria attain values, which permit the existence of an unbalanced ecosystem only. The highest measured figure of BOD_5 at the sampling point reached 150 mg/l (26 June 2002). Considering this value the source would belong to V Water Quality Class (> 15 mg/l) – very heavily polluted water, but subsequent monitoring showed stable BOD_5 values corresponding to II Water Quality Class. The reason for such high values of BOD_5 could be the location of dung near the sampling sites.

Monitoring – COD_{Cr}

The course of COD_{Cr} values in leachate (sampling from the surface of leachate pond) within the reporting period is presented in the following graph (Figure 11).

The COD_{Cr} values fluctuated between 387 and 2400 mg/l in leachate collected from the surface of the leachate pond. The mean COD_{Cr} value was 1299.80 mg/l. COD_{Cr} at this sampling point showed the highest rates of water pollution of all sampling sites and monitoring wells.

The development of the COD_{Cr} in groundwater (monitoring wells A,B and sampling point C) is displayed in the following graph (Figure 12).

COD_{Cr} values at monitoring well A oscillated in the reporting period between 10 and 19.2 mg/l. The mean value of COD_{Cr} reached 11.31 mg/l. With regard to sample A, it was not possible to determine COD_{Cr} value on 1 September 2008, since the sample could not be collected due to drought. On 27 December 2002 the COD_{Cr} could not be determined, because it was impossible to collect the sample as the monitoring well was frozen. Monitoring well B showed COD_{Cr} values ranging from 10 to 151 mg/l and the mean COD_{Cr} figure was 17.5 mg/l. At sampling point C the COD_{Cr} values ranged from 10 to 337 mg/l and the average value was 50.75 mg/l.

The Methodological Guide of the Ministry of Environment of the Czech Republic – „Soil and Groundwater Contamination Criteria" (1996)

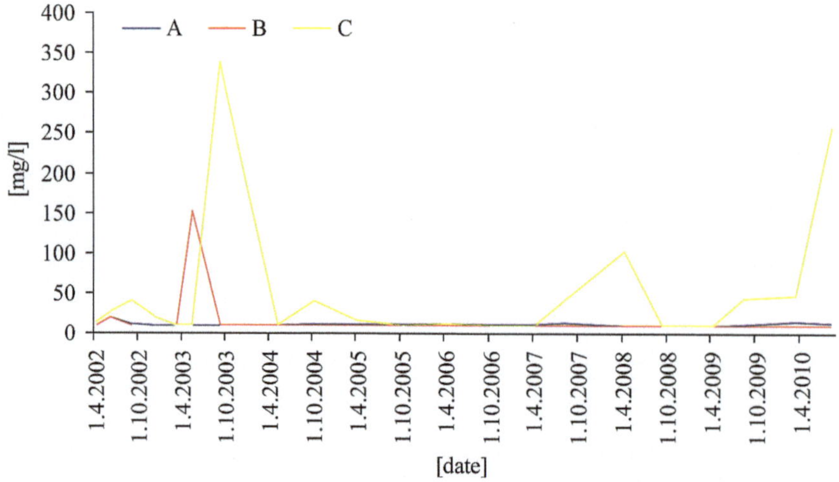

Figure 12. COD_{Cr} groundwater – A,B,C

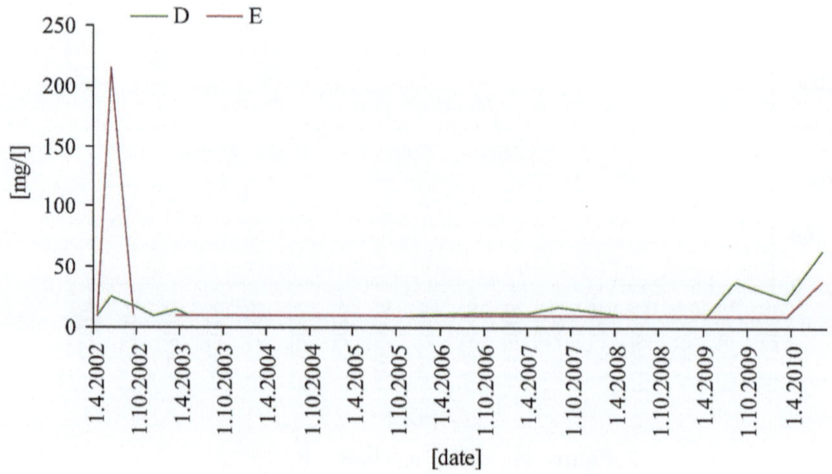

Figure 13. COD_{Cr} surface water – D, E

does not stipulate parameters for COD_{Cr} in groundwater.

The development of COD_{Cr} values for surface water at sampling points D and E is shown in the following graph (Figure 13).

COD_{Cr} values at sampling point D in the reference period fluctuated between 10 and 64.5 mg/l. The average COD_{Cr} figure reached 16.83 mg/l. With regard to sample D, it was not possible to determine the value of COD_{Cr} on 1 September 2008, since it could not be sampled due to drought. The COD_{Cr} values at sampling point E ranged from 10 to 214 mg/l and the mean value reached 21.7 mg/l. It was not possible to determine the value of COD_{Cr} on 27 December 2002, because the sample could not be collected as the sampling point was frozen.

When comparing the measured parameters for sampling points D and E (surface water) with the parameters set forth in the Czech National Standard ČSN 75 7111 "Classification of Surface Water Quality", it may be stated that sampling point D – relying on average values – is classified in II Water Quality Class (<25 mg/l). Thus, the water is slightly polluted – surface water condition that has been affected by human activities; however, water quality criteria attain values that enable the existence of a rich, balanced and sustainable ecosystem. The highest measured figure reached 64.5 mg/l (30 August 2010) at this sampling point. Considering this value, the source would belong to V Water Quality Class (> 60 mg/l). The reason for such high values of COD_{Cr} could be the agricultural intervention near the sampling site.

Sampling point E is classified according to the mean measured values to II Water Quality Class (< 25 mg/l). Thus, the water is slightly polluted – surface water condition that has been affected by human activities; however, water quality criteria attain values that enable the existence of a rich, balanced and sustainable ecosystem. The highest measured value of COD_{Cr} at the sampling point reached 214 mg/l (26 June 2012). Considering this figure, the source would belong to V Water Quality Class (> 60 mg / l) – very heavily polluted water, but subsequent monitoring recorded stable COD_{Cr}, values corresponding to II Water Quality Class. The reason for such high values of COD_{Cr} could be the presence of dung near the sampling site. Higher value (39.9 mg/l), exceeding the parameters of II Water Quality Class was also measured on 30 August 2010. At that moment this source would belong to III Water Quality Class > (45 mg /l) – Polluted water. The reason for higher values of COD_{Cr} could be the agricultural activity in the vicinity of the sampling point.

Monitoring – conductivity

Conductivity has been measured at the investigated landfill since 2005. The development of conductivity of leachate (sampling from the surface of leachate pond) within the examined period is presented in the following graph (Figure 14).

Conductivity values ranged from 80.2 to 1423.0 mS/m in leachate collected from the surface of the leachate pond (F). The mean value at the sampling point F was 944.60 mS/m. Of all sampling sites and monitoring wells this sampling point showed the highest conductivity values.

The course of conductivity in groundwater (monitoring wells A, B and sampling point C) is displayed in the following chart (Figure 15).

Conductivity values at monitoring well A ranged in the reporting period from 15.9 to 47.4 mS/m. The mean conductivity value was 36.26 mS/m. With regard to sample A, it was not possible to determine the conductivity value on 1 September 2008, since it could not be sampled due to drought. At monitoring well B the conductivity values fluctuated between 4.8 and 45.1 mS/m and the mean value was 37.49 mS/m. At sampling point C the conductivity values oscillated between 38.9 and 349 mS/m, whereas the average value reached 108.23 mS/m.

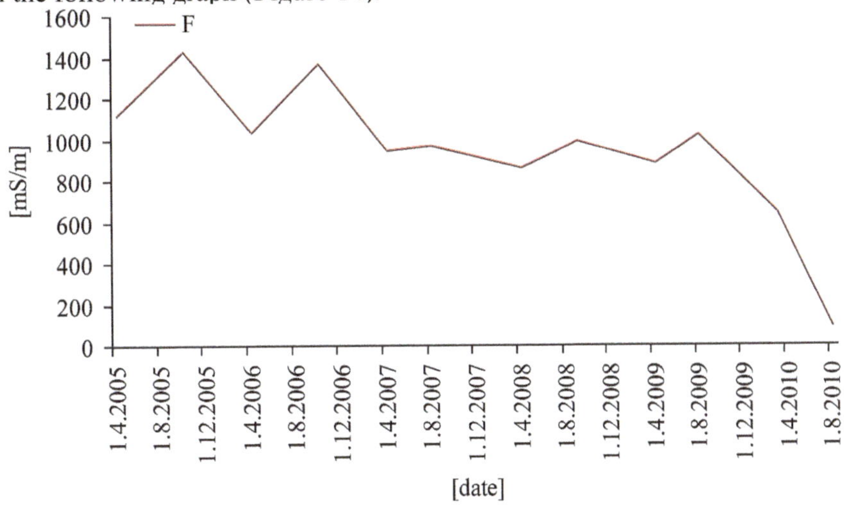

Figure 14. Conductivity leachate – F

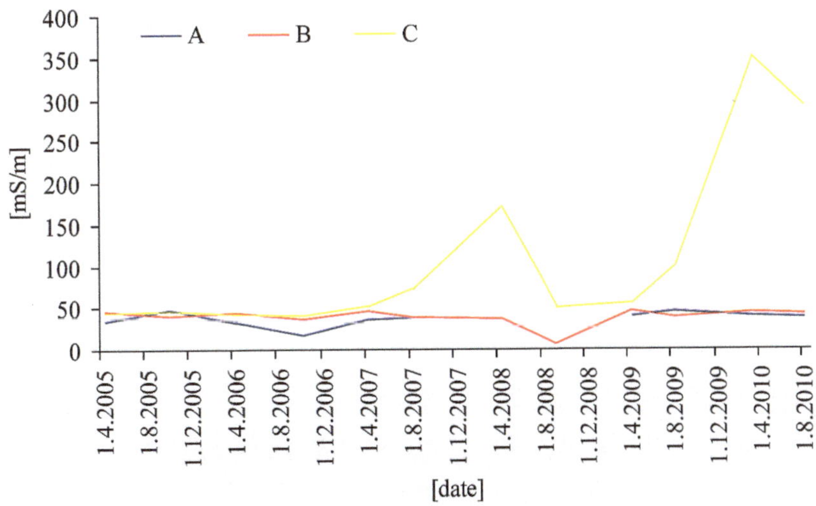

Figure 15. Conductivity groundwater – A, B, C

The Methodological Guide of the Ministry of Environment of the Czech Republic – „Soil and Groundwater Contamination Criteria" (1996) does not stipulate parameters for conductivity in groundwater.

The development of conductivity values in surface water at sampling points D and E is shown in the following graph (Figure 16).

Conductivity values at sampling site D ranged in the reporting period from 17.3 to 90.1 mS/m. The average conductivity value was 48.48 mS/m. With regard to sample D, it was not possible to determine the conductivity value on 1 September 2008, since it could not be sampled due to drought. At sampling point E the conductivity values ranged from 14.7 to 22.2 mS/m and the average value was 17.36 mS/m.

When comparing the measured parameters for sampling points D and E (surface water) with the parameters set forth in the Czech National Standard ČSN 75 7111 „Classification of Surface Water Quality", it may be stated that sampling point D – relying on average values – is classified in II Water Quality Class (<70 mS/m). Thus, the water is slightly polluted – surface water condition that has been affected by human activities; however, water quality criteria attain values that enable the existence of a rich, balanced and sustainable ecosystem. The highest measured figure at this sampling point reached 90 mS/m (30 August 2010). Considering this figure, the source would belong to III Water Quality Class (< 110 mS/m) – polluted water. The reason for such high conductivity value could be the agricultural intervention near the sampling site.

Sampling point E is classified into I Water Quality Class (<40 mS/m). That means unpolluted water – surface water condition that has not been substantially affected by human activities; water quality criteria do not exceed values consistent with normal natural background in surface streams.

Monitoring of the investigated parameters in time

At sampling point F the pH of leachate in the examined period slightly descended. At monitoring wells A, B and sampling point C (ground water) the pH mildly rose, while for surface water (sampling points D and E) the pH remained stable.

At sampling point F the BOD_5 values of leachate in the reporting period showed significant fluctuations. At monitoring wells A, B and sampling point C (groundwater) the BOD_5 values remained stable, except for the beginning of the examined period when the BOD_5 values were higher. Similarly, at sampling points D and E the BOD_5 values remained stable, except for the beginning of the period when the values increased.

At sampling point F the COD_{Cr} values for leachate in the reporting period showed fluctuations, but the overall progress had declining tendency. At monitoring well A the COD_{Cr} values showed stable development. At monitoring well B and sampling point C (groundwater) the COD_{Cr} values were stable, yet at the beginning and towards the end of the period the figures increased. As for surface water (sampling points D and E) the development of COD_{Cr} was stable, except

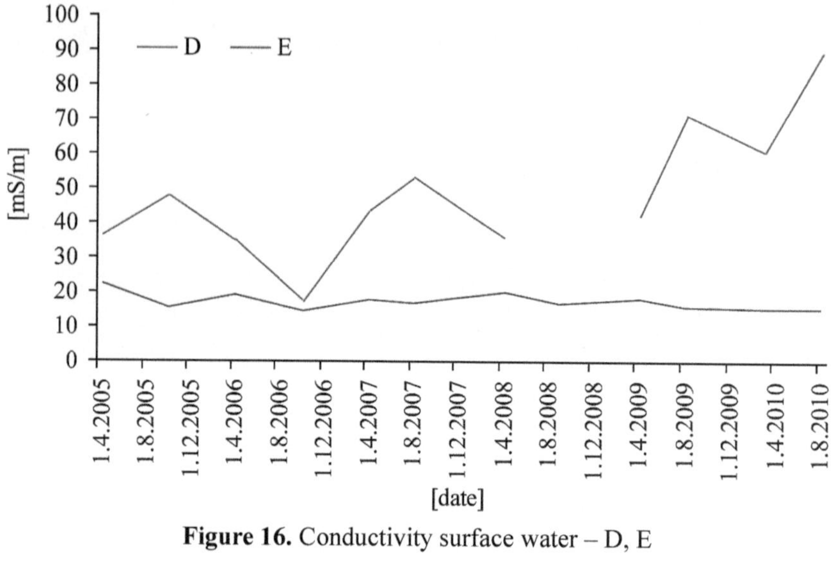

Figure 16. Conductivity surface water – D, E

higher figures at the beginning and at the end of the period.

At sampling point F (leachate) the conductivity showed high values at the beginning of the reporting period, subsequently there was a significant decrease. At monitoring wells for groundwater sampling (A, B) the course of conductivity remained stable, at sampling point C it had increasing tendency. At sampling points D and E (surface water) the progress of conductivity was as follows: at sampling point D oscillations were recorded, yet the values tended to increase; at sampling point E the conductivity curve remained stable.

Comparison of selected parameters with other countries

There are shown the range of pH, COD and conductivity values, that were investigated in selected countries [Baun and Christensen 2004] (Tables 2, 3 and 4).

COD values measured at the MSW landfill Štěpánovice reach similar values in Denmark. But the range of specific conductivity values indicates lowest values of all examined countries.

Leachate – relying on average values – reached the highest values in all observed parameters. Groundwater samples were collected at two monitoring wells and the sampling site (A, B, C). The Guide of the Ministry of Environment does not set forth values for the parameters BOD_5, COD_{Cr} and conductivity. The pH showed slightly acid values at all sampling points. Samples of surface water were collected from two sampling sites – D and E. The pH of surface water was slightly acid to neutral. BOD_5 values were stable over the examined period. Sampling point D belongs to – according to the average values – II Water Quality Class – slightly polluted water and sampling point E to

Table 2. Typical concentrations of pH in landfill leachate [Baun and Christensen, 2004]

Parameter	Range	Country*	Ref.**
pH	4.5–9.0	D	a, b
	4.5–8.6	DK	c, d
	4.5–8.2	USA	e, f
	6.4–8.0	UK	g
	7.2–8.4	HK	h
	5.9–7.0	N	f
	7.8–8.4	F	i
	5.82–8.83	**CZE – MSW Štěpánovice**	

* D, Germany; DK, Denmark; F, France; HK, Hong Kong; N, The Netherlands; UK, United Kingdom; USA, United States of America; CZE, Czech Republic.

** [a] Ehrig [1988], [b]Ehrig [1983], [c] Kjeldsen and Christophersen [2001], [d]Jørgensen and Kjeldsen [1995], [e]Krug and Ham [1997], [f]Johansen and Carlson [1976], [g] Rabinson [1995], [h]Ch *et al.*[1994], [i]Clement and Thomas [1995]

Table 3. Typical concentrations of COD in landfill leachate [Baun and Christensen, 2004]

Parameter	Range	Country*	Ref.**
CHSK	500–60 000	D	a,b
	16–2300	DK	c,d
	50–62 000	USA	e,f
	<10–33 700	UK	g
	147–1670	HK	h
	110–9425	N	f
	400–8000	F	i
	10–2400	**CZE – MSW Štěpánovice**	

* D, Germany; DK, Denmark; F, France; HK, Hong Kong; N, The Netherlands; UK, United Kingdom; USA, United States of America; CZE, Czech Republic.

** [a] Ehrig[1988], [b]Ehrig [1983], [c] Kjeldsen and Christophersen [2001], [d]Jørgensen and Kjeldsen [1995], [e]Krug and Ham [1997], [f]Johansen and Carlson [1976], [g] Rabinson [1995], [h]Ch *et al.*[1994], [i]Clement and Thomas [1995]

Table 4. Typical concentrations of specific conductivity in landfill leachate [Baun and Christensen, 2004]

Parameter	Range	Country*	Ref.**
Specific conductivity (μS/cm)	190–8900	DK	c,d
	50–35 000	USA	e
	503–18 400	UK	g
	2500–12 000	HK	h
	655–3380	N	f
	5000–18 200	F	i
	4.8–1423	**CZE – MSW Štěpánovice**	

* D, Germany; DK, Denmark; F, France; HK, Hong Kong; N, The Netherlands; UK, United Kingdom; USA, United States of America; CZE, Czech Republic.

** [a] Ehrig[1988], [b]Ehrig [1983], [c] Kjeldsen and Christophersen [2001], [d]Jørgensen and Kjeldsen [1995], [e]Krug and Ham [1997], [f]Johansen and Carlson [1976], [g] Rabinson [1995], [h]Ch *et al.*[1994], [i]Clement and Thomas [1995]

IV Water Quality Class – heavily polluted water. COD_{Cr} values remained stable over the reporting period. Sampling point D belongs to II Water Quality Class – slightly polluted water, as well as sampling point E. The average conductivity value at sampling points D and E in the reporting period remained constant. In line with the Czech National Standard ČSN 75 7221 "Classification of Surface Water Quality" sampling point D belongs to II Water Quality Class – slightly polluted water and sampling site E to Class I water quality – unpolluted water.

The authors believe that the fluctuations occurring with regard to the values of certain samples at the sampling points were not caused by the operation of the landfill itself, but rather were a result of the intense agricultural activity nearby the landfill.

Together with standard monitoring, biological monitoring with the use of bioindicators has been taking place in the surroundings of MSW landfill Štěpánovice since 2007 up to nowadays (still ongoing). Within the monitoring protected species have been found and identified, such as e.g. *Epipactis helleborine, Juniperus communis, Polygala chamaebuxus*. Also lichens e.g. *Cladonia arbuscula, Hypogymnia physodes* and *Xanthoria parietina* have been recorded.

In the rainwater reservoir (Figure 3) the occurrence of *Triturus vulgaris* was recorded. It is necessary to mention that its occurrence may reflect correct operation of the landfill considering the sensitivity of *Triturus vulgaris* to the environment. This species is legally protected and in accordance with the Regulation No. 395/1992 Coll. belongs to highly endangered species. Its presence in the rainwater reservoir water proves clarity of this water and shows that no contamination of water by leachate from the landfill or from drained water tank takes place.

CONCLUSION

The landfill leachate pollution reached the high values in observed parameters: pH, BOD_5, COD_{Cr} and conductivity. In line with the Czech "Classification of Surface Water Quality" sampling point belongs to Class II – slightly polluted water, and to Class I – unpolluted water. The fluctuations occurring with regard to the values of certain samples were a result of the intense agricultural activity nearby the landfill.

Acknowledgement

This study was supported by the city of Klatovy and the Technical Services of the city of Klatovy. We would like to express our great appreciation to the management of the landfill Štěpánovice. Namely, we are very grateful to Ing. Vladimír Král, Ph.D. and his colleagues for their assistance and their willingness to provide their time so generously.

REFERENCES

1. Baun D.L., Christensen T.H. 2004. Speciation of heavy metals in landfill leachate: a review. Waste Management and Research 22, 3–23.
2. Bagchi A. 1994. Design, construction, and monitoring of landfill. Willey, Canada.
3. Bila D.M., Montava˜o A.F., Silva, A.C., Dezotti M. 2005. Ozonation of a landfill leachate: evaluation of toxicity removal and biodegradability improvement. Journal of Hazardous Materials 117, 235–242.
4. Buket Yenigu N., Elfeki A.M.M., Gehrels J.C., van den Akker C., Hensbergen A.T., Dekking F.M. 2005. Reliability assessment of groundwater monitoring networks at landfill sites. Journal of Hydrology 308, 1–17
5. El-Fadel M., Bou-Zeid E., Chahine W. and Alayli B. 2002. Temporal variation of leachate quality from pre-stored and baled municipal solid waste with high organic and moisture content. Waste management 22, 269–282.
6. Godson R.H., Moore J. 1995. Subtitle D Groundwater monitoring statistics at a Greenfield Landfill Site in Alabama. In: Tappi International Environmental Conference; 909 International environmental conference International Environmental Conference Proceeding, 1995, pp. 909–915. Technical Association of the Pulp and Paper Industry.
7. Heron G., Bjerg P.L., Gravesen P., Ludvigsen L. and Christensen T.H. 1998. Geology and sediment geochemistry of a landfill leachate contaminated aquifer (Grinsted, Denmark). Journal of Contaminant Hydrology 29, 301–317.
8. Kerndorff H., Schleyer R., Milde G. and Plumb R.H. 1992. Geochemistry of groundwater pollutants at German waste disposal sites, in: Lesage, S., Jackson, R.E. (Eds.), Groundwater Contamination and Analysis at Hazardous Waste Sites. pp. 245–272. Marcel Dekker, New York 1992.
9. Kulikowska D., Klimiuk E. 2008. The effect of landfill age on municipal leachate composition. Bioresource Technology 99, 5981–5985.
10. Kurniawan T.A., Lo W.-H. and Chan, G.Y.S. 2006. Degradation of recalcitrant compounds from sta-

bilized landfill leachate using a combination of ozone-GAC adsorption treatment. Journal of Hazardous Materials 137, 433–455.

11. Laitinen N., Luonsi A. and Vilen J. 2006. Landfill leachate treatment with sequencing batch reactor and membrane bioreactor. Desalination 191, 86–91.
12. Lee G.F., Jones-Lee A. 1993. A groundwater protection strategy for lined landfills. Environmental Science and Technology 28, 584A–585A.
13. Lo I.M.-C. 1996. Characteristics and treatment of leachates from domestic landfills. Environment International 22, 433–442.
14. Massing H. 1994. Impacts of leakage from urban solid waste deposits on groundwater quality. Water Science and Technology 29, 239–244.
15. Mato R.R.A.M. 1999. Environmental implications involving the establishment of sanitary landfills in five municipalities in Tanzania: the case of Tanga municipality. Resources, Conservation and Recycling 25, 1–16.
16. Mikac N., Cosovic B., Ahel S.A. and Toncic Z. 1998. Assessment of groundwater contamination in the vicinity of a municipal waste landfill (Zagreb, Croatia). Water Science and Technology 37, 37–44.
17. Riediker S., Suter J.F.M. and Giger W. 2000. Benzene and naphthalene sulfonates in leachates and plumes of landfills. Water Resources 34, 2069–2079.
18. Salem Y., Hamouri K., Djemaa R. and Alois K. 2008. Evaluation of landfill leachate pollution and treatment. Desalination 220, 108–114.
19. Statom R.A., Thyne G.D. and McCray J.E. 2004. Temporal changes in leachate chemistry of municipal solid waste landfill cell in Florida, USA. Environmental Geology 45, 982–991.
20. Vaverková M., Adamcová D. 2014. Can vegetation indicate a municipal solid waste landfill's impact on the environment? Polish Journal of Environmental Studies 23(2), 501–509.
21. Wu J.J., Wu Ch-Ch., Ma H.-W. and Chang Ch-Cha. 2004. Treatment of landfill leachate by ozone-based advanced oxidation processes. Chemosphere 54, 997–1003.

5

INFLUENCE OF CHARCOAL CALCINATION IN FIELD CONDITIONS ON HEAVY METAL CONTENT IN PLANTS AND IN THE IMMEDIATE VICINITY OF THE RETORTS

Kalina Orłowska[1], Monika Mierzwa-Hersztek[2], Michał Kopeć[3]

[1] Department of Agricultural and Environmental Chemistry, Faculty of Agriculture and Economics, University of Agriculture in Krakow, 21 A. Mickiewicza Ave., 30-120 Kraków, Poland, e-mail: okaja1@wp.pl

[2] Department of Agricultural and Environmental Chemistry, Faculty of Agriculture and Economics, University of Agriculture in Krakow, 21 A. Mickiewicza Ave., 30-120 Kraków, Poland, e-mail: monika6_mierzwa@wp.pl

[3] Department of Agricultural and Environmental Chemistry, Faculty of Agriculture and Economics, University of Agriculture in Krakow, 21 A. Mickiewicza Ave., 30-120 Kraków, Poland, e-mail: rrkopec@cyf-kr.edu.pl

ABSTRACT

One of the ways of charcoal producing is its calcination in charcoal kilns called retorts in field conditions. In the charcoal production wood of deciduous trees is being subjected to the process of dry distillation. The process affects the surrounding environment. During the process a lot of pollutants are emitted to natural environment. One of them is ash, which contains heavy metals. The paper aimed at determining the effect of charcoal calcination on heavy metal content in the soil and plant material in the immediate vicinity. Charcoal calcination causes the alkalinization of the soil to 10 metres. In the closest vicinity, the calcination process increases Mn, Cd, Ni and Cu concentrations in the soil material. The influence of charcoal production on the concentrations Fe, Zn, Pb, Cr in soil material cannot be determined unanimously. The effect of the process on heavy metal content in plant material cannot be determined unanimously.

Keywords: charcoal, dry distillation, heavy metals, effect.

INTRODUCTION

Charcoal is one of the basic products produced during the wood "dry distillation" process. It is a light, black substance of low density, whose main factor (80%) is carbon [Encyklopedia... 2008]. It is used in everyday life, in metallurgical industry and for manufacturing carbonaceous electrodes, it was once used for producing activated carbon for water purification [Dębowski 1987]. Developing industry brought a necessity for developing a fast and easy way of its manufacturing, therefore, the Department of Forest Utilization and Forest Engineering, SGGW developed a technology of charcoal calcination in field conditions [Laurow 1996, 1997].

The best raw material for charcoal production is wood of deciduous trees. During a process of dry distillation occurring in special charcoal kilns called retorts organic matter is exposed to thermal energy at simultaneously limited access of atmospheric air [Rzedkowski 1999]. Finally, the percentage of final products is as follows: charcoal (35%), water (26.6%), gases (15.8%), mainly carbon dioxide, carbon oxide, ethylene, methane and tar sediment (8.11%), soluble tar (5.9%), acetic acid (6%), methanol (2.1%) and acetone (0.2%) [Rzedkowski 1999].

Initially, calcination was regarded as safe for the environment, because it generates pollution, the same as wood used for heating houses. The first investigations revealed that the only hazard was emission of temperature, CO and CO_2 into the air. Subsequent research pointed to the content of suspended and falling particulate matter and CO, SO_2, CH_2O and $C_2H_4O_2$ [Laurow 1996].

It was established that the harmful effect of the process is perceptible only in the close proximity of calcination site, not exceeding 50 m [Rzedkowski 1999].

The authors of charcoal calcination technology did not take into consideration possible emission of heavy metals to the environment, which might accumulate in the immediate vicinity. Therefore, a question arose whether charcoal production in field conditions influences heavy metal content in the soil and plants in direct vicinity of the installation.

METHODS

The soil material for analyses was collected in the area of active charcoal calcination area in the podkarpackie province, Komańcza Forest Inspectorate. The area around the calcination place was fresh mountain forest site, on three sides surrounded by closed forest, whereas to the south by grasses with some bushes. A small watercourse was flowing nearby. The installation for charcoal calcination in field conditions was constructed of 12 active retorts and auxiliary infrastructure, situated in a valley on three sides (to the north, east and west) surrounded by hills. Wood of deciduous trees (mainly alder and beech) was burned in the retorts. The set of retorts selected for the research was active during the whole vegetation period of the year when soil and plant material were sampled for analyses. Generally the set of retorts was operating for minimum of 5 years.

The soil material was collected in points by means of Egner's stick from the top layer of the soil cover (0–10 cm) in September 2011, from 4 cardinal directions (north, east, south and west) at the distance of 10.20 and 30 meters from the retort set. Approximately 1 kg of soil was collected as an average sample from the area of about 0.5 m^2 at a given distance. The directions of sample collection were established according to compass, where the reference point was the extreme retort to the given geographical direction. The soil material was dried in natural conditions and sifted through a sieve with 1mm mesh. Plant material from grass communities was collected into linen bags at the same time as the soil material. It was gathered from four cardinal directions, at the distance of 10, 20 and 30 meters from the retort set, in the area of about 1m^2. The material was dried and then ground.

Granulometric composition, pH in soil and water and soil with 1 mol·dm^{-3} KCl

solution suspension, electrolytic conductivity and total content of heavy metals were assessed in the soil material with methods commonly applied in environmental chemistry. Heavy metal concentrations were determined in the plant material [Ostrowska et al. 1991]. The paper presents average values from two replications and four geographical directions (n = 8) depending on the distance from the extreme retort.

RESULTS AND DISCUSSION

The soil material from the area under the influence of charcoal calcination sites was mostly classified to medium and heavy loams. Proportion of floatable particles in the soils classified to medium loams ranged from 41% to 69%, while in heavy loams from 53% to 70%.

The soil reaction of the material collected closest to the retorts ranged from 4.03–6.71 in soil and water suspension and from 3.34–6.01 in soil and KCl suspension (Table 1). The changes of maximum and minimum values and values of mean pH in the investigated area may evidence an alkalinizing effect of charcoal calcination in field conditions on the soil environment in the immediate retort vicinity, however the factor varies widely. In the research of Rzadkowski et al. [1999], the soil revealed the highest pH at the distance of 25 m from the active retorts and the value was decreasing with increasing distance from the installation. This distribution of soil pH values is caused by dispersion in the environment of one of pyrolysis side-effects (ash). The best raw material for charcoal production are deciduous trees, which after burning contain about 0.34–05.2% of ash (data for beech and oak tree) and reveal pH from 11.47–11.71, therefore, wood ash causes soil alkalinization [Ciesielczuk et al. 2011]. According to Arvidsson, Lundkvist [2003] and Park and others [2005] ash which comes from the burned trees is contributing to a large extent to the alkalinization of soil and it influences the amount of alkaline cations in the soil. Changeability of pH with the distance may result from the frequency of works around the retorts together with ash granulometric fraction translocation with ash fallout.

Changes occurred in the minimum, maximum and mean value of electrolytic conductivity in the soil material originating from the distance of less than 20 m from the installation in relation to the value characterizing the material collected at 30 meters from the retorts (Table 1). Average value of electrolytic conductivity ranged from 0.071– 0.114 mS·sm^{-1}.

Charcoal calcination in field conditions influences heavy metal content in soil material (Table 2). Average content of Mn, Cd, Ni and Cd in the soil material decreases with the distance from the retort set. Mn concentration decreases from 822.8 to 506.88 mg·kg^{-1} of soil. Cd content fluctuates from 0.25 to 0.07 mg·kg^{-1} of oil, whereas Ni ranges from 40.98 to 26.94 mg·kg^{-1} of soil. Cu concentration diminishes from 22.39 to 17.35 mg ·kg^{-1} of soil. An opposite tendency than in previously noted for pH and electrolytic conductivity was observed for Fe, Zn, Pb and Cr, because at 20 m from the set of operating retorts an increase in heavy metal concentrations occurred in relation to their content in the soil material collected at 10 and 30 m from the operating retorts. Obtained results did not corroborate the relationships observed by Rzadkowski [1999]. In his research charcoal calcination most probably affected Fe concentrations, because with growing distance from the retorts its content in the top layer of the soil material decreased up to the distance of about 50 m. In the research of the same author Mn and Zn content was gradually increasing to the distance of 50 m. Pb content declined to 25 m, then grew to the distance of 50 m and again decreased at longer distances. Considering Cu, its slight increase to the distance of 25 m was followed by the decline of its content in the soil material collected 50m from the retorts. Increasing content of this element was observed in material collected at longer distances. Heavy metal concentrations in the soil material might have been also affected by the ash from wood burning, since it contains great amount of these metals. According to Cieślelczuk, Kusz and Nemś [2001] ash from beech tree contains 45197 mg·kg^{-1} Mn, 1284.2 mg·kg^{-1} Zn, 231.8 mg·kg^{-1} Cu, 187.2 mg·kg^{-1} Ni, 52,40 mg·kg^{-1} Pb, 24.45 mg·kg^{-1} Cr and 1.30 mg·kg^{-1} Cd. Such high amounts of heavy metals in the ash, which is freely emitted outside the installation during retort unloading, may to a great extent affect the direct vicinity. On the other hand, the above mentioned authors state, after Wirsz and Matwiejew [2005] that the literature data reveal

Table 1. Minimum, maximum and man values of pH and electrolytic conductivity

Distance from retorts [m]	pH						Electrolytic conductivity [mS · cm^{-1}]		
	H_2O			KCl					
	minimum	maximum	mean ± SD	minimum	maximum	mean ± SD	minimum	maximum	mean ± SD
10	4.37	6.71	5.21±1.03	3.34	6.01	4.20±1.22	0.048	0.131	0.088±0.03
20	3.99	4.81	4.48±0.35	3.07	3.57	3.35±0.21	0.041	0.114	0.071±0.03
30	4.02	6.50	5.42±1.09	3.00	5.73	4.25±1.13	0.068	0.239	0.114±0.08

Explanation: SD – standard deviation.

Table 2. Content of total forms of elements in the soil material from the topsoil (0–10 cm) collected at 10, 20 and 30 m from the set of retorts

Element	Distance from retorts [m]								
	10			20			30		
	Content [mg · kg^{-1} d.m.]								
	minimum	maximum	mean ± SD	minimum	maximum	mean ± SD	minimum	maximum	mean ± SD
Fe	15900.0	21187.5	19025.0±2379.8	17900.0	22537.5	20090.6±3141.9	13800.0	21518.8	18817.2±3452.7
Mn	568.88	995.25	822.80±202.50	317.50	929.75	724.38±270.44	167.63	1181.63	506.88±463.93
Zn	54.94	73.52	65.26±9.11	67.19	80.51	71.06±8.06	40.81	73.59	55.25±13.20
Cd	0.04	0.49	0.25±0.26	0.05	0.34	0.13±0.18	0.05	0.09	0.07±0.04
Pb	13.60	25.26	18.67±4.59	18.26	26.14	22.07±6.62	15.84	23.59	22.02±6.94
Ni	34.81	46.87	40.98±6.34	22.23	46.41	37.77±12.26	21.19	37.29	26.94±7.37
Cu	19.96	25.19	22.39±3.06	13.02	28.75	22.04±7.40	9.90	23.66	17.35±5.77
Cr	15.92	22.66	20.50±3.06	19.36	22.71	21.10±1.99	17.11	22.33	19.25±4.04

Explanation: SD – standard deviation.

a higher content of chromium (30.8–133.3 mg/kg d.m.) but much lower manganese concentrations (68.7–398.0 mg/kg d.m.) in ash from wood briquettes. Also lower Zn concentration was registered, on the level of 9.75–47.7 mg/kg d.m. In the presented investigations, the material from the charcoal calcination site had the highest contents of Fe and Mn, then Zn, Ni, Pb, Cu and Cr but much lower quantities of Mn and Cd.

High values of standard deviation evidence a considerable diversification in the content of the analyzed elements in the soil material collected at the same distance from the operating retorts set. It is due to the specificity of the research object, because field experiments favour a considerable diversification of elements in soil. It may additionally testify an unequal effect of charcoal calcination in field conditions. It is most probably also caused by specific character of the terrain and possible tunnel effect, but also by prevalence of northern and southern winds [Michnia and Paczos 1972].

Charcoal calcination also influences heavy metal concentrations in the plant material from the immediate neighbourhood (Table 3). For Fe, Zn, Cd, Pb, Ni, Cu and Cr a tendency for their content diminishing was noted in the plant material collected 20 meters from the retorts in comparison with the contents in plants gathered at 10 and 30 meters. Only for Mn a tendency for its increasing concentration in the plant material became visible with increasing distance from the calcination site and it is opposite to the relationship noted for this element content in the soil material. In case of Fe, Ni, Cu and Cr the highest concentrations were registered in the soil material from the immediate vicinity of the retorts. On the other hand, the highest contents, as compared with the other sampling places, were assessed at the distance of 30 meters from the installation for Zn, Cd and Pb. The tendency observed for a majority of selected metals, i.e. their decreased content in the material collected at 20 m from the installation, has not been confirmed by a generally accepted principle that the lower pH, the more heavy metal forms available to plants. Soil reaction is one of the main factors conditioning the form in which heavy metals occur in the soil environment and therefore their availability to the local vegetation [Kabata-Pendias and Pendias 1999]. Decreasing pH of the soil material from slightly acid and acid, such as in this case to 4.48 and 3.35 should cause an increase in bioavailable metal forms in soil [Kwiatkowska-Malina, Maciejewska 201]. However, no such pH effect was observed. The situation might have been affected by organic matter content in soil, since it influences the strength of impact on heavy metal solubility. In soil enriched with organic matter (in this case charcoal) bioavailability of heavy metal forms and their toxicity to plants decrease [Kwiatkowska-Malina, Maciejewska 2009, 2011]. Confirmation of this thesis requires additional investigations of the area.

CONCLUSIONS

1. Charcoal calcination in field conditions had an alkalinizing effect and caused a decrease in electrolytic conductivity value in soil material in the immediate vicinity – 10 meters from the installation.

Table 3. Content of total forms of elements in the plant material collected at 10, 20 and 30 m from the set of retorts

Element	Distance from retorts [m]								
	10			20			30		
	Content [mg · kg^{-1} d.m.]								
	minimum	maximum	mean ± SD	minimum	maximum	mean ± SD	minimum	maximum	mean ± SD
Fe	1310.50	7358.50	3918.00±2592.59	538.85	2905.00	1328.61±1041.66	675.85	7616.50	3171.66±3024.65
Mn	196.90	399.50	287.26±79.22	264.70	554.25	399.94±111.92	292.80	1181.50	664.58±358.18
Zn	37.68	51.54	44.16±5.41	24.39	45.28	35.43±9.59	38.49	78.39	49.99±18.39
Cd	0.20	0.54	0.33±0.14	0.11	0.26	0.17±0.07	0.32	1.15	0.73±0.52
Pb	1.24	4.69	2.89±1.32	0.80	2.43	1.37±0.75	1.28	10.89	4.18±4.22
Ni	4.02	15.53	9.24±4.47	3.44	7.20	4.90±1.55	5.78	16.84	8.77±5.05
Cu	8.43	11.84	10.03±1.47	3.02	9.91	6.05±2.69	5.14	6.73	6.11±1.30
Cr	3.57	14.53	7.86±4.75	2.51	5.71	3.65±1.36	3.43	10.63	6.47±3.25

Explanation: SD – standard deviation.

2. Calcination process influences an increase in the content of total forms of Mn, Cd, Ni and Cu in the soil material in the close proximity of the retorts.
3. It cannot be unanimously stated if charcoal calcination immediately affects heavy metal concentrations in plants.

REFERENCES

1. Arvidsson H., Lundkvist H. 2003. Effects of crushed Wood Ash on soil chemistry on Young Wood Norway spruce stands. Forest Ecology and Management, 176, 121–132.
2. Ciesielczuk T., Kusza G., Nemś A. 2011. Nawożenie popiołami z termicznego przekształcenia biomasy źródłem pierwiastków śladowych gleb. Ochrona Środowiska i Zasobów Naturalnych, 49, 219-227.
3. Dębowski Z. 1987. Krajowe możliwości produkcji i regeneracji węgli aktywnych dla potrzeb oczyszczania wody. Wydawnictwo PZITS 521/2-3, (32-33), 29-32.
4. Encyklopedia dla wszystkich. Chemia. 2008. Wydawnictwo Naukowo-Techniczne, Warszawa, ss. 404.
5. Kabata-Pendias A. i Pendias H. 1999. Biogeochemia pierwiastków śladowych. Wyd. Nauk. PWN, Warszawa, 236-239.
6. Kwiatkowska-Malina J., Maciejewska A. 2009. Wpływ materii organicznej na pobieranie metali ciężkich przez rzodkiewkę i facelię. Ochrona Środowiska i Zasobów Naturalnych, 40, 217-223.
7. Kwiatkowska-Malina J., Maciejewska A. 2011. Pobieranie metali ciężkich przez rośliny w warunkach zróżnicowanego odczynu gleby i zawartości materii organicznej. Ochrona Środowiska i Zasobów Naturalnych, 49, 43-51.
8. Laurow Z. 1996. Wpływ zwęglania drewna w retortach stalowych na środowisko przyrodnicze. Przegląd Techniki Rolniczej i Leśnej, 12,
9. Laurow Z. 1997. Produkcja węgla drzewnego w warunkach polowych. Biblioteczka leśniczego, 83, Wydawnictwo Świat, Warszawa, ss. 18.
10. Michna E., Paczos S. 1972. Zarys klimatu Bieszczadów Zachodnich. Lubelskie Towarzystwo Naukowe, Wydawnictwo Polskiej Akademii Nauk, ss. 73.
11. Ostrowska A., Gawliński S., Szczubiałka Z. 1991. Metody analizy i oceny właściwości gleby i roślin. Wydawnictwo IOŚ, Warszawa, ss. 325.
12. Park B.B., Ruth D.Y., James M.S., Don K.L., Lawrence P.A. 2005. Wood ash effects on plant and soil in a willow bioenergy plantation. Biomass and Bioenergy 28, 355-365.
13. Rzadkowski S. Wawrzoniak J., Wójcik J. 1999. Ocena wpływu wypalania drewna w Bieszczadach na środowisko leśne. Wyd. Instytut Badawczy Leśnictwa, ss.19.
14. Wisz J., Matwiejew A. 2005. Biomasa – badania w laboratorium w aspekcie przydatności do energetycznego spalania, Energetyka, 9, 631-636.

6

APPLICATION OF SOKOLKA COMPOST TO PROTECT PESTICIDE GRAVEYARD AREA

Katarzyna Ignatowicz[1], **Jacek Piekarski**[2]

[1] Department of Technology in Engineering and Environmental Protection, Białystok University of Technology, Wiejska 45a, 15-351 Białystok, Poland, e-mail: k.ignatowicz@pb.edu.pl

[2] Department of Water-Sludge Technology and Waste Utilization, Koszalin University of Technology, Śniadeckich 2, 75-543 Koszalin, Poland, e-mail: jacek@wbiis.tu.koszalin.pl

ABSTRACT

The paper presents the results of the researches in sorption process. Selected waste materials were checked to build a barrier for pesticide migration from existing graveyards. Due to primary researches Sokolka compost was selected for further researches. The sorption barrier was designed by the surface and underground water run-off. The line of the barrier was made around the centre of graveyard, 0.5 meter width and 4.0 meters deep up to the non-permeable ground layer. GC analyses have shown the presence of pesticides in piezometer before the barrier and no pesticides in water outside sorption barrier. Due to researches it was found that prevention barrier was successful.

Keywords: sorption, pesticide, compost.

INTRODUCTION

Useless plant protection means and the packaging are one of the most dangerous chemical contaminants for environment. The problem is serious in former Soviet Union countries: Armenia, Belarus, Moldova, Kazakhstan, Kyrgyzstan, Uzbekistan, Ukraine, and Russia, but also in Albania, Romania, Turkey, and Poland. [Sofia 2005; Feng 2003; Ignatowicz 2007a, 2007b, 2007c, 2008c; Li 2012; Siłowiecki 2003] Landfills with outdated and useless plant protection means are the largest threat for natural environment resulting from agriculture chemization. For many years, burial sites and water quality in the vicinity have been one of the most difficult problems to solve, which have made enormous harm to a man and natural environment. Large amounts of useless plant protection means in 1960's were one of the reasons to build waste repositories in Poland. Such a situation intensified in 1970's, when part of means considered as harmful, was withdrawn from use. Burial sites were most often constructed from concrete circles insulated using tar or adhesive; old military objects, such as bunkers or fortifications were also adapted. No hydrological recognition of an area was made during the waste repository localization – usually, remote sites were selected. The problem attracted some attention in 1990's, when burial sites were considered as objects harmful for natural environment due to leakages of outdated pesticides into the environment, which resulted in the environment pollution. Therefore, activities aiming at cataloguing the waste repositories have been undertaken. According to Polish Geological Institute, the number of waste repositories and burial sites filled with outdated pesticides amounted to 284 in 2003 in Poland [Siłowiecki 2003]. Ministry of Environment estimated (before 2009) that about 13.9 Mg of harmful substances are deposited. According to cataloguing carried out in 2009 by Regional Directorates of Environment Protection and the data updated and published by Ministry of Agriculture in July 2010, that number increased almost 200-fold up to 2567.81 Mg. In West-Pomerania province, there are 19 burial sites with 1010.9 Mg of deposited wastes, whereas in Podlasie province – 10 waste repositories, including 5 intended for liquidation and containing 18.41 Mg of pesticides. The construction of waste repositories gets worse and worse every year, which makes a considerable threat to natural environment contamination and also may be a harm for inhabitants in the vicinity, because contaminants penetrating the environment are invisible. Ground and surface waters as well as soils become polluted as a consequence of pesticide wastes penetration from leaky burial sites. Leakage, when reaches the water-carrying layers, is transported along the underground water streams and may be seized by surface waters. Therefore, activities aiming at eliminating the burial sites are undertaken, although they should be made in reliable way and in accordance to the current state of knowledge. Having no external funds, owners of waste repositories refuse to eliminate them at their own expense, which reaches even hundreds of thousand PLZ. Every object should be considered individually, thus examination of natural waters quality is performed to evaluate their contamination and transport of chemical contaminants in the vicinity. In the case of corrosion and damage of waste repository, a continuous supply of contaminants into the open waters is and will be observed for many years. [Biziuk 2009; Ignatowicz 2007b; Li 2012].

It should be remembered that even if the burial site is removed, the effects of storing toxic substances will be noticeable for many years, both in soil and waters. Therefore, it is necessary to search for means of reducing the pesticide migration within the environment and introducing new solutions.

METHODOLOGY

The research aimed at developing the new, low-cost technology for reducing toxic substances penetration from waste repositories into environment. All this inspired the author to carry out the study on applying the sorption process on selected waste materials that can be used as a barrier for chemicals penetration into the environment in order to reduce their migration from existing burial sites and waste repositories. The wastes can be also used for land reclamation by filling the pits made after dump storage liquidation, thus creating conditions appropriate for energetic plant species grown on contaminated areas.

The study has been carried out since 2000 and included several stages:

1) Qualitative and quantitative evaluation of pesticide remains in water and soil environments in the vicinity of burial ground;
2) Evaluation of different waste materials usefulness for sorption of pesticides stored within burial grounds;
3) Actual-scale applied research in working burial ground.

RESULTS

Qualitative and quantitative determination of pesticide remains in environmental elements adjacent to burial grounds

Analyses of plants, soils (up to 30 m distance from burial ground), natural waters, including drinking water for neighboring inhabitants (up to 2 km from burial ground) were performed in the direct vicinity of an outdated pesticides store. Material for study was collected in spring during thawing and heavy rainfalls; other seasons – during storms. Not filtered water samples were subject to determinations for 47 biologically active substances representing various chemical groups of pesticide remains: organochlorine, organophosphorus, pyrethroids, nitrophenols, triazines, and phenoxyacids, being components of the most often stored agents. The water samples were also subject to determine heavy metals that were also sometimes constituents of plant protection means (e.g. in organomercuric pesticides): aluminum, arsenic, boron, barium, beryllium, bismuth, cadmium, cobalt, chromium, copper, iron, gallium, mercury, lithium, manganese, molybdenum, nickel, lead, selenium, strontium, thallium, vanadium, and zinc [Ignatowicz 2007a, 2007b, 2007c].

Pesticide concentrations in all samples were determined in accordance with methodology obligatory in a certified laboratory by means of gas chromatograph AGILENT6890 applying ECD1 and NPD2 columns. Moreover, after sample digestion according to EPA 3015 procedure using microwave digester Mars 5, metal concentrations were determined by means of emission atomic spectrophotometry combined with inductively coupled plasma (ICP-AES), although mercury was analyzed with a help of CV-AAS. [Bizuk 2009; Feng 2003; Spadotto 2003; Li 2012, Wiater 2007].

The largest amounts of biologically active substances remains were found in soil samples. Analyzed soil samples contained 10 out of 16 tested organochlorine substances: pirimicarb, o,p-DDT, DDE, p,p-DDT, α-HCH, γ-HCH, δ-HCH, HCB, methoxychlor, and endrine [Ignatowicz 2007b, 2007c, 2008a]. These are the most durable and hardly-degradable compounds. The results of soil analysis show that quite a large studied area is contaminated with pesticides originating most likely from non-hermetic chambers of dump or pesticide wastes buried in ground. The presence of insecticides, pyrethroids, and herbicides was found. The achieved contents of pesticides indicate the spreading of pollution within the soil. Buried grounds contained largest amounts of pesticides from organochlorine insecticides group, which is also reflected by the presence of the toxins and their metabolites in the environment surrounding the burial ground in Folwarki Tylwickie. Occurrence of p,p'DDT isomer at concentrations from 0.11 to 0.63 μg/dm^3 in waters and 136 μg/kg in soils, as well as g-HCH (lindan), which is the hardest degradable of all HCH metabolites, (0.15-9.0 μg/dm^3) and 13μg/kg in soil, is very characteristic. These extremely toxic substances are characterized by long decomposition time and are difficult to biodegrade. Decomposition time of 95% DDT in the environment is from 4 up to 30 years, while that for g-HCH is from 3 to 10 years [Biziuk 2009; Ignatowicz 2007b, 2007c, 2008a].

Evaluating the usefulness of various natural and waste materials for sorption of pesticides dumped in burial grounds

The studies under static conditions were carried out according to methodology by Chemviron Carbon [Atkin 2006; Ignatowicz 2011a; Spadotto 2003]. They aimed at plotting the adsorption isotherms, due to which a comparison of sorption quantity of different adsorbents on different sorbents can be possible. The analysis of the sorption process was performed on a base of achieved results (applying Statistica software); Freundlich, Langmuir, BET, Huttig, and Dubinin-Raduszkiewicz isotherms were estimated as non-linear models by means of the least squares method with a help of Gauss-Newton or Lovenberg-Marquandt algorithm [Atkins 2006; Ignatowicz 2008b, 2009a, 2011a; Spadotto 2003].

Subsequent study stage, when finding out which pesticide substances migrate into the environment, consisted in a series of experiments aiming at evaluating the usefulness of natural sorbents in reducing this emission. Characterization of applied waste sorbents indicates that they meet conditions for sludge used for natural, including agricultural purposes. [Ignatowicz 2008a, 2008b, 2009a, 2009b, 2011a].

Applied studies in working burial ground

The results achieved in stages I–II of the study allowed for designing the sorption barrier localized near the burial ground on a line of surface and underground waters outflow, which had to prevent from pesticide migration. The barrier design was described by Ignatowicz [2011b]. the applied study was carried out in 2009 and in 2012 in Folwarki Tylwickie village (Zabłudów commune, Podlasie province). The burial ground is localized in a direct vicinity of the cultivated field. That burial ground was classified into the 1st category of threat.

Figure 1. The photo of well's outcrop [Ignatowicz, 2011b]

Water samples were collected from the installed piezometers since the barrier mounting during the heavy rainfalls and snow thawing. Mean values of the results achieved in 2009–2010 were presented by Ignatowicz [2011b]. Minimum and maximum values of achieved results in 2010–2014 were presented in Table 1. The GC analyses revealed that despite of pesticides presence in piezometer just before the barrier, they were absent in waters behind it, thus suggesting that the protective shield satisfied its function.

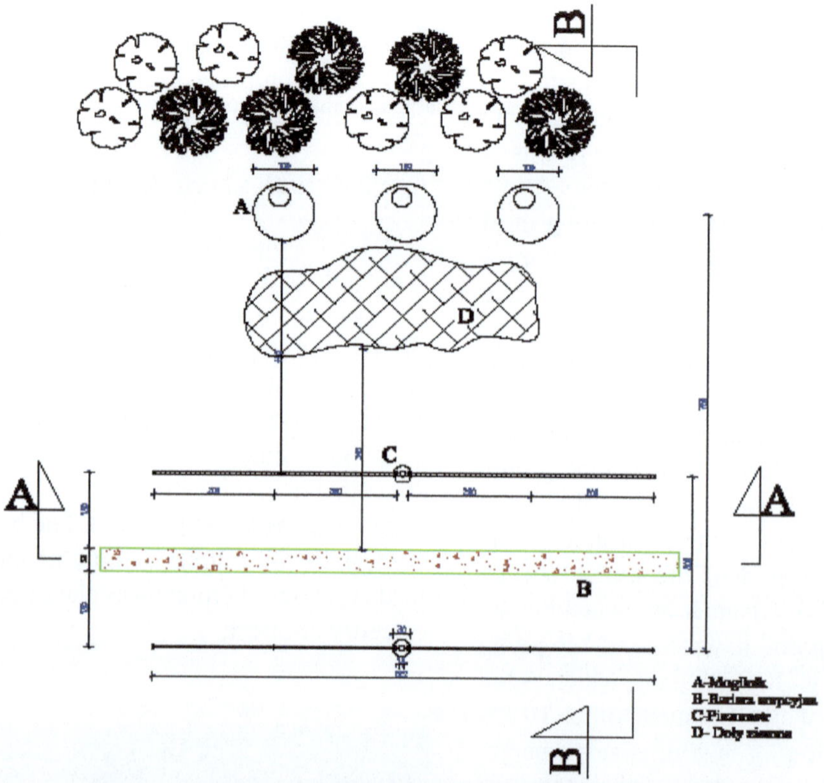

Figure 2. The view of the graveyards area [Ignatowicz 2011b]

Table 1. Concentration of pesticides marked in water samples 2010–2014

Pesticide		Water [μg/dm³] Piezometers "before"	Water [μg/dm³] Piezometers "behind"	Pesticide		Water [μg/dm³] Piezometers "before"	Water [μg/dm³] Piezometers "behind"
H	Atrazine	nd	nd	OC	HCB	nd	nd
	Chlorpropham	nd			α - HCH	0,92-0,94	
	Dichlorprop	nd			β - HCH	0,25-0,29	
	Dinoseb	nd			γ - HCH	9,00-9,89	
	DNOK	nd			δ - HCH	0,06-0,07	
	MCPA	nd			Methoxychlor	0,22-0,28	
	Mecoprop	nd		OP	Bromfenvinfos	nd	
	Simazine	nd			Chlorfenvinfos	nd	
	2,4 - D	nd			Chlorpyrifos	nd	
IC	Carbaryl	nd			Chlorpyrifos - methyl	nd	
	Carbofuran	nd			Diazinon	nd	
	Pirimicarb	nd			Dimethoate	nd	
	Propoxur	nd			Fenthion	nd	
OC	Chlorfenson	nd			Fenitrotion	nd	
	p,p' – DDD	1,12-1,89			Heptenophos	nd	
	o,p' – DDT	0,20-0,25			Izofenphos	nd	
	p,p' – DDE	0,52-0,62			Methidation	nd	
	p,p' – DDT	0,18-0,27			Parathion	nd	
	Dieldrine	0,12-0,18			Thiometon	0,28	
	DMDT	nd			Triazophos	0,94	
	endrine	0,22-0,25		P	Cypermethrin	nd	
	α - β Endosulfan	0,65-0,71			Deltamethrin	nd	
	Endosulfan - sulfate	nd			Fenpropathrin	nd	

nd – not detected.

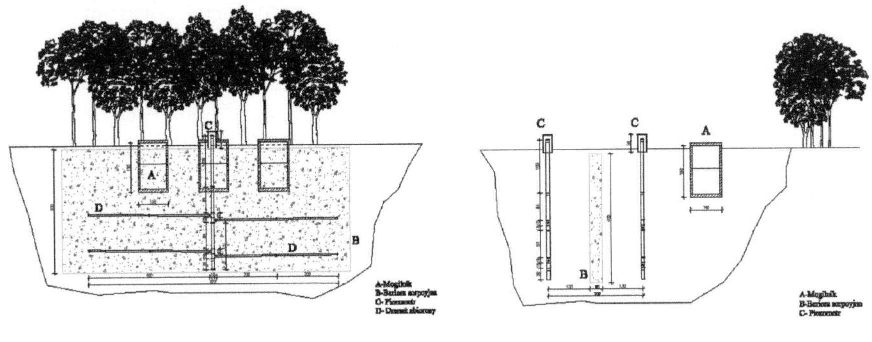

Figure 3. The sectional view of graveyard and sorption barrier. [Ignatowicz 2011b]

CONCLUSION

The research is a comprehensive cycle of studies around active burial grounds along with a scientific description of a new low-cost technology for reducing penetration of toxic substances from waste dumps into environment, as well as technology of reclamation of closed objects. The research persisted for several years, which allowed for repeating and confirming the achieved results. The results were given to the owners and leaseholders of areas with orchards and crop and horticultural plantations in the vicinity of burial grounds. The cooperation with owners of horticultural and agricultural farms neighboring the pesticide burial grounds, including the owner of Cultivation and Orchard Farm "WIBO", Mr. Witold Kulesza from Stelmachów, it was possible to set up the sorption barrier made of compost from Sokółka around the real object. The barrier was installed in Folwarki Tylwickie, where active burial ground consisting of three tanks built of concrete well circles isolated with roofing paper and cement as well as non-hermetic ground pits, still works. It is from the 1st category of threat. Analyzed tanks are of the following capacities: 1.26, 1.8, and 1.8 m^3. According to the cataloguing in 2009, the burial ground holds almost 2.3 Mg of outdated pesticides, while estimated weight of dangerous pesticide wastes to neutralize amounts to 43.3 Mg. The burial ground functions at the depth of about 1.6 m, while underground waters can be found at 2 m level. It is situated among undergrowth directly in the cultivated field and near protected areas and compact building. Stabilized sewage sludge and its composts applied in the study can be also used to fill the ground pits formed after burial grounds close down, and to reclamation of the surrounding area. Due to their beneficial sorption properties, it would additionally prevent surrounding environment from migration of pesticide remains and would make costs of the object liquidation lower.

In future, the research can find its application to reduce the spread of other toxic substances within the environment during an ecological disaster or a poison leakage.

Acknowledgments

Financial support for this research was provided by Ministry of Science and Higher Education within the project N N304 274840.

REFERENCES

1. 8th International HCH and Pesticides Forum Sofia, Bułgaria 26-28.05.2005.
2. Atkins P.W. 2006. Physical chemistry, 8th ed. Oxford University Press.
3. Biziuk M., Kosikowska M. 2009. Methods of determination of pesticides residues – a review Ecological Chemistry and Engineering S, 16 (32), 207–220.
4. Feng K., Yu B.Y., Ge D.M., Wong M.H., Wang X.C., Cao Z.H. 2003. Organo-chlorine pesticide (DDT and HCH) residues in the Taihu Lake Region and its movement in soil-water system – I. Field survey of DDT and HCH residues in ecosystem of the region. Chemosphere, 50(6), 683–687.
5. Ignatowicz K. 2007a. Content of harmful substances in natural environment elements near pesticide tomb. Oceanological and Hydrobiological Studies, Vol. XXXVI, 3, 163–170.
6. Ignatowicz K. 2007b. Evaluation of pesticide remains and heavy metals concentrations near burial grounds. Polish Journal of Environmental Studies, 16 (3B), 177–181.

7. Ignatowicz K. 2007c. Określenie wielkości migracji pozostałości pestycydów i metali ciężkich z mogilników do wód naturalnych. Rocznik Ochrona Środowiska, 9, 281–293.

8. Ignatowicz K. 2008a. Sorption process for migration reduction of pesticides from graveyards. Archives of Environmental Protection, 34(3), 143–149.

9. Ignatowicz K. 2008b. Zastosowanie sorpcji na odpadowych materiałach naturalnych do ograniczenia migracji pestycydów z mogilników. Przemysł Chemiczny.

10. Ignatowicz K. 2008c. Zanieczyszczenie wód podziemnych i powierzchniowych środkami ochrony roślin w pobliżu mogilnika pestycydowego. Gaz, Woda i Technika Sanitarna, 9, 13–15.

11. Ignatowicz K. 2009a. Selection of sorbent for removing pesticides during water treatment. Journal of Hazardous Materials, 169, 953–957.

12. Ignatowicz K. 2009b. Sewage sludge compost applying for migration reduction of pesticide from graveyards. Polish Journal of Environmental Studies. Series of Monographs, 6, 49–54.

13. Ignatowicz K. 2011a. A mass transfer model fort he adsorption of pesticide on coconut shell based activated carbon. International Journal of Heat and Mass Transfer, 54, 4931–4938.

14. Ignatowicz K. 2011b. Application of Sokółka compost to pratect pesticide graveyard area. Rocznik Ochrona Środowiska, 13, 1201–1221.

15. Li W., H. Yang, Q. Gao, H. Pan, H. Yang, 2012. Residues of Organochlorine Pesticides in Water and Suspended Particulate Matter from Xiangshan Bay, East China Sea. Bull Environ Contam Toxicol 89, 811–815.

16. Siłowicki A. 2003. Iwentaryzacja odpadów środków ochrony roślin Mat. IOR, projekt w ramach Projektu GEF w Polsce.

17. Spadotto C.A., Hornsby A.G., 2003. Soil sorption of acidic pesticides: modeling pH effects. Journal Environmental Quality, 32, 949–995.

18. Wiater J., 2007. Assessment of manurial value of composts made from municipal wastes. Pol. J. Environ. Stud., 16(2A), 707–710.

7

SOIL POLLUTION OF SELECTED PAHS AS A FACTOR AFFECTING THE PROPERTIES OF HUMIC ACIDS

Bożena Dębska[1], Magdalena Banach-Szott[2], Ewa Rosa[3]

[1] Department of Environmental Chemistry, University of Technology and Life Sciences in Bydgoszcz, 6 Bernardyńska Str., 85-029 Bydgoszcz, Poland, e-mail: debska@utp.edu.pl

[2] Department of Environmental Chemistry, University of Technology and Life Sciences in Bydgoszcz, 6 Bernardyńska Str., 85-029 Bydgoszcz, Poland, e-mail: mbanach@utp.edu.pl

[3] Department of Environmental Chemistry, University of Technology and Life Sciences in Bydgoszcz, 6 Bernardyńska Str., 85-029 Bydgoszcz, Poland, e-mail: ewa.rosa@o2.pl

ABSTRACT

It is well-known that the properties of humus soil substances (including humic acids) are soil-type-specific. However, one shall consider the fact that properties of organic matter of soil can be modified by farming system (crop rotation, fertilisation) as well as other external factors, including pollutants; PAHs. The objective of the paper is to determine the effect of a single-time pollution of soils with high rates of PAHs on the properties of humic acids. The research was performed with the samples of soils representative for the Kujawy and Pomorze Region (*Phaeozems, Luvisol, Haplic Arenosols, Fluvisols*). Soil samples were polluted with selected PAHs; fluorene, anthracene, pyrene and chrysene at the amount corresponding to 100 mg PAHs · kg^{-1}. Treatments, i.e., soils + PAHs, were incubated for 180 and 360 days at the temperature of 20–25 °C and at constant moisture of 50% of field water capacity. Humic acids were extracted from the soil samples prior to and after 180 and 360 days of incubation. The following analyses were performed for separating humic acids: elemental composition, UV-VIS and IR spectrophotometric analyses, susceptibility to oxidation. Results demonstrated that a single introduction of fluorene, anthracene, pyrene and chrysene at very high rates into soils affects the properties of humic acids. There was mostly recorded a decrease in coefficients of absorbance $A_{2/6}$ and $A_{4/6,}$ an increase in the parameter defining the susceptibility of humic acids to oxidation. There were also noted changes in the pattern of spectra in infrared and the values of the parameter defining the degree of internal oxidation of the humic acids molecules.

Keywords: soils, PAHs, humic acids, elemental composition, UV-VIS, FTIR.

INTRODUCTION

Polycyclic aromatic hydrocarbons (PAHs) are considered, due to a strong carcinogenic effect and mutagenic properties to represent the group of pollutions most hazardous to human health. Interestingly, soil is the component of the natural environment in which, finally, most hydrophobic organic pollution gets accumulated, including polycyclic aromatic hydrocarbons (PAHs). The PAHs presence in the environment is the source of infection and, as such, it poses a direct threat to all the living organisms in the contaminated area [14, 19]. As reported in the literature [10, 11, 16, 17, 20, 21, 23], an important parameter which determines the immobilization of PAHs by soils is the quantity and the quality of organic matter.

The main factor of organic matter is made up by humus substances being most common organic compounds found in the nature. The compounds participate in all the processes which occur in soil and affect its physical, chemical and biological properties. They are the main link in the sequestration of carbon and releasing CO_2 to the air, as well as the detoxication of the environment.

It is well-known that the properties of organic matter of soils (including humic acids) are specific for a given type of soil [5, 8, 9, 15, 18]. However, one shall note that the properties of organic matter of soils can be modified for example by the method of land cultivation (crop rotation, fertilisation) as well as by other external factors, including pollution by PAHs [4, 5, 8, 9, 15].

The aim of the paper is to determine the effect of a single pollution of soils with high rates of PAHs on the properties of humic acids.

MATERIAL AND METHODS

The research was performed drawing on the soils under agricultural use based on the soils sampled in the Kujawy and Pomorze Province, in the northern part of Poland. Their basic properties are listed in Table 1.

Soil samples were polluted with selected PAHs (fluorene, anthracene, pyrene and chrysene), at the amount corresponding to 100 mg PAHs · kg^{-1}. They were not polluted and PAHs-polluted soil samples were incubated for 180 and 360 days at the temperature of 20–25 °C and fixed moisture (50% of field water capacity). After a specific incubation time, entire samples were destroyed. The incubation was performed in 3 replications for each sampling date.

The properties of humic acids were determined for the samples of soils unpolluted with PAHs and 180 and 360 days after adding PAHs.

Extraction of humic acids

Humic acids (HAs) were extracted from soil samples prior to the incubation (additionally non-polluted with PAHs), and after 180 and 360 days of incubation. Humic acids were extracted and purified using the following procedure:

- decalcification (24 h) with 0.05 M HCl (1:10, *w/v*); after centrifugation the residue was washed with distilled water till neutral,
- extraction (24 h) of the remaining solid with 0.5 M NaOH (1:10, w/v), with occasional mixing, followed by centrifugation,
- precipitation (24 h) of humic acids from the resulting alkaline extract with 2 M HCl to pH=2 and centrifugation,
- purification of the resulting humic acids as follows: the humic acid residue was treated with a mixture of HCl/HF (950 cm^3 H_2O, 5 cm^3 concentrated HCl, 5 cm^3 concentrated HF) over a 24 h period, followed by centrifugation. This procedure was repeated three times. The humic acids residue was treated with distilled water until a zero reaction to chloride was achieved [7].

The preparations were lyophilised and powdered in agate mortar. Ash content in the humic acids preparations was lower than 2%.

Characteristics of humic acids

The separated humic acids were analysed for:

- elemental composition (Perkin Elmer 2400 CHN analyser, Table 2). The H/C, O/C atomic ratios and ω (internal oxidation degree) were calculated according to the formula:

$$\omega = (2O + 3N - H) : C,$$

where: O, N, H, C – content in atomic %.

- UV-VIS absorption spectra (Perkin Elmer UV-VIS Spectrometer, Lambda 20, Table 3). VIS spectra were obtained from 0.02 % humic acid solutions in 0.1 M NaOH and UV-spectra after fivefold dilution. Absorbance was measured at 280 nm (A_{280}), 465 nm (A_{465}) and 665 nm

Table 1. Soil types and basic properties of soil [4, 5]

Sample	Soil	TOC	Nt	TOC/Nt	pH in H_2O	Particle size distribution		
						2-0.05 mm	0.05-0.002 mm	<0.002 mm
		[g · kg^{-1}]				[%]		
Ph2	*Phaeozems*	25.37	1.91	13.28	5.9	67	20	13
Lu5	*Luvisol*	11.74	1.03	11.40	5.9	75	19	6
Hp8	*Haplic Arenosols*	5.72	0.42	13.62	6.4	93	4	3
Fl9	*Fluvisols*	34.02	2.69	12.65	6.4	55	29	16
Fl10	*Fluvisols*	8.17	0.63	12.97	6.3	82	12	6

Explanations: TOC – Total Organic Carbon, Nt – total nitrogen. TOC and Nt determined with the analyser Vario Max CNS (Elementar, Germany).

Table 2. Elemental composition (atomic %) of humic acids and the values of atomic ratio and the degree of internal oxidation

Sample	C	H	N	O	H/C	O/C	ω
	in atomic %						
Ph2 - 0	35.63	41.40	2.72	20.25	1.16	0.568	0.204
Ph2 - 180	35.36	41.40	2.72	20.52	1.17	0.580	0.220
Ph2 - 360	35.81	40.90	2.58	20.72	1.14	0.579	0.231
Lu5 - 0	35.93	40.93	2.56	20.59	1.14	0.573	0.220
Lu5 - 180	37.18	39.88	2.46	20.50	1.07	0.552	0.230
Lu5 - 360	36.33	41.28	2.41	19.98	1.14	0.550	0.163
Hp8 - 0	35.51	42.86	2.66	18.97	1.21	0.534	0.086
Hp8 - 180	35.23	42.61	2.61	19.55	1.21	0.555	0.122
Hp8 - 360	35.01	44.10	2.60	18.29	1.26	0.522	0.008
Fl9 - 0	33.99	42.62	2.70	20.69	1.25	0.609	0.202
Fl9 - 180	34.76	42.23	2.88	20.13	1.21	0.579	0.184
Fl9 - 360	34.32	42.99	2.62	20.07	1.25	0.585	0.146
Fl10 - 0	35.84	41.23	2.58	20.35	1.15	0.568	0.201
Fl10 - 180	35.50	41.78	2.45	20.27	1.18	0.571	0.172
Fl10 - 360	35.39	42.66	2.33	19.62	1.21	0.554	0.101

Explanations: 0, 180, 360 – stands days of incubation.

Table 3. Spectral properties of humic acids

Sample	A_{280}	A_{465}	A_{665}	$A_{2/4}$	$A_{2/6}$	$A_{4/6}$	ΔA_{665u}
Ph2 - 0	4.69	0.826	0.165	5.69	28.42	5.00	65.83 %
Ph2 - 180	5.37	1.052	0.221	5.11	24.29	4.75	71.88 %
Ph2 - 360	3.70	0.705	0.150	5.26	24.70	4.70	60.13 %
Lu5 - 0	5.00	1.06	0.236	4.73	21.16	4.47	67.52 %
Lu5 - 180	4.49	1.11	0.270	4.04	16.65	4.13	70.29 %
Lu5 - 360	4.10	0.839	0.195	4.89	21.05	4.30	70.52 %
Hp8 - 0	4.35	0.843	0.183	5.16	23.83	4.61	71.54 %
Hp8 - 180	3.78	0.931	0.216	4.05	17.52	4.32	74.71 %
Hp8 - 360	3.05	0.629	0.131	4.85	23.27	4.80	66.11 %
Fl9 - 0	3.87	0.598	0.099	6.47	38.95	6.02	74.40 %
Fl9 - 180	3.31	0.555	0.099	5.95	33.29	5.59	84.89 %
Fl9 - 360	2.68	0.408	0.071	6.58	37.75	5.75	74.93 %
Fl10 - 0	3.95	0.661	0.126	5.98	31.29	5.23	76.09 %
Fl10 - 180	4.78	0.871	0.178	5.49	26.88	4.89	78.23 %
Fl10 - 360	2.75	0.532	0.095	5.16	29.00	5.62	69.38 %

Explanations: 0, 180, 360 – stands days of incubation.

(A_{665}) and was used to calculate coefficient values:

$A_{2/4}$ – 280 nm and 465 nm absorbance ratio,
$A_{2/6}$ – 280 nm and 665 nm absorbance ratio,
$A_{4/6}$ – 465 nm and 665 nm absorbance ratio [12];

- susceptibility to oxidation with H_2O_2 by the measurement of HAs solution absorbance (0.02% HAs in 0.1 M NaOH and 1.5% H_2O_2, Table 3) at 665 nm. The susceptibility to oxidation was calculated from the following formula:

$$\Delta A_x = ((A_0 - A_u)/A_0)\cdot 100\ \%$$

where: x – wavelength (665 nm),
A_0 – initial absorbance (prior to adding H_2O_2),
A_u – absorbance after oxidation [8];

- infrared spectra (Perkin-Elmer FT-IR Spectrometer, Spectrum BX) over 400–4400 cm^{-1}

were obtained for HAs (3 mg) in KBr (800 mg); deconvolution was applied, with a filter generating the bands of $g = 4$ narrower, and using the process of smoothing, for which the length parameter was $l = 80\%$ [3].

The results were processed with the analysis of variance with the use of the Tukey test at the significance level of $\alpha = 0.05$, for the two-factor experiments without reps, applying the ANALWAR 5.2 statistics software (Author: Prof. Franciszek Rudnicki), with the following factor design: factor I – time; factor II – soil type. The results of the analysis of variance for parameters, for which a significant LSD value was recorded for factor I given in Table 4.

RESULTS AND DISCUSSION

One of the basic parameters characteristic for the structure of humic acids (HAs) is their elemental composition. The elemental composition of the humic acids analyzed is presented in atomic percentage (Table 2).

The content of carbon in the molecules of humic acids ranged from 34.32 to 37.18%, hydrogen from 39.88 to 44.10%, nitrogen from 2.33 to 2.88% and oxygen accounted for 18.29–20.72%. The basic factor differentiating the content of respective elements was the sampling site. Introducing selected PAHs (chrysene, pyrene, anthracene, fluorene) into soil samples only slightly modified the degree of internal oxidation (Tables 2 and 4).

Humic acids (HAs) from *Luvisol*, *Haplic Arenosols* and *Phaeozems* isolated from the soil samples after 180 days of incubation showed higher values of the degree of internal oxidation, as compared to the HAs of unpolluted soils. As for the HAs of *Fluvisols* (sample Fl9, Fl10) a reverse dependence was observed. After 360 days of incubation, only the humic acids of *Phaeozems* showed higher values of the degree of internal oxidation, as compared to the HAs of initial soil. With that in mind, one can assume that introducing chrysene, pyrene, anthracene and fluorene to soils modifies the properties of humic acids connected with their degree of oxidation, however the direction of changes depends on soil properties.

Spectrometric parameters in the UV-VIS range are important parameters used for characterising the properties of humic acids and evaluating the transformations of organic matter in the soils. Based on years' research some relationships between the values of absorbance at the wavelengths of 280, 465 and 665 nm and the structure of humic acids were demonstrated. The values of absorbance are used to calculate the ratios of absorbance $A_{2/4}$, $A_{2/6}$, and $A_{4/6}$. Coefficients $A_{2/4}$, $A_{2/6}$ and $A_{4/6}$ are important indices of organic materials humification advancement and the characteristics of the humus substances produced, as well as changes in the properties of humic acids which occur due to various anthropogenic factors [8, 9, 12]. Among the humic acids of the initial soils (additionally non-polluted with PAHs), the highest values of $A_{2/4}$, $A_{2/6}$ and $A_{4/6}$ were recorded for the HAs isolated from the *Fluvisols* with the texture of fine sand (sample no Fl9, Table 3).

The lowest values of the coefficients of absorbance were recorded for the HAs of the *Luvisol* (Lu5). Providing the soils with very high rates of PAHs (100 mg · kg^{-1}) decreased value $A_{2/4}$, $A_{2/6}$ and $A_{4/6}$ of humic acids isolated after 180 days of incubation (Table 4).

Table 4. Results of analysis of variance of selected parameters of humic acids

Factor		ω	$A_{2/4}$	$A_{2/6}$	$A_{4/6}$	ΔA_{665u}
Mean - Factor I	0	0.183	5.606	28.73	5.066	71.076
	180	0.186	4.928	23.726	4.736	76.000
	360	0.130	5.348	27.154	5.034	68.214
Mean - Factor II	Ph2	0.218	5.353	25.803	4.817	65.947
	Lu5	0.204	4.553	19.620	4.300	69.443
	Hp8	0.072	4.687	21.540	4.577	70.787
	Fl9	0.177	6.333	36.663	5.787	78.073
	Fl10	0.158	5.543	29.056	5.247	74.567
LSD - Factor I		0.053	0.512	2.025	0.324	5.340
LSD - Factor II		0.083	0.799	3.162	0.505	8.335

Explanations: 0, 180, 360 – stands days of incubation.

In the soils, between 180 and 360 days of incubation, there occur processes leading to the formation of humic acids with higher values $A_{2/4}$, $A_{2/6}$ and $A_{4/6}$, as compared to HAs after 180 days of incubation. Despite the increase in the values of the above ratios of absorbance, after 360 days of incubation, the HAs of the soils polluted with PAHs recorded, in general, still lower values $A_{2/4}$, $A_{2/6}$ and $A_{4/6}$, as compared to the HAs of non-polluted soils.

The structure of humic acids can be also indirectly inferred from the changes in the value of absorbances which occur as a result of the reaction of oxidation with hydrogen peroxide of humic acids [8]. According to Gonet [8], the aliphatic part of the HAs molecules showed to be more susceptible to oxidation, while the aromatic part is more resistant. The changes in the value of absorbance at 665 nm of humic acids which occur as a result of oxidation with H_2O_2 are given in Table 3. The values of parameter ΔA_{665u} (susceptibility to oxidation determined at the wavelength of 665 nm) recorded for the HAs of the initial soils (with no PAHs added) ranged from 65.83 (Ph2) to 76.09% (Fl10). Highest values of ΔA_{665u} were recorded for the humic acids isolated from the soil samples additionally polluted with PAHs after 180 days of incubation. For the HAs of *Phaeozems* (Ph2), *Haplic Areonosols* (Hp8) *and Fluvisols* (Fl10), the values of ΔA_{665u} after 360 days of incubation were significantly lower, as compared to the HAs of non-polluted soils. Humic acids of *Fluvisols* with the texture of fine sand (sample no Fl9) after 360 days of incubation showed value ΔA_{665u} similar to the HAs of non-polluted soil. On the other hand, the HAs of *Phaeozems* demonstrated the lowest values ΔA_{665u} after 360 days of incubation. Therefore, introduction of chrysene, pyrene, fluorene and anthracene to soils resulted in changes in the structure of humic acids, which must have increased the share of unsaturated aliphatic structures, which, in turn, decreased the resistance of the molecules of humic acids to oxidation [6, 8].

The changes in the structure of the molecules of humic acids obtained once PAHs were introduced into soils are confirmed by the spectra produced in the IR range. Spectra of HAs *Phaeozems* and *Fluvisols* (Fl9, Fl10) isolated from the samples of soils additionally polluted with high values of PAHs, after 180 days of incubation, as compared to the HAs of the original soil, demonstrated a higher band intensity of the frequency of about 1510 cm^{-1}, 1460 cm^{-1}, 1420 cm^{-1}, 1260 cm^{-1} and the peaks in the range 1130–1000 cm^{-1} (Figure 1, due to a similar pattern of HAs spectra, the present paper includes spectra for a single variant only).

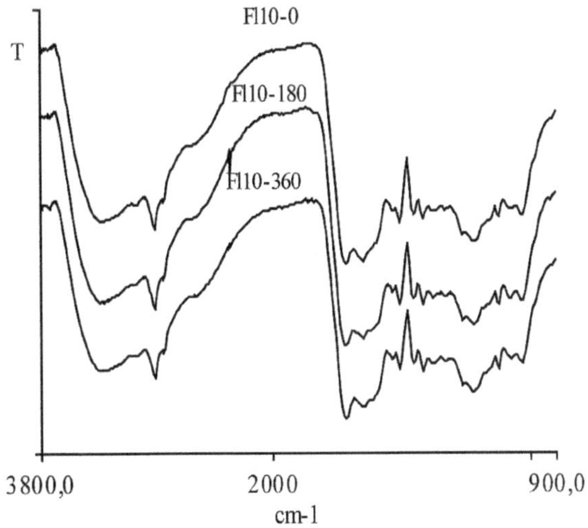

Figure 1. FT-IR of humic acids of *Fluvisol*

The band at 1520–1500 cm^{-1} is characteristic for aromatic compounds (C=C). The band in the range of 1460–1440 cm^{-1} corresponds to the deformation vibration of groups $-CH_3$ and $=CH_2$, the band 1420–1400 cm^{-1} points to the presence of groups O-C-O in esters and C=O in carboxyl acids, and OH groups. In the 1280–1200 cm^{-1} range there occur bands of C–O groups of phenols, carboxyl acids, esters and ethers, and the bands in the range 1130 cm^{-1} and 1080–1030 cm^{-1} point to the presence of polysaccharides. They can also be related to the presence of groups OCH_3 which occur in lignins and OH in alcohols [8, 3, 15, 18].

The intensity of the above absorption bands decreased from the 180 to 360th day of incubation. However, the humic acids isolated after 360 days of incubation, as compared to the HAs of the original soils, show a slightly higher share of simple aromatic structures (a higher intensity of the 1510 cm^{-1} band), as well as higher band in the range of 1460–1030 cm^{-1}. Spectra of HAs of *Luvisol* and *Haplic Arenosols* isolated from the samples of soils additionally polluted with high values of PAHs, after 180 and 360 days of incubation, as compared to the HAs of the original soil showed a similar pattern (Figure 2).

The spectrum pattern of the humic acids investigated confirms, on the one hand, the effect of PAHs on the properties of humic acids, and, on the other hand, points that the process of transformations of pollutants introduced into soil can differ and it also depends on soil properties.

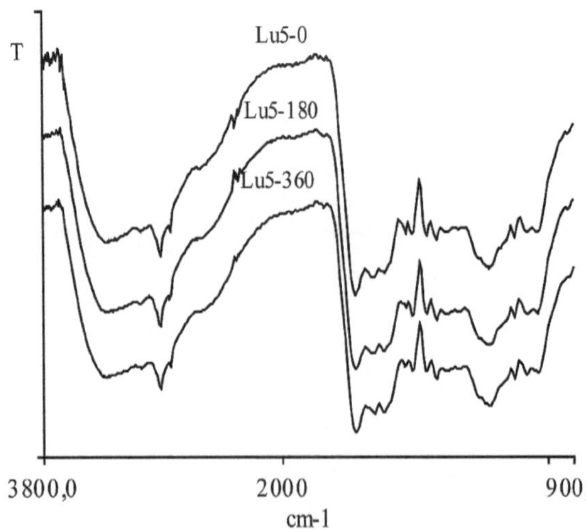

Figure 2. FT-IR of humic acids of *Luvisol*

The present results (elemental composition, spectrometric parameters in the UV-VIS and IR range then) coincide with earlier literature reports [1, 2, 11, 13, 17, 20, 22] which stressed a considerable share of organic matter, including humic acids, in the PAHs sorption processes. However, one shall also remember that the compounds formed in the process of PAHs decomposition can be bound by the fractions of organic matter (humic acids, fulvic acids, humins).

CONCLUSION

In the humic acids isolated from the samples of soils additionally polluted with PAHs at the dose 100 mg · kg^{-1}, as compared with the HAs of the non-polluted soils, the following were observed:

- a change in the parameter ω – the degree of internal oxidation;
- an increase in the intensity of absorption bands in the 1520–1000 cm^{-1} range, higher in all the HAs spectra isolated after 180 days than after 360 days. The band connected with an increase in the share of simple aromatic and aliphatic structures in the molecules of humic acids;
- an increase in the susceptibility to oxidation recorded for the humic acids isolated after 180 days of incubation.

The results point to the need of considering while investigating the properties of soils in the areas exposed to the effect of PAHs, the role of those compounds in the organic matter properties development.

Acknowledgements

The research has been made as part of N N310 3123 34 research project, financed by the Ministry of Science and Higher Education.

REFERENCES

1. Barančiková C., Gergelová Z. 1995. Soil parameters influencing of PCBs sorption. [In:] W.J. den Brink, R. Bosman, F. Arendt (Eds), Contaminated Soil'95. Kluwer Academic Publishers, Dordrecht/Boston/London, 357-358.
2. Chiou C.T. 1989. Theoretical considerations of the partition uptake of nonionic organic compounds by soil organic matter. [In:] B.L. Sawhney, K. Brown (Eds), Reactions and movement of organic chemicals in soils. SSA Special Publication, Madison, 22, 1-29.
3. Cocozza C., Miano T. 2002. Structural resolution of metal-humic acids interactions through deconvolution FT-IR spectroscopy. [In:] G. Davis, E.A. Ghabbour (Eds), Proc. 11th IHSS Meeting. IHSS, Boston, USA, 264-266.
4. Dębska B., Banach-Szott M., Rosa E., Drąg M. 2011. Effect of soil pollution with PAHs on organic matter properties. Pol. J. Soil Sci., 44(2): 97-103.
5. Dębska B., Drąg M., Tobiasova E. 2012. Effecf of post-harvest residue of maize, rapeseed, and sunflower on humic acids properties in various soils. Pol. J. Environ. Stud., 21(3): 603-613.
6. Drąg M., Dębska B., Dziamski A. 2007. Properties of humic substances od forest and meadow soil in the area of the Wierzchlas Reserve. Humic Subs. in Ecos. 7, 141-151.
7. Dziadowiec H., Gonet S.S. 1999. Methodological quide to the study of soil organic matter. PTG, Warszawa, 66 pp. (in Polish).
8. Gonet S.S. 1989. Properties of humic acids of soils of varied fertilisation. Dissertation, UTA Bydgoszcz, 55 pp. (in Polish).
9. Gonet S.S., Debska B. 1999. Properties of humic acids produced during decomposition of plant residues in soil. Rostlinna Vyroba, 45(10): 455-460.
10. Guo X., Luo L., Ma Y., Zhang S. 2010. Sorption of polycyclic aromatic hydrocarbons on particulate organic matters. J. Hazard. Mater., 173(1-3), 130-136.
11. Kim S.J., Kwon J.H. 2010. Determination of partition coefficients for selected PAHs between water and dissolved organic matter. Clean - Soil Air Water, 38(9): 797-802.
12. Kumada K. 1987. Chemistry of soil organic matter. Developments in Soil Science, Japan Sc. Soc. Press Tokyo, Elsevier, Amsterdam, 241 pp.

13. Maliszewska-Kordybach B. 1995. The persistence of carcinogenic micropollutant – pyrene – in soils of different organic matter content. Arch. Ochr. Środ., 2: 183-190.

14. Maliszewska-Kordybach B. 1999. Persistent organic contaminants in the environment; PAHs as a case study. [In:].C. Block J, V.V. Goncharuk, P Baveye. (Eds), Bioavalibility of organic xenobiotics in the environment. NATO ASI series, Kluver Academic Publischers, Dordrecht, 3-37.

15. Orlov D.S. 1986. Humus acids of soils. A.A. Balkema (Ed.), Rotterdam, 378 pp.

16. Pan B., Xing B.S., Liu W.X., Tao S., Lin X.M., Zhang X.M., Zhang Y.X., Xiao Y., Dai H.C., Yuan H.S. 2006. Distribution of sorbed phenanthrene and pyrene in different humic fractions of soils and importance of humin. Environ. Pollut., 143(1): 24-33.

17. Plaza C., Xing B., Fernandez J.M., Senesi N., Polo A. 2009. Binding of polycyclic aromatic hydrocarbons by humic acids formed during composting. Environ. Pollut., 157(1): 257-263.

18. Tan K.H. 1998. Principles of soil chemistry. Marcel Dekker Inc., New York, Basel, Hong Kong, 521 pp.

19. Wang W., Huang M., Kang Y., Wang H., Leung A., Cheung K., Wong M. 2011. Polycyclic aromatic hydrocarbons (PAHs) in urban surface dust of Guangzhou, China: Status, sources and human health risk assessment. Sci. Total Environ., 409: 4519-4527.

20. Weissenfelts W.D., Klewer H-J., Langhoff J. 1992. Adsorption of polycyclic aromatic hydrocarbons (PAHs) by soil particles: influence on biodegradability and biotoxicity. Appl. Microbiol. Biot., 36(5): 689-696.

21. Yang Y., Zhang N., Xue M., Tao S. 2010. Impact of soil organic matter on the distribution of polycyclic aromatic hydrocarbons (PAHs) in soils. Environ. Pollut., 158(6): 2170-2174.

22. Zaccone C., Gallipoli A., Cocozza C., Trevisan M., Miano T.M. 2009. Distribution patterns of selected PAHs in bulk peat and corresponding humic acids from Swiss ombrotrophic bog profile. Plant Soil, 315(1-2): 35-45.

23. Zhang Y., Li F., Wang, J. 2011. Determination of retardation effect of SOM on aqueous leaching of polycyclic aromatic hydrocarbons using confocal laser scanning microscope. Int. J. Environ. Res., 5(4): 999-1008.

8

EVALUATION OF THE CONTENT OF HEAVY METALS IN FLUVISOLS OF FLOODPLAIN AREA DEPENDING ON THE TYPE OF LAND USE

Mirosław Kobierski[1]

[1] Department of Soil Science and Soil Protection, University of Technology and Life Sciences in Bydgoszcz, Bernardyńska 6/8, 85-029 Bydgoszcz, Poland, e-mail: kobierski@utp.edu.pl

ABSTRACT

The aim of the research was the evaluation of the potential contamination with heavy metals in Fluvisols, used as grasslands and arable soils of Vistula River floodplain in the area of the Chełmiński and Nadwiślański Complex of Landscape Parks. The indicators proposed by Håkanson allow to evaluate the potential ecological risk of the contamination with heavy metals associated with the accumulation of one metal or a combination of multiple metals. The mean total content of Cd, Pb, Ni Cu, Zn, Mn, as well as Fe in Fluvisols at the depth of 120–150 cm was assumed as the content of the local geochemical background and it was: 1.0 $mg \cdot kg^{-1}$, 22.8 $mg \cdot kg^{-1}$, 26.9 $mg \cdot kg^{-1}$, 1.4 $mg \cdot kg^{-1}$, 60.8 $mg \cdot kg^{-1}$, 591 $mg \cdot kg^{-1}$, and 17.6 $g \cdot kg^{-1}$, respectively. The values of the indicators such as contamination factor (CF), enrichment factor (EF) revealed higher levels of the accumulation of heavy metals in the soils of grasslands, which shows that the method of their use has a significant effect on the total metal content. Contamination with heavy metals in the surface layer of the investigated Fluvisols was found, and in terms of the content with cadmium a moderate and considerable potential ecological risk was reported. Due to the fact that no unfavourable effect of trace elements on the riverside environment was proved and that the floodplain areas are under agricultural use, to evaluate the contamination with metals, the limit values for the soils of agricultural land were assumed as stipulated in the Regulation of Minister of the Environment of September 9, 2002. According to that criterion, the soils studied do not qualify as contaminated with metals. Only in one of the soil sampling points the total content of zinc was higher than the one determined as the maximum for agricultural land soils, namely 350 $mg \cdot kg^{-1}$. A significantly positive correlation was noted between the content of C org and the total content of metals as well as very numerous interactions between metals. The results of cluster analysis confirm that the method of use determines the concentration of metals in the surface layer of the Fluvisols studied.

Keywords: heavy metals, floodplain, potential ecological risk index, enrichment factor

INTRODUCTION

The areas within the Chełmiński and Nadwiślański Complex of Landscape Parks undergo annual flood events. Every year the Vistula River is affected by flooding owing to the number of tributaries and shallow riverbeds that involve overflowing during spring floods or seldom periods of heavy rain in summer. The natural floodplain is limited with the floodbank and it makes up wasteland grown with riparian woodlands and it remains under agricultural use as grasslands and arable land. In many places the right bank of the Vistula River is regularly eroded and the amount of the deposited sediments on the floodplain depends on the reach of the flood as well as local landscape-related conditions. The sedimentation and erosion processes observed in the river channel take a form of a loss of fine sediments at the foot of the riverbanks during strong currents, especially spring floods. More coarse sediments, namely gravels and sand are transported by the river and deposited within the riverbed. However, during flood recessions, the fine sediment (silt and clay fraction) that is transported along the river is accumulated on the floodplain [Kordowski 2003]. The fluvial deposits, rich in humic and mineral colloids, act as a collector of water pollution in the river catchment, and as such they can constitute an essential source of chemical substances [Wyżga, Ciszewski 2010, Skorbiłowicz, Samborska 2014]. It is difficult to define unambiguously how soil properties change during overflowing in the periods of flooding as well as to what extent they affect the content of metals [Macklin, Klimek 1992, Singh et al. 2005]. An increased total content of metals in Fluvisols of floodplains comes mostly from the properties of sedimentation material but also from agricultural use [Kobierski, Piotrowska 2010]. Anthropogenic heavy metals deposited into soils as a result of human activities are more mobile than lithogenic and pedogenic metals the origins of which are difficult to distinguish [Kabata-Pendias 2004]. The accumulation of metals in soils increases with the decreasing particle size but the main factors controlling the behaviour of trace metals are organic matter contents, iron and manganese hydroxides, pH and redox potential [Harmsen 2007, Van Gestel 2008]. Physical and chemical properties of Fluvisols can reflect the processes which occur in riverside environment as well as give a possibility of interpreting the changes which occur in the entire catchment since they are a resultant of geological, pedogenic and anthropogenic factors [Saint-Laurent et al. 2010].

The aim of the study was to characterize the Fluvisols of the floodplain of the Grudziądz Basin on the evaluation of the total content of Cd, Pb, Ni Cu, Zn, Mn, and Fe depending on the type of land use.

MATERIALS AND METHOD

The research covered the floodplain areas between the Vistula riverbed and the flood embankment within the Grudziądz Basin (the Lower Vistula River). The riverside environment was evaluated in 12 study areas (zones). The samples from Fluvisol profiles numbered I.19, I.24, I.26, I.28, I.30, I.32 were taken from grasslands, while samples II.20, II.21, II.23, II.25, II.29, II.31 – from arable land. For the purpose of this study, soil samples were taken from layers: 0–30 cm, 60–90 cm, 120–150 cm, in which the basic properties as well as the total content of Cd, Pb, Ni, Cu, Mn and Fe were determined. The soil analyses were performed according to the commonly applied methods:

- texture, using the areometric method [PN-ISO-11277: 2005],
- pH in 1 mol·dm^{-3} KCl solution (1:2,5 soil-solution ratio), using the potentiometric method [ISO 10390: 2005].

The content of total organic carbon was assayed with analyser vario Max CN Elementar Analysensysteme GmbH (Hanau, Germany). The total content of metals was assayed after the mineralization in the mixture of acids HF and $HClO_4$ [Crock, Severson 1980]. The precision of this procedure was confirmed with reference materials TILL-3 for metals determination. The concentration of metals was defined using atomic absorption spectrometry (AAS, Philips 9100, Cambridge, UK). Soil properties were treated with standard statistics and statistical tests (ANOVA and the Tukey test). Pearson's correlation analysis was also done between the soil properties and total content of metals variables. The statistical analyses were made using Statistica 7.0 (StatSoft Inc, Tulsa, USA).

Potential ecological risk index

The potential ecological risk index *RI* is used to assess the ecological risk associated with the accumulation of one metal or a combination of multiple metals in surface soils or sediments, which was proposed by [Håkanson 1980]. The value of *RI* can be calculated with the following formulas:

$$C^i_f = C^i / C^i_b$$

$$C_{deg} = \Sigma\, C^i_f$$

$$E^i_r = T^i \times C^i_f$$

$$RI = \Sigma\, E^i_r$$

where: C^i_f – the contamination factor;
C^i – the content of metal in topsoil;
C^i_b – the background concentration of metal;
C_{deg} – contamination degree;
E^i_r – the potential risk of a single metal;
T^i – the toxic-response factor for a given heavy metal;
RI – the sum of potential risk of asingle metal caused by the overall contamination.

In order to unify the assessment results, the local background contents of heavy metals were used as the reference values for calculating the indices. As the content of the geochemical background for respective elements their mean total content in the 120–150 cm layer was assumed. The sum of the value of C^i_f for respective metals is expressed by the degree of contamination (C_{deg}) of the ecosystem (Table 1).

Table 1. Contamination factor C^i_f and degree of contamination categories C_{deg}

C^i_f classes	C^i_f and C_{deg} terminologies	C_{deg} classes
$C^i_f < 1$	Low contamination C^i_f and C_{deg}	$C_{deg} < 8$
$1 \le C^i_f < 3$	Moderate C^i_f and C_{deg}	$8 \le C_{deg} < 16$
$3 \le C^i_f < 6$	Considerable C^i_f and C_{deg}	$16 \le C_{deg} < 32$
$C^i_f \ge 6$	Very high C^i_f and C_{deg}	$C_{deg} \ge 32$

Based on the Håkanson [1980] approach, the toxic-response factors for Zn is 1; Cu – 2; Pb, Cu, Ni – 5; Cd – 30. Potential ecological risk index (*RI*) is a methodology developed by Håkanson, who proposed five categories of E^i_r, and four categories of *RI*.

The following terminologies was used to describe the risk factor: E^i_r<40, low potential ecological risk; 40≤E^i_r<80, moderate potential ecological risk; 80 E^i_r<160, considerable potential ecological risk; 160≤E^i_r<320, high potential ecological risk; and E^i_r ≥320, very high ecological risk. The potential ecological risk index (*RI*) was defined as the sum of the risk factors: *RI*<150, low ecological risk; 150≤*RI*<300, moderate ecological risk; 300≤*RI*<600, considerable ecological risk; and *RI*>600, very high ecological risk.

Enrichment factor

A common approach to estimating the human activity impact on soils is to calculate a normalized enrichment factor (*EF*) for metal concentrations above background levels. The values of the enrichment factor (*EF*) were calculated according to the formula:

$$EF = [C_n/C_{nFe}]/[B_n/B_{nFe}]$$

where: C_n – total content of metal;
C_{nFe} – total content of Fe as the reference element;
B_n – content of metal for the geochemical background;
B_{nFe} – content of Fe in geochemical background [Martin, Meybeck 1979].

Such elements as Ti, Al, Li, Zr as well as Fe get slightly anthropogenically accumulated and they do not participate actively in geochemical cycles. Neither do they have substantial anthropogenic sources. Based on *EF* value the enrichment categories were determined: <2 – deficiently to minimal enrichment, 2–5 moderate enrichment, 5–20 – considerable enrichment, 20–40 very high enrichment, >40 extremely high enrichment [Sutherland et al. 2000]. For the calculations it was assumed that the content of metals at the depth of 120–150 cm corresponds to the content of the local geochemical background.

RESULTS AND DISCUSSION

A varied texture (from medium sand to clay loam) as well as the results of the analyses of basic properties were characteristic for Fluvisols (Table 2, 3). The content of clay fraction ranged from 0 to 31%, and at the depth of 120–150 cm of a few soil profiles sand and sandy loam were found. The reaction was neutral and alkaline and the average content of organic carbon in the plough horizon of arable soils was 12.4 $g \cdot kg^{-1}$ and it was twice as low than the one identified in the 0–30 cm layer of grasslands. In the soils of permanent grasslands the content of organic matter increases naturally. Hoffmann et al. [2009] point to a significant influence of the depositional environment and sedimentary facies on the total organic carbon content (TOC). They showed that TOC increases with increasing clay fraction content. The soil (0–20 cm depth) sampled in a frequent flood zone demonstrates a lower content of organic carbon [Gervais-Beaulac et al. 2013]. The loss of biomass (organic matter) during the flood recession results in a net loss of organic carbon to the subsurface soils. Successive flooding triggers Fluvisols impoverishment and changes the pedogenic processes and soil development.

The total content of cadmium, lead, nickel, copper, zinc, manganese and iron in layer 0–30 cm in the soils ranged from 56.1 to 384 mg $Zn \cdot kg^{-1}$, from 9.9 to 41.5 mg $Cu \cdot kg^{-1}$, from 418 to 1455 mg $Mn \cdot kg^{-1}$, from 17.2 to 58.8 mg $Pb \cdot kg^{-1}$, from 22.2 to 49.9 mg $Ni \cdot kg^{-1}$ as well as from 1.22 to 3.23 g $Fe \cdot kg^{-1}$ (Table 4, 5). A significantly higher total content of cadmium, lead, copper and zinc was recorded in the surface layer of grassland soils, as compared with the content of those metals in arable soils (Table 6).

According to the Regulation of Minister of the Environment of September 9, 2002, the total content of Cd, Pb, Ni, Cu and Zn found in the

Table 2. Basic properties of grassland soils

Profile No.	Layer	pH	Corg	Sand	Silt	Clay fraction	Texture acc.	
							PTG*	USDA
		1 M KCl	g·kg⁻¹	%			2008	
I.19	1	7.03	21.1	72	19	9	gpdr	FSL
	2	7.42	11.1	56	29	15	gl	SL
	3	7.48	8.7	55	31	14	gl	SL
I.24	1	7.11	30.4	28	48	24	gz	L
	2	7.30	10.9	27	52	21	pyi	SiL
	3	7.31	9.3	30	51	19	pyi	SiL
I.26	1	6.93	28.9	49	35	16	gz	L
	2	7.29	9.5	22	59	19	gl	SL
	3	7.41	5.7	32	55	13	pyi	SiL
I.28	1	6.81	39.1	39	47	14	gz	L
	2	7.31	6.9	35	48	17	gz	L
	3	6.76	4.5	22	53	25	pyi	SiL
I.30	1	6.67	34.7	20	53	27	pyi	SiL
	2	7.08	6.5	38	48	14	gz	L
	3	7.23	1.9	70	24	6	gpdr	FSL
I.32	1	6.68	16.2	55	31	14	gl	SL
	2	7.06	16.1	20	52	28	gi	L
	3	7.20	9.9	17	62	21	pyi	SiL

Explanation: 1 – 0–30 cm, 2 – 60–90 cm, 3 – 120–150 cm
* Textural classes are based on the Polish System of Soil Classification [PTG, 2008].
gpdr/FSL – fine sandy loam, gl/SL – sandy loam, gz/L – loam, gi/CL – clay loam, pyi/SiL – silt loam

Table 3. Basic properties of arable soils

Profile No.	Layer	pH KCl	Corg	Sand	Silt	Clay fraction	Texture acc.	
							PTG	USDA
		1 M KCl	g·kg⁻¹	%			2008	
II.20	1	7.32	18.2	30	56	14	pyi	SiL
	2	7.33	9.0	22	58	20	pyi	SiL
	3	7.66	6.9	56	36	8	gl	SL
II.21	1	7.20	11.4	28	52	20	pyi	SiL
	2	7.57	4.5	47	40	13	gz	L
	3	7.35	5.2	32	52	16	pyi	SiL
II.23	1	7.40	8.4	62	28	10	gl	SL
	2	7.30	2.9	26	57	17	pyi	SiL
	3	7.67	3.5	49	45	6	gpdr	FSL
II.25	1	7.46	8.4	39	46	15	gz	L
	2	7.61	6.7	39	50	11	gz	L
	3	7.55	0.7	95	5	0	plśr	MS
II.29	1	6.79	13.7	51	35	14	gz	L
	2	6.64	3.7	53	36	11	gl	SL
	3	6.67	3.6	21	50	29	gi	CL
II.31	1	7.18	14.2	27	42	31	gi	CL
	2	7.43	3.8	55	33	12	gl	SL
	3	7.45	1.2	85	10	5	pgdr	FSL

Explanation: see Table 2.
plśr/MS – medium sand, pgdr/LFS – loamy fine sand

Table 4. Total metal content of grassland soils

Profile No.	Layer	Cd	Pb	Ni	Cu	Zn	Mn	Fe
		mg·kg^{-1}						g·kg^{-1}
I.19	1	^{A}**1.8**	22.3	18.9	10.3	^{A}**159**	418	12.9
	2	^{A}**1.4**	36.5	22.9	9.6	49.0	715	18.3
	3	^{A}**1.2**	39.3	21.2	7.2	79.9	538	14.4
I.24	1	^{A}**2.2**	43.8	^{A}**42.5**	29.8	^{A}**212**	1455	32.3
	2	^{A}**1.1**	25.1	37.7	19.1	82.5	945	27.2
	3	0.9	21.8	25.4	16.5	69.5	549	23.2
I.26	1	^{A}**1.1**	36.2	27.8	22.4	^{A}**155**	831	20.1
	2	^{A}**1.1**	25.8	^{A}**42.8**	21.6	82.0	1188	30.9
	3	1.0	20.7	32.1	16.9	62.7	790	23.4
I.28	1	^{A}**2.9**	^{A}**58.6**	^{A}**44.1**	^{A}39.8	$^{A|B}$**384**	1177	24.6
	2	^{A}**1.3**	24.9	^{A}**35.1**	18.0	70.6	756	20.9
	3	^{A}**1.2**	21.8	^{A}**43.1**	16.5	80.4	1083	30.8
I.30	1	^{A}**3.3**	^{A}**58.8**	^{A}**49.9**	^{A}41.5	^{A}**238**	1085	26.9
	2	^{A}**1.3**	22.8	28.7	10.7	55.2	568	15.3
	3	^{A}**1.1**	18.0	19.6	2.6	33.0	285	7.7
I.32	1	^{A}**3.1**	^{A}**56.0**	^{A}**49.9**	^{A}36.8	^{A}**277**	1035	28.1
	2	^{A}**1.7**	37.4	^{A}**46.7**	24.6	^{A}**130**	1232	29.4
	3	^{A}**2.1**	37.3	^{A}**36.9**	19.4	^{A}**137**	925	21.4

Explanation: 1 – 0–30 cm, 2 – 60–90 cm, 3 – 120–150 cm
Above the limit for soils: protected by law – **A**, agricultural land – **B**

Table 5. Total metal content of arable soils

Profile No.	Layer	Cd	Pb	Ni	Cu	Zn	Mn	Fe
		mg·kg^{-1}						g·kg^{-1}
II.20	1	^{A}**2.0**	39.5	^{A}**36.2**	19.7	^{A}**156**	1071	26.7
	2	^{A}**1.1**	35.9	^{A}**39.2**	18.5	96.4	894	27.3
	3	^{A}**1.1**	24.2	16.0	8.0	42.7	424	14.0
II.21	1	0.6	24.3	31.0	20.1	56.1	626	21.1
	2	0.6	17.3	24.7	10.9	47.0	474	17.7
	3	0.6	19.6	32.4	14.4	56.9	801	23.2
II.23	1	0.8	17.2	22.2	9.9	68.8	432	12.2
	2	0.7	21.2	^{A}**36.9**	17.5	60.8	875	25.4
	3	0.5	16.7	24.6	8.7	45.4	468	14.0
II.25	1	1.0	20.1	28.6	12.9	58.9	654	19.4
	2	1.0	20.2	31.9	12.5	51.4	662	20.3
	3	0.7	9.3	7.4	1.9	6.2	122	3.4
II.29	1	^{A}**1.4**	26.8	27.6	16.0	89.1	510	12.6
	2	^{A}**1.1**	18.9	25.5	6.9	47.6	332	13.1
	3	1.0	28.9	^{A}**49.4**	23.9	92.5	876	30.2
II.31	1	^{A}**1.8**	37.7	^{A}**37.1**	22.1	^{A}**133.5**	832	22.4
	2	^{A}**1.4**	20.7	25.1	10.7	42.2	417	11.1
	3	^{A}**1.2**	15.5	14.5	1.1	22.9	231	5.0

Explanation: 1 – 0–30 cm, 2 – 60–90 cm, 3 – 120–150 cm
Above the limit for soils: protected by law – **A**

Table 6. Results of statistical analysis; mean total content of metals (for Anova, the Tukey test)

Metal	Layer	Significant level (p)	I grasslands	II arable soils
			mean content	
Cd	1	0.02	2.4	1.3
	2	–	1.3	1.0
Pb	1	0.03	45.9	27.6
	2	–	28.7	22.4
Ni	1	–	38.8	30.4
	2	–	35.6	30.5
Cu	1	0.03	30.1	16.8
	2	–	17.3	12.8
Zn	1	0.004	237	93.7
	2	–	78.2	57.6
Mn	1	–	1000	687
	2	–	901	609
Fe	1	–	24.1	19.1
	2	–	23.7	19.1

Explanation: 1 – 0–30 cm, 2 – 60–90 cm

0-30 cm layer of the soils of the areas protected by law (group A), should not exceed 1.0 mg·kg^{-1}, 50.0 mg·kg^{-1}, 35.0 mg·kg^{-1}, 30.0 mg·kg^{-1}, 100.0 mg·kg^{-1} respectively [Dz. U. No 165, item 1359]. In all the soil samples of the surface layer of the grasslands as well as three arable soils a higher total content of Cd than the maximum for the soils of group A was recorded. In the 0–30 cm layer of grasslands the contents of Pb and Cu were higher than the maximum in three samples; Ni in four samples, and Zn in all the samples. In the surface layer of arable soils the content of Cd higher than the maximum was noted in 3 samples and Ni_t and Zn – in two samples. Due to the fact that no unfavourable effect of trace elements on the riverside environment was shown as well as that the floodplain areas are under agricultural use, to evaluate the contamination with metals, the references for the soil of agricultural land (group B) were assumed. According to that criterion the soils do not qualify as contaminated with Cd, Pb, Cu and Zn. Only in one of the soil sampling points (I.28) the total content of zinc was higher than the one defined as the maximum for the soils of agricultural land, namely 350 mg·kg^{-1}. The mean total content of Cd, Pb, Ni Cu, Zn, Mn, as well as Fe in the Fluvisols studied at the depth of 120–150 cm was assumed as the content of local geochemical background and it was 1.0 mg·kg^{-1}, 22.8 mg·kg^{-1}, 26.9 mg·kg^{-1}, 11,4 mg·kg^{-1}, 60.8 mg·kg^{-1}, 591 mg·kg^{-1}, and 17.6 g·kg^{-1} respectively (Table 7).

Table 7. Basic statistics data of total metal content at the depth of 120–150 cm

Metal	Min.	Max.	Mean	SD	CV
	mg·kg^{-1}				%
Cd	0.5	2.1	1.0	0.41	41.0
Pb	9.3	39.3	22.7	8.71	38.4
Ni	7.4	79.4	26.9	12.33	45.8
Cu	1.1	23.9	11.4	7.52	66.0
Zn	6.2	137	60.7	34.91	57.5
Mn	122	1083	591	302.9	51.3
Fe	3.4	30.8	17.6	9.24	52.5

Explanation: SD – standard deviation, CV – coefficient of variation

Martin and Whitfield [1983] report on the mean total content of metals in soils globally being: Cd – 0.35 mg·kg^{-1}, Pb – 35 mg·kg^{-1}, Ni – 50 mg·kg^{-1}, Cu – 30 mg·kg^{-1}, Zn – 90 mg·kg^{-1}, Mn – 1000 mg·kg^{-1} as well as Fe 40 g·kg^{-1}. While Kabata-Pendias and Pendias [2001] claim that the mean total content of metals in Fluvisols of Poland is: Cd – 0.3 mg·kg^{-1}, Pb – 39 mg·kg^{-1}, Ni – 9 mg·kg^{-1}, Cu – 22 mg·kg^{-1}, Zn – 84.5 mg·kg^{-1}, Mn – 1085 mg·kg^{-1}.

Value C_{deg} pointing to moderate contamination was recorded in four samples of soil of the surface layer of pastureland, and in study zone no. I.28 a considerable contamination (C_{deg}>16) (Table 8) was noted, whereas the value of the degree of contamination C_{deg} demonstrated that in the samples of the 0–30 cm layer of two research areas of arable soils (II.20, II.31) moderate contamination was identified (Table 9).

The indices E^i_r and *RI* suggested by Håkanson [1980] facilitating the evaluation of the potential ecological risk of contamination show a moderate potential ecological risk for Cd in three samples of the surface layer of arable soils (II.20, II.29) as well as two grasslands (I.19, I.24), while in three others (I.28, I.30, II.31) a considerable potential ecological risk (Table 8, 9) was recorded. The value of potential ecological risk index *RI,* associated with the accumulation of a combination of multiple metals, was 150, which points to a low ecological risk. In three samples taken from the 0–30 cm layer of grasslands (I.28, I.30, I.32) the value of *RI* was higher than 120. The results of *EF* analysis indicate that the Fluvisols showed low to moderate enrichment with metals. The enrichment factor of respective metals assumed slightly lower values for arable soils, as compared with the soils of grasslands. Values *EF* calculated

Table 8. Values of indices for grasslands

Profile No.	Layer	Cd_t			Pb_t			Ni_t			Cu_t			Zn_t			C_{deg}	RI
		EF	C_f^i	E_r^i	EF	C_f^i	E_r^i	EF	C_f^i	E_r^i	EF	C_f^i	E_r^i	EF	C_f^i	E_r^i		
I.19	1	2.4	1.8	54	1.3	1.0	5.0	1.0	0.7	3.5	1.2	0.9	1.8	3.6	2.6	2.6	7.0	66.9
	2	1.3	1.4	42	1.5	1.6	8.0	0.8	0.8	4.0	0.8	0.8	1.6	0.8	0.8	0.8	5.4	56.4
I.24	1	1.2	2.2	66	1.0	1.9	9.5	0.9	1.6	8.0	1.4	2.6	5.2	1.9	3.5	3.5	11.8	92.2
	2	0.7	1.1	33	0.7	1.1	5.5	0.9	1.4	7.0	1.1	1.7	3.4	0.9	1.4	1.4	6.7	50.3
I.26	1	1.0	1.1	33	1.4	1.6	8.0	0.9	1.0	5.0	1.7	2.0	4.0	2.2	2.5	2.5	8.2	52.5
	2	0.6	1.1	33	0.6	1.1	5.5	0.9	1.6	8.0	1.1	1.9	3.8	0.8	1.3	1.3	7.0	51.6
I.28	1	2.1	2.9	87	1.8	2.6	13.0	1.2	1.6	8.0	2.5	3.5	7.0	4.5	6.3	6.3	16.9	121.3
	2	1.1	1.3	39	0.9	1.1	5.5	1.1	1.3	6.5	1.3	1.6	3.2	1.0	1.2	1.2	6.5	55.4
I.30	1	2.1	3.3	99	1.7	2.6	13.0	1.2	1.9	9.5	2.4	3.6	7.2	2.6	3.9	3.9	15.3	132.6
	2	1.5	1.3	39	1.2	1.0	5.0	1.2	1.1	5.5	1.1	0.9	1.8	1.0	0.9	0.9	5.2	52.2
I.32	1	1.9	3.1	93	1.5	2.5	12.5	1.2	1.9	9.5	2.0	3.2	6.4	2.9	4.6	4.6	15.3	126.0
	2	1.0	1.7	51	1.0	1.6	8.0	1.0	1.7	8.5	1.3	2.2	4.4	1.3	2.1	2.1	9.3	74.0

Explanation: 1 – 0–30 cm, 2 – 60–90 cm; EF – enrichment factor, Cfi – contamination factor, Eir – potential ecological risk, Cdeg – contamination degree, RI – potential ecological risk index

Table 9. Values of indices for arable soils

Profile No.	Layer	Cd_t			Pb_t			Ni_t			Cu_t			Zn_t			C_{deg}	RI
		EF	C_f	Er	EF	C_f	Er	EF	C_f	Er	EF	C_f	Er	EF	C_f	Er		
II.20	1	1.3	2.0	60	1.1	1.7	8.5	0.9	1.3	6.5	1.1	1.7	3.4	1.7	2.6	2.6	9.3	81.5
	2	0.7	1.1	33	1.0	1.6	8	0.9	1.5	7.5	1.0	1.6	3.2	1.0	1.6	1.6	7.4	52.8
II.21	1	0.5	0.6	18	0.9	1.1	5.5	1.0	1.1	5.5	1.5	1.8	3.6	0.8	0.9	0.9	5.5	33.5
	2	0.6	0.6	18	0.8	0.8	4.0	0.9	0.9	4.5	0.9	1.0	2.0	0.8	0.8	0.8	4.1	29.3
II.23	1	1.1	0.8	24	1.1	0.8	4.0	1.2	0.8	4.0	1.2	0.9	1.8	1.6	1.1	1.1	4.4	34.9
	3	0.5	0.7	21	0.6	1.0	5.0	0.9	1.4	7.0	1.1	1.5	3.0	0.7	1.0	1.0	5.6	36.5
II.25	1	0.9	1.0	30	0.8	0.9	4.5	1.0	1.0	5.0	1.0	1.1	2.2	0.9	1.0	1.0	5.0	42.7
	3	0.9	1.0	30	0.8	0.9	4.5	1.0	1.2	6.0	0.9	1.1	2.2	0.7	0.8	0.8	5.0	43.0
II.29	1	1.9	1.4	42	1.6	1.2	6.0	1.4	1.0	5.0	1.9	1.4	2.8	2.0	1.5	1.5	6.5	57.3
	2	1.5	1.1	33	1.1	0.9	4.5	1.3	0.9	4.5	0.8	0.6	1.2	1.0	0.8	0.8	4.3	44.0
II.31	1	1.4	1.8	54	1.3	1.6	8.0	1.1	1.4	7.0	1.5	1.9	3.8	1.7	2.2	2.2	8.9	75.0
	2	2.2	1.4	42	1.4	0.9	4.5	1.5	0.9	4.5	1.5	0.9	1.8	1.1	0.7	0.7	4.8	53.5

Explanation: see Table 8.

for Cd, Zn, Cu and Pb in most cases indicate the anthropogenic origin of those heavy metals.

The values of contamination factor revealed higher levels of accumulation of heavy metals in grassland soils, which shows that the method of their use has a significant effect on the total content of metals. Józefowska et al. [2014] report, based on their research, on different types of land use not affecting the content of heavy metals and only the accumulation index of lead was higher in grassland soils than in arable land soils.

In statistics, Ward's method is a criterion applied in hierarchical cluster analysis. Figure 1 is the result of the cluster analysis and two main groups have been distinguished in the dendrogram (Figure 1).

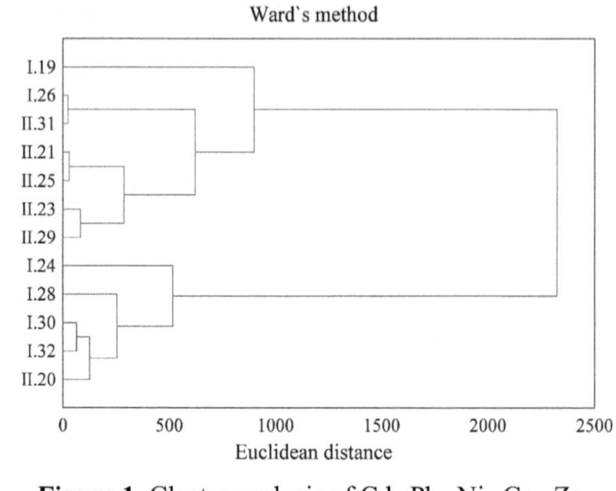

Figure 1. Cluster analysis of Cd_t, Pb_t, Ni_t, Cu_t, Zn_t, Fe_t, Mn_t in the surface horizon of Fluvisols

The results of the analysis show that the method used determines the content of heavy metals in the surface layer of the Fluvisols under study. A similar total content of metals was reported for the samples (1 cluster) taken from 5 arable soils and 2 pastures (II.23, II.29, II.21, II.25, II.31 and I.26, I.19) as well as samples (2 cluster) taken from 4 pastures and one arable soil (I.24, I.28, I.30, I.32 and II.20). In the Fluvisols investigated a significantly positive correlation between the content of C org and the total content of metals was found. The soil samples with a higher content of the clay fraction also contained a significantly higher total content of Pb, Ni, Cu, Zn, Mn and Fe (Table 10). Between the total content of Cd, Pb, Cu and Zn and the content of Corg higher values of the coefficient of correlation than with the clay fraction were noted, unlike Ni, Mn and Fe. A significantly positive correlation between the content of organic carbon and the content of metals confirms a significant role of humus in bonding metals [Van Gestel 2008]. Numerous interactions were identified between the total content of respective metals and some of them are very high dependencies with the value of the coefficient of correlation of $r>0.90$ ($p<0,05$).

Table 10. Significant correlation coefficient at $p < 0.05$

Soil properties	Cd_t	Pb_t	Ni_t	Cu_t	Zn_t	Mn_t	Fe_t
C org	0.75	0.81	0.45	0.77	0.88	0.59	0.40
Clay fraction		0.49	0.80	0.65	0.35	0.74	0.80
Cd_t		0.87	0.54	0.73	0.87	0.54	0.34
Pb_t			0.66	0.84	0.88	0.69	0.54
Ni_t				0.87	0.63	0.89	0.90
Cu_t					0.86	0.83	0.77
Zn_t						0.65	0.49
Mn_t							0.93
Fe_t							

Apart from chemical and metallurgical industries, municipal waste is responsible for most significant river contaminations which result in soil degradation in their close vicinity, both locally and regionally. In the Fluvisols of the Vistula floodplain areas the content of heavy metals is usually elevated, as compared with the content of the geochemical background and fluvial deposits can contain more mobile forms of metals [Czarnowska et al. 1995, Dąbkowska-Naskręt et al. 2000, Kobierski, Piotrowska 2010]. Skorbiłowicz and Samborska [2014] demonstrate that the arable land in the Vistula river catchment are characterized by a low content of Cu, whose average value was 8.2 ±1.5 mg·kg^{-1}, while the average nickel content was 10.6 ±5.1 mg·kg^{-1}. An excessive content of heavy metals, both indispensable to plants and not playing metabolic functions, can have an unfavourable effect on plants due to the capacity of metals for bio-accumulation. As a result of the alkalization of the soil environment and the higher the content of humus substances and the clay fraction in soil, the bioavailability of metals can get limited [Kabata-Pendias, Pendias 2001].

CONCLUSIONS

1. The total content of metals in Fluvisols depends on the use of soils. A higher total content of metal was found in the soils of grasslands.
2. Compliant with Regulation of Minister of the Environment of September 9, 2002, the total content of cadmium and zinc found in the 0–30 cm layer of all the pastureland investigated was higher from the maximum for the areas protected by law (land group A). A higher total content of Pb and Ni than the maximum for the soils of the protected areas was reported also in a few soil samples in the surface horizon of Fluvisols. No unambiguous, unfavourable, effect of trace elements was found in the riverside environment and so to evaluate the contamination the limit values specified in Regulation for agricultural land were assumed. According to this criterion the Fluvisols of the Grudziądz Basin, besides one soil sampling point, do not qualify as contaminated with the metals under the study.
3. The values of enrichment factor calculated for Cd, Pb, Ni, Cu and Zn were slightly higher for the soils of grasslands and point to the anthropogenic accumulation of metals, which concerns especially zinc for which in the surface horizon of grasslands the value of EF ranges from 1.9 to 4.5.
4. Potential ecological risk demonstrated that overall risk caused by the content of metals ranges from low to considerable. The value of potential ecological risk index was found to be low, and varied between 29.3 to 132.6 indicating that the Fluvisols were not contaminated with heavy metals.

Acknowledgements

The studies were financed by the NCN (National Science Centre), Poland, under research project No. 2716/B/P01/2011/40.

REFERENCES

1. Crock J.G., Severson R. 1980. Four reference soil and rock samples for measuring element availability in the western energy regions. U.S. Geological Survey circular 841, 16.
2. Czarnowska K., Broda D., Chojnicki J., Turemka E. 1995. Heavy metals in illuvial soils of the Vistula valley. Roczn. Glebozn. 46, 3/4, 5–18 [in Polish].
3. Dąbkowska-Naskręt H., Kobierski M., Różański S. 2000. Trace elements distribution and mobility in alluwial soils of the Wisła River Valley, Poland. Mengen und Spurenelemente, 20. Arbeitstagung. Jena, 152–157.
4. Dziennik Ustaw 2002. Rozporządzenie Ministra Środowiska w sprawie standardów jakości gleby oraz jakości ziemi. Nr 165, poz. 1359. [in Polish]
5. Gervais-Beaulac V., Saint-Laurent D., Berthelot J.S., Mesfioui M. 2013. Organic carbon distribution in alluvial soils according to different flood risk zones. J. Soil Sci. Environ. Manage., 4, 8, 169–177.
6. Håkanson L. 1980. An ecological risk index for aquatic pollution control – a sedimentological approach. Water Research 14, 975–1101.
7. Harmsen J. 2007. Measuring bioavailability: from a scientific approach to standard methods. J. Environ. Qual., 36, 5, 1420–1428.
8. Hoffmann T, Glatzel S, Dikau R. 2009. A carbon storage perspective on alluvial sediment storage in the Rhine catchment. Geomorphology 108, 127–137.
9. ISO 10390: 2005. Soil quality – Determination of pH.
10. Józefowska A., Miechówka A., Gąsiorek M., Zadrożny P. 2014. Content of zinc, lead and cadmium in selected agricultural soils in the area of the Śląskie and Ciężkowickie foothills. J. Ecol. Eng., 15, 1, 74–80.
11. Kabata-Pendias A. 2004. Soil-plant transfer of trace elements – an environmental issue. Geoderma 122. 143–149.
12. Kabata-Pendias A., Pendias H. 2001. Trace elements in soils and plants. Boca Raton. London. New York. Washington DC, 1–331.
13. Kobierski M., Piotrowska A. 2010. Profile distribution of heavy metals and enzymatic activity in Fluvisols of the Vistula River Valley, Poland. Fresenius Environ. Bulletin 19, No 2b, 303–311.
14. Kordowski J. 2003. Internal structures and granulometry of lower Vistula valley overbank deposits in the Toruń and Unisław Basins. Przegląd Geograficzny 75, 4, 601–621.
15. Macklin M.G., Klimek K. 1992. Dispersal, storage and transformation of metal contaminated alluvium in the Upper Vistula basin, southwest Poland. Appl. Geogr., 12, 7–30.
16. Martin J.-M., Meybeck M. 1979. Elemental massbalance of material carried by major world rivers. Marine Chemistry 7, 3, 173–206.
17. Martin J.-M., Whitfield M. 1983. The significance of the river input of chemical elements to the ocean. [In:] Trace metals in seawater. Wong C.S., Boyle E., Bruland K.W., Goldberg E.D. (Editors), Plenum, New York. 265–296.
18. PN-ISO–11277: 2005. Soil quality – Determination of particle size distribution in mineral soil material – Method by sieving and sedimentation.
19. Saint-Laurent D., St-Laurent J., Lavoie L., Drouin A., Ghaleb B. 2010. Floodplain sedimentation rates, soil properties and recent flood history in Southern Québec. Global. Planetary Change 70, 76–91.
20. Singh K., Malik A., Sinha S., Singh V., Murthy R. 2005. Estimation of source of heavy metal contamination in sediments of Gomti river (India) using principal component analysis. Water Air Soil Pollut., 166, 321–341.
21. Skorbiłowicz E., Samborska A. 2014. Content of copper and nickel in soils of Vistula River catchment. J. Ecol. Eng., 15, 2, 53–59.
22. Sutherland R.A., Tolosa C.A, Tack F.M.G., Verloo M.G. 2000. Characterization of selected element concentration and enrichment ratios in background and anthropogenically impacted roadside areas. Arch. Environ. Contam. Toxicol., 38, 428–438.
23. Van Gestel C.A.M. 2008. Physico-chemical and biological parameters determine metal bioavailability in soils. Sci. Total Environ., 406, 385–395.
24. Wyżga B., Ciszewski D. 2010. Hydraulic controls on the entrapment of heavy metal-polluted sediments on a floodplain of variable width, the upper Vistula River, southern Poland. Geomorphology 117, 272–286.

9

THE GEOLOGICAL CONDITIONING OF HYDROCARBON EMISSIONS RESULTING FROM SOIL CONTAMINATION

Ewa J. Lipińska[1]

[1] Subcarpathian Voivodeship Inspector for Environmental Protection Voivodeship Inspectorate for Environmental Protection in Rzeszów, Gen. M. Langiewicza 26 Str., 35-101 Rzeszów, Poland, e-mail: wios@wios.rzeszow.pl

ABSTRACT

Synchronization economy of oil mining and mineral waters is associated with planning the functions of spa treatment. Environmental protection of the spa areas also applies to preserve their technical and cultural heritage. This article attempts to determine the places of natural and anthropogenic hydrocarbon pollution substances. Their presence in the soil affects the quality of the environment. As a result, maps are produced showing directions of research: (1) the natural background of biodiversity, and (2) potential anthropogenic pollution. They are assessed in the context of the health and human life, protection of the environment and the possibility of damage to the environment. Research is conducted in communes of the status of the spa – for special protection.

Keywords: geology, hydrocarbon substances, mineral water, pollution, soil, environment.

INTRODUCTION

Substance is any chemical element and its compounds, except for radioactive substances, modified micro-organisms and genetically modified organisms. The substances are released, subjected to various forms of emissions and gradual partial inclusion in the geochemical cycle. These processes are varied and complex. They also have various forms: physical, chemical and biological (or hydrous, atmospheric or colloidal).

In the administrative practice, it is difficult to explain what level of substances in the environment is considered as *natural – initial – normal.* It is difficult to determine the natural range of their occurrences. The reason is different concentrations of their geochemical background in time and space.

Emission is the direct or indirect release of substances to the ground from individual or diffuse sources [Directive 2010]. Natural processes in the environment usually lead to a degradation or assimilation of these substances. Their volume, speed, emissions, short- and long-term toxicity must be taken into account when being entered into the environment. A self-cleaning of the environment is following and the restoration of its natural value but its usefulness is diminished. There is no risk of damage to the environment.

When the processes of the self-cleaning of the environment are too weak to overcome the influence of the substance with the forces of nature, a damage to the environment appears.

Damage to the environment is a negative measurable change in the condition or function of natural elements, evaluated in relation to the initial state [the Act 2007].

Soil is the top layer of the earth's crust situated between the bedrock and the surface. It is composed of mineral particles, organic matter of water, air, and living organisms. Most of soil is anthropogenically changed, although to a differentiated degree. For this reason, the determination of the natural concentrations of substances in soils and the soil pollution is complex.

Pollution is a direct or indirect introduction – as a result of human activity – of a substance, which may be harmful to human health or the environment, dangerous to property, or which impair the environment, or interfere with other legitimate uses of the environment.

Partial results of the investigation of natural and antropogenic emissions of hydrocarbon impact on the environment is presented in the paper. The research covered areas of the communes with the status of the spa, i.e. Horyniec-Zdrój, Iwonicz-Zdrój, Rymanów and Solina. The maps of these communes show biodiversity and anthropogenic changes. The maps are made for the purpose of spatial planning.

METHODOLOGY

The object of research is the geological environment of communes of the status of the spa. Special rules of protection [e.g. the Act 2005 and 2007].

The aim of the study is to identify the geological conditions in areas of the communes of the status of spas, their soil and preparation of maps indicating the biodiversity, infrastructure and mining of mineral hydrocarbon materials.

The utilitarian aim is to identify the direct (potential, natural) and specific (anthropogenic) pollution in the areas of communes with the status of spa. The evaluation of the geological environment is made by analyzing data from state environmental monitoring. The object of study is a continuation and extension of the issues described in [Lipińska 2013].

THE BEHAVIOUR OF LIQUID MINERAL DEPOSITS OF POLISH CARPATHIANS

In Poland, Carpathians cover an area of ca 19,600 km^2, i.e. 6% of the total area of the country. They are subjected to various forms of conservation. Protection of *Natura 2000* in Polish Carpathians includes the Alpine biogeographical region, which covers an area of ca 10,050 km^2.

Carpathians are composed of tracks flysch. A large variation of sediment thickness and intense tectonic disturbances is characteristic of this area. The occurrence of mineral resources is directly related with the respective geological structure.

In the central and eastern part of Carpathians deposits of mineral water and hydrocarbon substances can be found (Figure 1). Carpathian oil reserves are estimated at ca 27 km^2 and natural gas at 16.5 km^2.

Oil, natural gas and water are liquid natural resources. When these fluids surface or are exploited, their place is filled with the expanding substances. These are substances that remain in the fluid, including rock, or substances that flow into the bed of the external areas. The nature of inflowing material determines the basic energy system deposits. It determines the behaviour of the reservoir and its size.

Five systems of energy deposits of hydrocarbon substances was established. The actual active deposits is more than one power system (which is a mixed energy system).

The energy system of the deposit is determined on the basis of: (1) geological information, (2) the behaviour of the respective bores in the operating conditions, (3) the overall behaviour of the reservoir.

In Figure 2 examples of natural emissions of hydrocarbon were presented.

THE RESEARCH PROBLEM – GEOLOGY, SOILS, BIODIVERSITY IN COMMUNES OF WITH THE STATUS OF THE SPA

The commune of Horyniec-Zdrój

The commune of Horyniec-Zdrój lies within two geological units. The area is composed of Miocene deposits. These are mostly clay, covered by arrears clay, glacial sands and silty sands. Subsequently, sedimentary rocks of Cretaceous, as well as Tertiary sandstones and limestone are to be found. Single boulders, which are reminiscent of post-glacial are also found.

Mineral deposits are mainly represented by peat deposits, mineral deposits, deposits of marl, sand, sandstone and limestone. Peat mining area has an area of ca 550 ha and resources of about 380,000 m^3. The mining area of mineral deposits is ca 180 ha. No deposits of hydrocarbon materials are reported.

The variability of soil and soil quality is a product of geological diversity of the commune of Horyniec-Zdrój. In this area podsoils, alluvial soil and brown soil evolved. In some places there are humus and peat.

In Figure 3, the biodiversity of the area of the commune of Horyniec-Zdrój is shown, as well as the distribution of mineral water wells and the limits of mining areas, against the backdrop of protected areas.

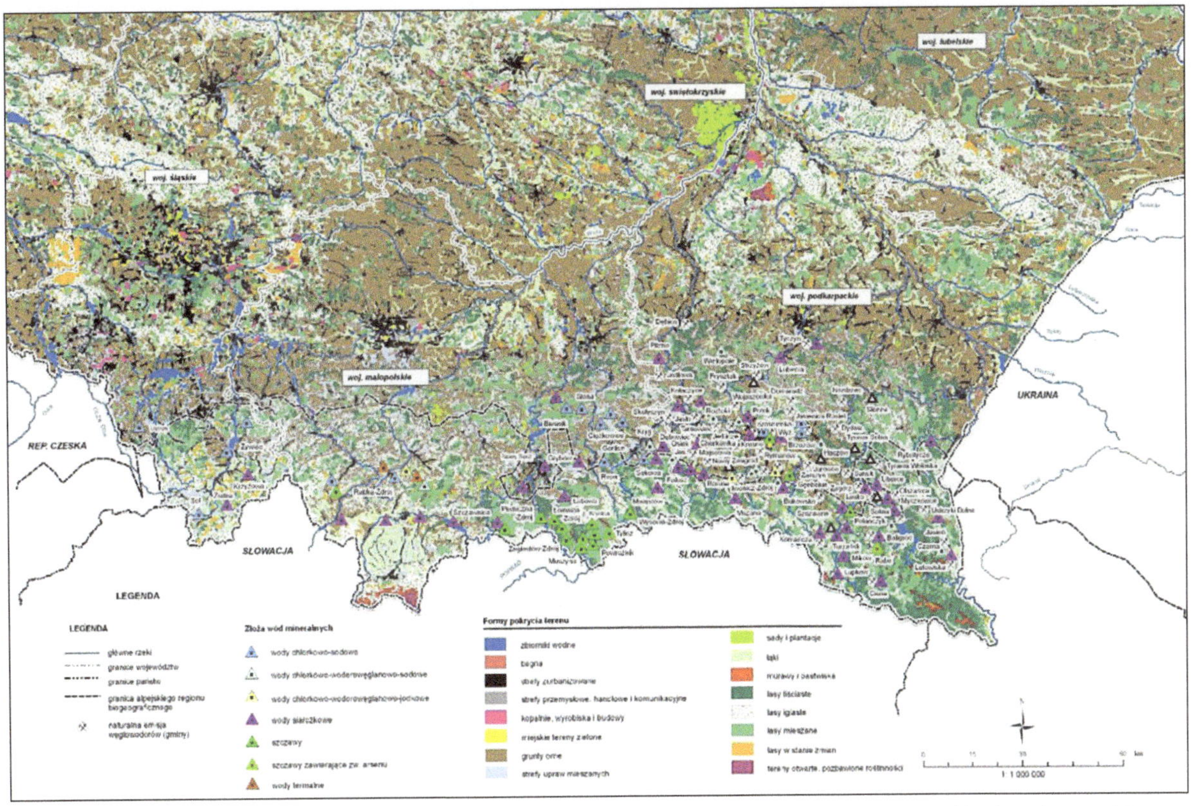

Figure 1. Polish Carpathians – forms of biodiversity in the area of the Alpine biogeographical region and location of deposits of mineral

Figure 2. Well known *kopanka* with wooden housing; area of the village Turze Pole, the commune of Brzozów, Brzozów county, Subcarpathian voivodeship

Figure 3. *Kopanka*; substitute area, the commune of Brzozów, Brzozów county, Subcarpathian voivodeship

The commune of Iwonicz-Zdrój

The commune of Iwonicz-Zdrój lies in the eastern part of Polish Carpathians. This mountain range, extending for ca 40 km long and 5 km wide, are built of works of Paleogene flysch and Cretaceous. Hard and resistant to weathering sandstones dominate there.

The occurrence of mineral resources results from the geological structure. These are deposits of mineral water, oil and natural gas. Mineral waters are mined from two deposits. Two deposits of oil and natural gas are documented and expoited.

The beds of oil and gas are in sandstones. The variability of soil in this area is the result of its diversified relief. Soil weathering are character-

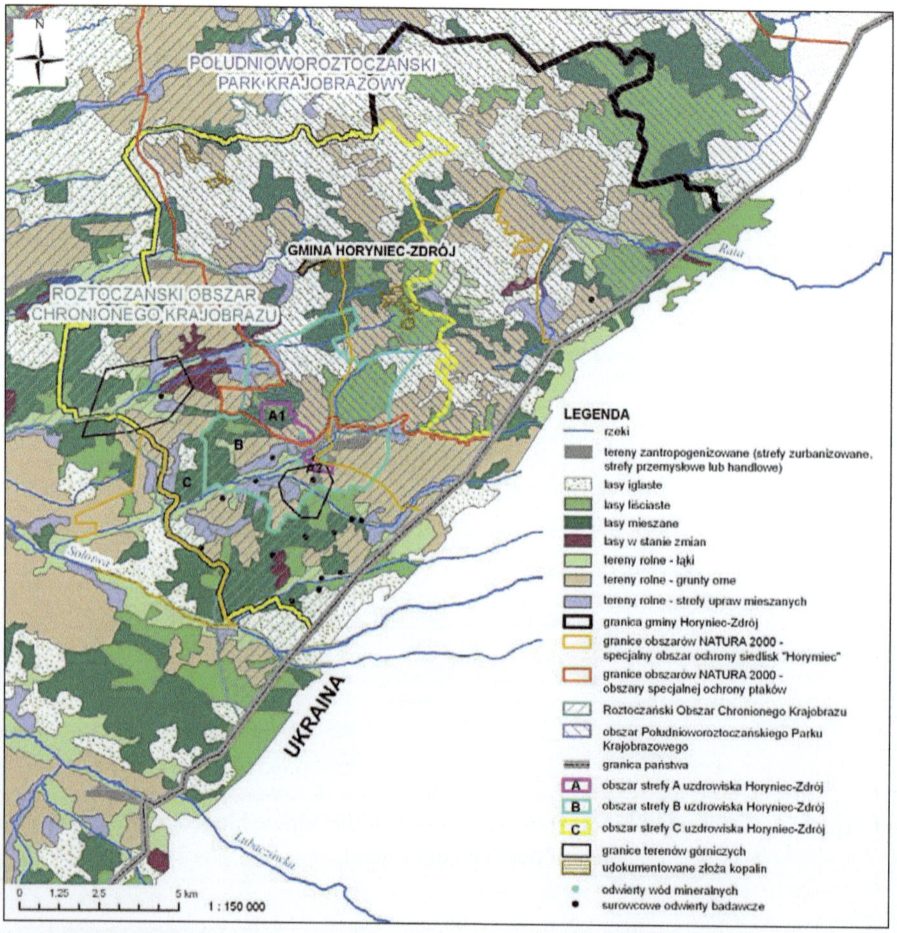

Figure 4. The commune of Horyniec-Zdrój – the distribution of biodiversity, mineral deposits, drilling research documented mineral deposits, the boundaries of mining; the Subcarpathian voivodeship

ized by granulometric composition aluminum, dust and silt. Brown soils dominate while alluvial soils occur locally.

The biodiversity of the area of the commune of Iwonicz-Zdrój is shown in Figure 4, location of mineral water intakes in relation to the areas of the occurrence of mineral deposits, active and liquidated drilling of oil and natural gas on the background of protected areas.

The commune of Rymanów

The commune of Rymanów lies in the eastern part of Polish Carpathians. The mountain range extends for ca 40 km long and 5 km wide. It is built of songs Paleogene flysch and Cretaceous. Upper Cretaceous sandstones are represented by layered slate. Tracks flysch sediments are covered by Quaternary sediments.

Natural resources are the raw materials of rock, mineral deposits, oil and natural gas. Soil variability is the result of diverse geological relief. Soil with a particle size of aluminum, dust and silt dominate. Brown and alluvial soils were produced.

In Figure 5 the biodiversity of the area of the commune of Rymanów is shown, as well as the location of sources and mineral drilling, oil drilling, natural resource exploration wells, the limits of the mining areas on the background of protected areas.

The commune of Solina

The commune of Solina lies within Eastern Carpathians. Sandstone layers with variable thickness of sediments dominate. Tracks flysch sediments are covered with quaternary deposits. The occurrence of mineral resources results from the geological structure. There are documented deposits of oil. Currently, wells are dismantled and the area is being reclaimed.

Exploitation of standstone was completed in 1991. Mineral waters are found in three horizons aquifers. This area is characterized by variability of soil due to the diversity of its relief. There exist soil particle size of aluminum, dust and silt. Brown soils and alluvial soils dominate. In Figure 6 the biodiversity of the commune of So-

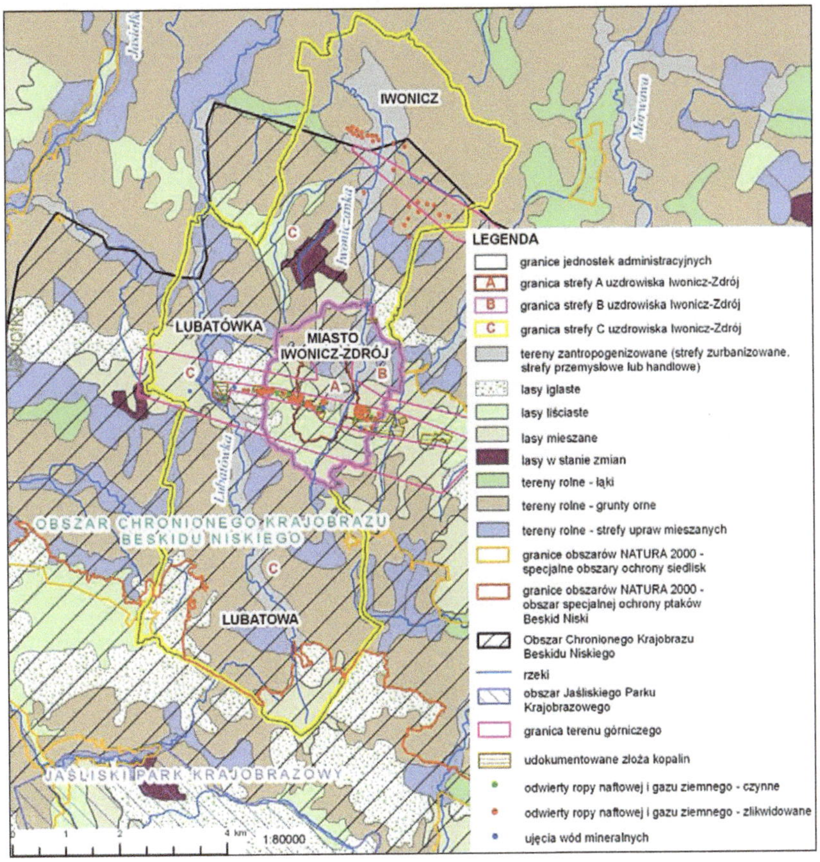

Figure 4. The commune of Iwonicz-Zdrój – the biodiversity, mineral deposits and hydrocarbon materials, liquidated and non-liquidated wells, the limits of the mining area; the Krosno county, Subcarpathian Voivodeship

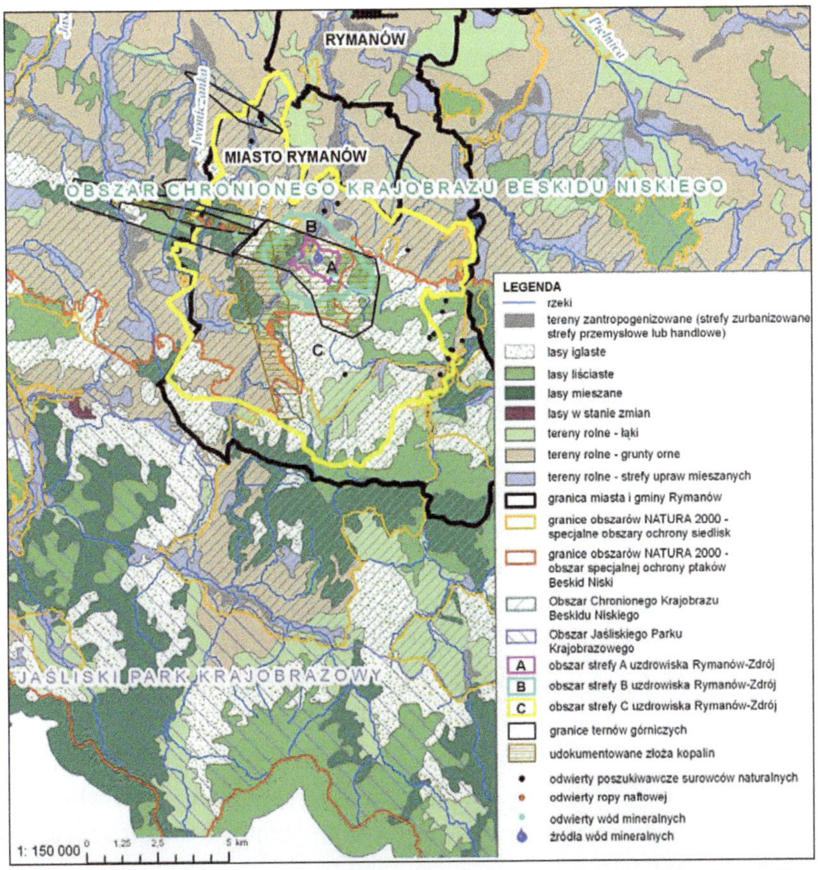

Figure 5. The commune of Rymanów – the distribution of biodiversity, mineral deposits and hydrocarbon materials, liquidated and non-liquidated wells, the limits of mining areas; the Krosno county, Subcarpathian Voivodeship

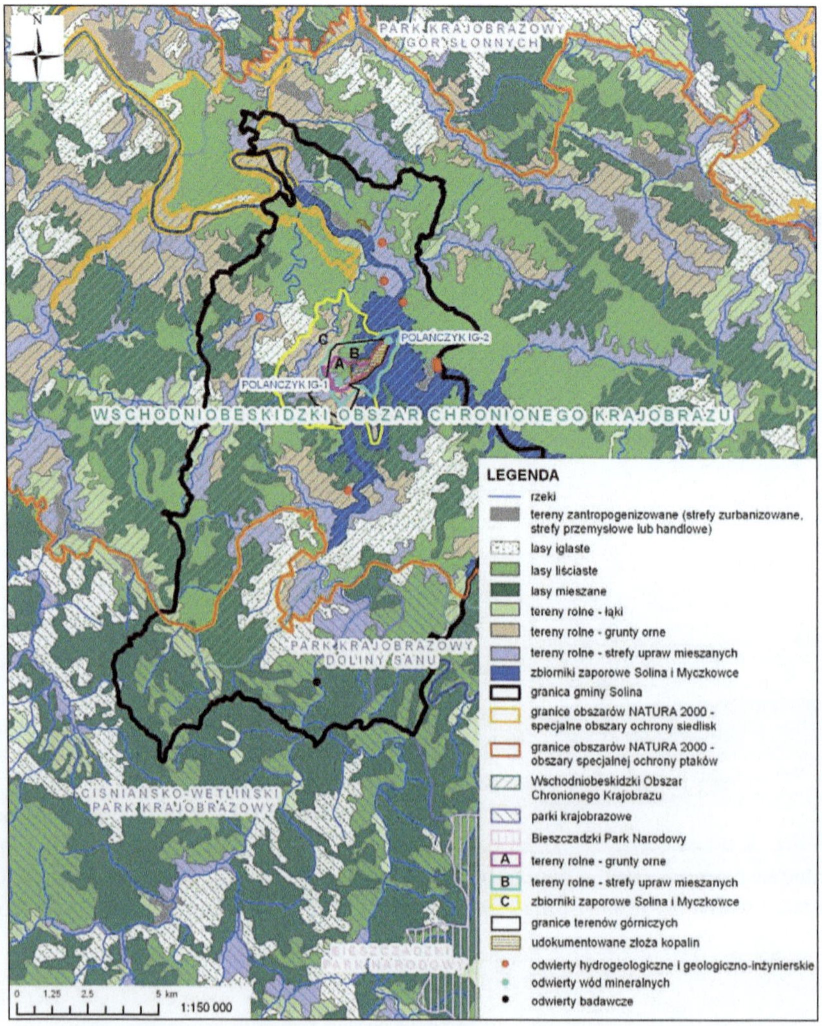

Figure 6. The commune of Solina – the biodiversity, mineral deposits, hydrogeological drilling, the limits of the mining area; the Lesko County, Subcarpathian Voivodeship

lina is shown, as well as the distribution of mineral water wells, the limits of the mining area, including the limits of the documented mineral deposits, against the background of the protected areas.

CONCLUSIONS

The proposed concept of studies seems satisfactory. The research goals specified in the *Methodology* section were reached. Postulates and conclusions are as follows.

1. The study is partial, reduced to the observation of only a certain part of the analysed statistical collectivity, geological environment of the health resorts. This requires a careful technical and organizational preparation of study.
2. Mineral resources whose extraction can contaminate soil were identified.
3. The studies expose information about warning, the indication of places, where natural and anthropogenic pollution currently exist and may occur in the future on the maps.
4. Maps have been elaborated, in which the area can be considered bi-directionally: (1) as a natural geochemical background, (2) as a possible anthropogenic pollution.
5. The research is addressed to the administration of communities with spa. The maps provide the original material for the development of the study of conditions and directions of the spatial development of the communes and the environment.
6. The owners of the technical infrastructure should find indicators on what work should be done to eliminate the direct and potential risks, eliminate their sources and reclaim the contaminated area to acquire hydrocarbon materials.

7. The result of the co-occurrence of mineral deposits and those of hydrocarbon materials is the landscape and nature of spontaneous flora species richness, the composition of which often finds the occurrence of rare and vulnerable to extinction in the neighbouring areas. This wealth affects the behaviour and enrichment of the local biodiversity.
8. An important scientific contribution to the research is an interdisciplinary range of issues raised, concerning, inter alia, environmental engineering, environmental protection, geology and drilling.
9. A discussion of the impact of impurities remains, due to the size limitation of this paper within the scope of subsequent analyses.

REFERENCES

1. Directive 2010/75/EU of the European Parliament and of the Council of 24 Nov. 2010 on industrial emission (integrated pollution prevention and control) (Recast) (Text with EEA relevance) L 334/17.
2. Lipińska E.J. 2013. Distribution and environmental impact of emissions from co-occurring deposits of hydrocarbon materials from mineral deposits in the municipality of Iwonicz-Zdrój. Scientific Association of Social Space and Environment. Rzeszów.
3. The Act of 13 April 2007 on the prevention of environment damage and its repair (Journal of Laws No. 75, item. 493).
4. The Act of 28 July 2005 on health resorts, spas and spa conservation areas and the municipalities spa (developed on the basis of Acts of 2012, pos. 651, 742).

10

CONCENTRATION OF URANIUM IN WATERS OF THE BIGGEST LAKES OF THE TYWA RIVER DRAINAGE BASIN

Jacek Kubiak[1], **Sylwia Machula**[1]

[1] Department of Hydrochemistry and Aquatic Biological Resources, West Pomeranian University of Technology in Szczecin, Kazimierza Królewicza 4, 71-550 Szczecin, Poland, e-mail: jkubiak@zut.edu.pl; smachula@zut.edu.pl

ABSTRACT

The paper presents the results of research of uranium concentrations in its different kinds – suspended and dissolved – in waters of the largest lakes located in the catchment area of the River Tywa – Strzeszowskie Lake, Dłużyna Lake, Długie Lake and Dłuzec Lake. Small (or the order of several 0,01 μg/l) variations in concentration of uranium in different lakes were noted. The study has also shown a seasonal variation – in a similar range – in concentrations of the above species of uranium, as well as total uranium. The content of dissolved uranium was highest in the autumn and winter, lower in the spring and summer. Overall, total uranium was found in greatest concentrations during the fall, in other seasons concentrations were lower and similar to each other. Suspended uranium was found in largest concentrations in autumn and summer, in lower ones in spring and winter. Concentrations of the different species of uranium during the study period showed a small variation – variation coefficient below 10% for total uranium and dissolved uranium, and about 25% for suspended uranium. The observed concentrations of uranium were typical of uncontaminated unpolluted water.

Keywords: Tywa River, Dłużec Lake, Długie Lake, Strzeszowskie Lake, Dołgie Lake, uranium, uranium concentration in water.

INTRODUCTION AND RESEARCH OBJECTIVES

Uranium is a heavy metal, generally regarded by the public as rare – however, in the environment it is in fact more common than such elements as cadmium or selenium [Bowen 1979]. Uranium occurs naturally in earth's crust at concentrations of approximately 4 ppm; such concentrations are similar to those characteristic of arsenic, boron or other lanthanides, and an order higher than some non-heavy metals, for example silver and mercury. Uranium is found in varying concentrations in all soils, rocks, minerals, as well as food and water (Kabata-Pendias, Kabata, 1999).

Tests of 3700 water samples collected in Ontario (Canada) in 1990–1995 showed the mean uranium content was 0.05–4.21 μg/l, and in treated potable water – 0.40 μg/l. Average uranium concentration in drinking water in New York was in the range of 0.03–0.08 μg/l. In five Japanese cities, the mean level of uranium in drinkable water was 0.9 μg/l. However, there have been cases of occurrence of uranium at much higher concentrations, especially in small supplies. For example, concentration of 700 μg/l was found in a small private water supply in Canada. Research conducted in Finland in waters used to supply drinking water to the populace, the median concentration was found at the level of 28 μg/l. In a study of 476 Norwegian groundwater samples, 18% of them had uranium concentrations in excess of 20 μg/l. Concentrations in excess of 20 μg/l have also been found in groundwater in New Mexico, USA and in central Australia [WHO 2011].

Uranium present in natural water reservoirs comes primarily from anthropogenic sources, such as: fertilization with phosphate fertilizers, processing of apatites, production of construction materials (cement), manufacturing pigments for glass and ceramics industry. The use of phosphate fertilizers is a major source of anthropogenic uranium (fertilizers contain up to 390 mg of U/kg, depending on production technology, raw materials used and the fertilization dose applied) because of high uranium content in phosphate ores (uranium in apatite: 30–120 mg/kg) [Zielinski et al. 1,997; Kratz and Schnug 2006]. Other anthropogenic sources include uranium mining waste, burning of coal and petroleum products, emissions from nuclear power plants and military uses of uranium. Also coal-fired power plants are one of its significant sources in the environment, as coal and peat are characterized by high uranium content, of up to 200 mg/kg [Bowen 1979]. Sources of uranium in the marine environment include atmospheric precipitation of terrigenous material, as well as inland waters inflow [Skwarzec et al. 2002].

CHARACTERISTICS OF THE RESEARCH AREA

Landscape of the Tywa River basin, as well as the whole West Pomeranian Lakeland was shaped in the last Scandinavian glaciation, during glacier stagnation in the Pomeranian stage [Kondracki 2001]. Southern parts of the Tywa River drainage basin are located in the Myślibórz Lake District; its northern parts belong to the Wełtyń Plain, while the estuarine segment is located within the Lower Oder Valley [Kępińska-Kasprzak et al. 1997, Raczyńska 1999, Kondracki 2001].

Tywa River – with a length of 47,9 km and catchment area of 264,5 km ^{2} – is a right tributary of the East Oder. SNQ reliable flow calculated for the section of the mouth of the river is 0.71 m^3/s. Tywa catchment area is located in the geographical terrain where the share of underground water and lakes water recharge is estimated at more than 60% of the total runoff. Average specific runoff of the Tywa River in the years 1961–1985 was 3,41 $l/s/km^2$ [Duda et al. 1991].

Lakes, especially in the middle section of the river, are the dominant component (2.8%) of the landscape of the Tywa River drainage basin [Kępińska-Kasprzak et al. 1997, Duda et al. 1991, Raczyńska 1999] – 30 lakes were counted in the Tywa catchment area (18 with the surface of less than 10 ha). Tywa runs through eight ribbon lakes; only 4 of them have the surface exceeding 50 hectares – the series of lakes: Strzeszowskie, Dołgie (Dłużyna), Długie (Swobnickie) and Dłużec, where the present study was conducted. The lakes are situated in the western part of the Myślibórz Lake District, within the Chojna-Moryń Lake Group [Piskorski, 1979, Kondracki 1994, 2000, Kubiak 2003]. The aquifers are located among forests, and are used for tourism and recreation purposes, as well as lake fishing. The lakes' morphometric characteristics are presented in Table 1, while the uses of the catchment area are shown in Table 2. A detailed description of the general limnological and geographical characteristics is given by Kubiak et al. [2009, 2013].

Table 1. Morphometric characteristics of the studied lakes

No	Indicator	Strzeszowskie Lake	Dołgie Lake	Swobnickie Lake	Dłużec Lake
1	Latitude	52°59.6'	53°00.5'	53°03.7'	53°07.3'
2	Longitude	14°36.9'	14°37.3'	14°38.6'	14°39.4'
3	Maximum length (m)	1900	1850	3900	2400
4	Maximum width (m)	1150	450	1200	500
5	Average width (m)	670	304	882	355
6	Elongation ratio	1.6	4.1	3.2	4.8
7	Maximum effective length (m)	1900	1700	3900	2200
8	Maximum effective width (m)	1150	450	1200	600
9	Shoreline length (m)	5750	4350	9000	5875
10	Shoreline development index	1.44	1.64	1.37	1.79
11	Water table surface (ha)	127.2	56.3	344	85.2
12	Volume (thous. m^3)	9499	1700	14485	5228
13	Maximum depth (m)	14.2	6.6	6.8	10.4
15	Average depth (m)	7.4	3.1	4.2	6.1
16	Depth indicator	0.52	0.47	0.62	0.58

Table 2. Uses of catchment areas of the studied lakes

Lake	Swobnica				Dłużec				Strzeszowskie				Dłużyna			
Usage	Direct drainage basin		Total drainage basin		Direct drainage basin		Total drainage basin		Direct drainage basin		Total drainage basin		Direct drainage basin		Total drainage basin	
	km²	%	km²	%	km²	%	km²	%	km²	%	km²	%	km²	%	km²	%
Forests	3.9	34.1	26.8	26.0	0.5	4.3	28.4	22.8	0.1	2.2	1.8	15.7	4.7	59.5	12.9	23.2
Grazing land	-	-	5.4	5.3	0.4	3.2	7.0	5.6	0.5	10.4	0.5	4.6	0.2	2.0	4.6	8.3
Waters	3.4	30.4	5.1	5.0	1.0	7.9	6.1	4.9	1.3	25.6	1.4	12.7	0.6	7.6	0.9	1.6
Arable land	3.8	33.3	62.7	60.9	9.5	79.5	79.1	63.4	2.6	52.6	6.8	60.8	2.4	29.9	36.1	65.2
Other	0.3	2.2	2.9	2.8	0.6	5.1	4.1	3.3	0.5	9.2	0.7	6.2	0.1	1.0	1.0	1.7
Total	11.3	100	102.9	100	12.0	100	124.7	100	5.0	100	11.3	100	7.9	100	55.4	100

The lakes Strzeszowskie, Dłużec near Banie and Dołgie are eumictic, periodically presenting tachymictic features, while the lake Długie (Swobnickie) is polymictic. Oxygen systems in these lakes are proof of their high trophic index [Kubiak et al. 2009].

Tywa River drainage basin is characterized by fairly significant differences in hypsometric terms. In the area of the Myślibórz Lake District, where the tested lakes are located, the landscape is hilly – however, the moraine hills of this lake district only in a few instances exceed the height of 100 m above sea level, with relative heights of 20 to 40 m. The Tywa drainage basin is situated at the average altitude of 57m above sea level, longitudinal slope from the source to the river mouth is 2,2‰, while the average slope is of 8‰ [Duda et al. 1991, Raczyńska 1999].

Surface lithology of the Tywa River catchment area is mainly made up of rock material left behind after the last ice age. Glacial till extends over most of the Tywa River basin, only its lower parts are covered with glacial sands. The majority of the Tywa drainage basin area is covered with podsolic soils (about 67% of the total surface of the basin), extending mainly over the central and estuarial part of the catchment area. Light to medium brown earths are present in the upper section of the drainage basin (about 27% of the total surface) [Duda et al. 1991].

The discussed catchment area is rich in natural plant communities; synanthropic species are only a small share of the plant life, pointing to low contamination of the Tywa River waters. The quality of water in the river makes it a habitat of salmonid fish; also eutrophication index has not been exceeded in water.

Given the low degree of contamination of lake waters in the Tywa River catchment area, as well as a limited number of available publications on the presence of uranium in the Polish lakes. The key objective of the present study was to determine the content of this metal in the waters of the largest lakes in the Tywa River catchment area.

METHODOLOGY

Tests of samples from the lakes Strzeszowskie. Dołgie (Dłużyna). Długie (Swobnickie) and Dłużec were conducted in all seasons. Every year from 2008 to 2010. Water samples were taken at varying depths in the deepest areas of the lakes. For measuring uranium contents ICP-MS spectrometer – Elan DRC from Perkin-Elmer company (detection of <0.1 μg/l) was used. It is a fast and exact procedure for determination of uranium concentration. Even of the order of ng/l in drinking water and mineral waters (Himiri et al. 2000. WHO 2011). Uranium was measured in samples both non-filtered and filtered through a fiberglass filter of 0.45 μm retention. After digestion with nitric acid; this procedure allowed the author to measure the total uranium content as well as the levels of dissolved uranium – the difference in these two concentrations was used for calculation of the suspended uranium concentration.

In order to determine the impact of metabolic processes on the presence of uranium in waters of the tested lakes studies of other hydro-chemical indicators were carried out in parallel. Taking into account the above impact and variability of concentrations of different species of uranium between the aquifers as well as their seasonal fluctuations statistical analysis of the obtained results was performed using the Statistica 10.0 software [StatSoft Inc. 2009].

PRESENTATION AND DISCUSSION OF THE RESULTS

Uranium is one of the most important – in geochemical terms – radioactive elements; it is found in waters usually in small amounts. The highest concentrations are known to occur in acidic waters of igneous rocks and in some types of sedimentary rocks. Concentrations of uranium in seawater vary between 1 and 4 μg/l and remain stable – groundwater uranium contents, which usually do not exceed μg/l, are equally stable. In mineral waters, uranium levels can reach 500 μg/l. In water supplied from uranium deposits the concentrations exceeding 1 mg/l are frequent – although in such areas concentrations to the order of 15 mg/l were also recorded. Average content of uranium in groundwater of temperate climate is 2.1 μg/l (Birke et al. 2009). Uranium levels in the flowing waters of Germany vary between 0.007 and 43.7 μg/l. with a median of 0.33 μg/l (Birke et al. 2009). Earlier publications stated that in the surface waters on the German, uranium concentrations were within the range of 0.007 to 52.7 μg/l (median 0.65 μg/l). At that time highest values were recorded around uranium mines and in areas known for intensive use of phosphate fertilizers [Birke et al. 2003].

The solubility and mobility of uranium compounds in the aquatic environment are a complex issue and depend on the ionic form in which uranium is present [Merkel et al. 2002]. Water migration of uranium is also dependant on oxidation-reduction conditions to a large extent; with an increase in Eh, uranium content in water grows as well. In an oxidizing environment. the element is present in the form of easily migrating uranyl ion. UO_2^{+2}. Uranium shows a marked tendency to binding, forming larger compounds, for example with ions: carbonates. fluorides. sulphates. phosphates and hydroxyl ions [Bernhard, 2004]. Most of the generated cationic, anionic or neutral complexes are very mobile in aqueous solution [Merkel, Sperling, 1998]. Hence, the ubiquitous nature of uranium, particularly in well-oxygenated water and subsurface groundwater. Water migration is restricted by water sorption processes; as soluble uranium compounds are readily adsorbed by silica gels. iron hydroxides and aluminum hydroxides as well as clay minerals [Bernhard, 2004]. In a reducing environment, the reduction of U (VI) to U (IV) results in removal of autogenic uranium from the water to the bottom sediments, as it is then only very weakly soluble [Skwarzec et al. 2002. Bernhard, 2004]. The uranium exchange in the lakes between the bottom sediments and water is described by Chappaz et al. [2010], who further explain that in the anaerobic environment in hypolimnion uranium is reduced to its stable insoluble species and precipitates down to the bottom sediments.

The uranium content in the Oder River waters around Gozdowice was at 1.5 μg/l [Birke et al. 2009]. Different studies conducted in 1999 and 2000 showed that in the Oder River water in the area of Krajnik uranium concentrations (U^{234} and U^{238}) were 1.89 and 1.58 μg/l respectively, while the suspended uranium contents of same isotopes were 0.39 and 0.35 μg/l [Pietrzak-Flis et al. 2004]. The quoted data refer to the waters of the River Oder adjacent to the Tywa River drainage area. In the above years analyses of concentrations of the two isotopes of uranium in waters of lakes Drawsko, Wądąg and Wigry were also carried out. In the Drawsko Lake recorded concentrations were at 0.31 μg/l (1999) and 1.18 μg/l (2000) for the dissolved uranium species. and – respectively for U^{234} and U^{238} – at 0.34 and 0.37 μg/l for suspended uranium. More uranium-rich were found to be the waters of the Wądąg Lake – concentrations of 1.32 and 1.52 μg/l of dissolved uranium and 0.81 and 0.39 μg/l of suspended uranium were recorded in the samples. These studies have also shown that only a negligible proportion of the total uranium amount is made up of the isotope U^{235} [Pietrzak-Flis et al. 2004]. In the Baltic Sea waters (1996–1998) uranium was present at concentrations of 6.11 μg/l in the area of the Gdańsk Deep and 3.96 μg/l in the region of the Bornholm Deep [Skwarzec et al. 2002].

The studies carried out clearly point to the conclusion that in waters of the largest lakes in the Tywa River drainage area uranium is present only at low concentrations (Table 3), similar to those naturally occurring in surface waters [Skwarzec 2002, Pietrzak-Flis et al. 2004]. In the Tywa River waters uranium contents varied within the range of 0.49–2.42 μg/l. The mean concentration of uranium was at 1.26 μg/l in the river water and at 1.19 μg/l in the lakes. In most instances, uranium levels in the studied lake waters were contained within the range of 1.0–1.2 μg/l – frequency of 57.7% of the samples.

Dissolved uranium in waters of all the studied lakes of the Tywa River drainage basin was most frequently measured at within the range

Table 3. Concentrations of different forms of uranium in waters of the studied lakes in the Tywa River drainage area

Lake	Indicator	N	Mean	Confidence interval	Confidence interval	Median	Min	Max	Range	Variation coeff.	Max. freq. range	Freq. (%)
All lakes: total	U dissolv.	272	0.88	0.87	0.89	0.89	0.59	1.18	0.59	10.4	0.8–0.9	48.5
	U total	272	1.19	1.17	1.20	1.19	0.77	1.65	0.80	9.8	1.0–1.2	57.7
	U susp.	272	0.31	0.30	0.32	0.30	0.10	0.58	0.48	28.7	0.3–0.4	42.3
	U dissolv/total (%)	272	74.5	73.7	75.2	75.0	58.3	90.3	32.0	8.5	70–75	31.9
	U susp./total (%)	272	25.5	24.7	26.2	25.0	7.5	41.7	34.2	24.7	20.–25	34.0
Dłużec	U dissolv.	46	0.87	0.85	0.88	0.88	0.75	1.00	0.25	6.7	0.8–0.9	54.3
	U total	46	1.16	1.13	1.19	1.17	1.00	1.42	0.42	7.7	1.1–1.2	41.3
	U susp.	46	0.30	0.29	0.32	0.30	0.20	0.42	0.22	20.4	0.2–0.3	52.1
	U dissolv/total (%)	46	74.1	72.5	75.7	73.1	61.6	82.6	21.0	7.4	70–75	34.8
	U susp./total (%)	46	26.2	24.6	27.7	26.8	17.4	38.4	21.0	19.9	25–30	32.6
Dołgie	U dissolv.	92	0.90	0.89	0.92	0.90	0.70	1.17	0.47	9.2	0.8–0.9	46.7
	U total	92	1.23	1.22	1.25	1.23	1.09	1.65	0.50	6.8	1.1–1.2	39.1
	U susp.	92	0.33	0.31	0.34	0.33	0.17	0.50	0.34	22.3	0.3–0.4	46.7
	U dissolv/total (%)	92	73.4	72.3	74.6	74.4	58.3	87.6	29.3	7.5	70.75	38
	U susp./total (%)	92	27.0	25.9	28.0	26.0	12.9	41.7	28.8	18.6	20–25	39.1
Swobnica	U dissolv.	45	0.86	0.82	0.90	0.84	0.65	1.18	0.60	15.1	0.8–0.9	28.9
	U total	45	1.13	1.09	1.17	1.12	0.77	1.45	0.68	12.5	1.1–1.2	40
	U susp.	45	0.27	0.25	0.30	0.30	0.10	0.46	0.37	30.4	0.2–0.3	42.2
	U dissolv/total (%)	45	76.4	74.3	78.5	77.2	59.2	90.3	31.1	9.2	75–80	37.8
	U susp./total (%)	45	23.6	21.5	25.7	22.8	9.7	40.8	31.1	29.7	20–25	37.8
Strzeszowskie	U dissolv.	89	0.88	0.86	0.90	0.89	0.59	1.11	0.52	9.9	0.8–0.9	57.3
	U total	89	1.18	1.15	1.20	1.18	0.81	1.52	0.71	10.6	1.0–1.21	31.5
	U susp.	89	0.30	0.28	0.33	0.30	0.11	0.58	0.47	35.3	0.2–0.3	42.7
	U dissolv/total (%)	89	74.8	73.3	76.2	75.0	60.9	89.7	28.8	9.2	70–75	28.1
	U susp./total (%)	89	24.5	23.0	26.1	25.0	7.5	39.1	31.6	29.5	20–25	25.8

of 0.8–0.9 μg/l (frequency 48.5%). with mean concentration of 0.88 μg/l and concentration extremes of 0.6 and 1.2 μg/l. Average suspended uranium contents was of 0.31 μg/l (maximum frequency range 0.2–0.3 μg/l. frequency of 42.3%) (Table 3). Dissolved uranium is very mobile in well oxygenated waters [Merkel. Sperling. 1998], hence the different variability of uranium concentrations – lower for total uranium and dissolved uranium (variation coefficient <10%). and higher for suspended uranium (variation coefficient ~ 25%). Overall, the element was present in the tested waters mainly in the form of dissolved uranium; on average nearly 75% of the total uranium content in the samples was its dissolved species (Tables 3 and 4).

The geographical differences in the occurrence of uranium are mainly due to the lithology variations of the catchment areas of individual aquifers and anthropogenic factors impacting the lakes and their basins. Maximum concentrations are usually observed in the surface waters in old mining districts. High uranium contents are found also in loess soils areas under intense agricultural exploitation (concentrations correlated positively with utilization of phosphate fertilizers) [Birke et al. 2009]. Ground water contains uranium usually in the range of 2 to 12 μg/1; the lowest uranium levels are recorded generally in the humid climate regions and the highest – in dry climates [Birke et al. 2009].

Consistent geological structure of the Tywa River drainage basin [Mikołajski, 1964] and evenly-spread human pressure in the area [Kubiak et al. 2013] mean that no significant statistical differences in uranium concentrations between the waters flowing in and out of the studied lakes occurred, and that only low variation (of the order of hundredths of a μg/l) in uranium concentrations in the different lakes were recorded. Similarly

Table 4. Concentrations of different forms of uranium in waters of the studied lakes in the River Tywa drainage area in different seasons

Season	Indicator	N	Mean	Confidence interval	Confidence interval	Median	Min	Max	Range	Variation coeff.	Max. freq. range	Freq. (%)
All lakes: total	U dissolv.	272	0.88	0.87	0.89	0.89	0.59	1.18	0.59	10.4	0.8–0.9	48.5
	U total	272	1.19	1.17	1.20	1.19	0.77	1.65	0.80	9.8	1.0–1.2	57.7
	U susp.	272	0.31	0.30	0.32	0.30	0.10	0.58	0.48	28.7	0.3–0.4	42.3
	U dissolv./total (%)	272	74.5	73.7	75.2	75.0	58.3	90.3	32.0	8.5	70–75	31.9
	U susp./total (%)	272	25.5	24.7	26.2	25.0	7.5	41.7	34.2	24.7	20–25	34.0
Spring	U dissolv.	73	0.87	0.85	0.88	0.9	0.7	1.0	0.3	7.0	0.8–0.9	58.9
	U total	73	1.15	1.13	1.17	1.2	0.8	1.3	0.5	8.3	1.1–1.2	45.2
	U susp.	73	0.29	0.27	0.31	0.3	0.1	0.5	0.4	31.2	0.2–0.3	45.2
	U dissolv./total (%)	73	75.2	73.8	76.7	75.0	60.9	90.3	29.4	8.4	70–75	30
	U susp./total (%)	73	24.9	23.4	26.3	25.0	9.7	39.1	29.4	25.1	20–25	42.5
Summer	U dissolv.	87	0.87	0.85	0.90	0.9	0.6	1.2	0.6	14.3	0.8–0.9	28.7
	U total	87	1.20	1.17	1.23	1.2	0.8	1.5	0.7	12.2	1.1–1.2	28.7
	U susp.	87	0.33	0.31	0.35	0.3	0.1	0.5	0.4	28.8	0.3–0.4	48.3
	U dissolv./total (%)	87	72.6	71.1	74.1	72.0	58.3	89.7	31.4	9.7	70–75	26.4
	U susp./total (%)	87	27.6	26.1	29.1	28.0	10.3	41.7	31.4	25.3	25–30	27.6
Epilimnion	U dissolv.	39	0.90	0.86	0.95	0.9	0.6	1.2	0.6	15.0	0.8+0.9	28.2
	U total	39	1.23	1.18	1.28	1.2	0.8	1.5	0.7	12.3	1.2–1.3	30.8
	U susp.	39	0.33	0.30	0.36	0.3	0.1	0.5	0.4	29.7	0.3–0.4	48.7
	U dissolv./total (%)	39	73.2	71.0	75.4	72.9	62.7	85.8	23.2	9.3	70–75	25.6
	U susp./total (%)	39	26.9	24.7	29.1	27.1	14.2	37.3	23.2	25.5	25–30	25.6
Metalimnion	U dissolv.	17	0.85	0.80	0.91	0.8	0.6	1.0	0.4	13.0	0.8–0.9	41.2
	U total	17	1.19	1.12	1.25	1.2	0.9	1.4	0.5	10.3	1.2–1.3	35.3
	U susp.	17	0.33	0.27	0.38	0.4	0.1	0.5	0.4	33.6	0.3–0.4	36.3
	U dissolv./total (%)	17	72.2	67.9	76.6	69.2	61.9	89.7	27.8	11.7	60–65	23
	U susp./total (%)	17	27.7	23.4	32.1	30.8	10.3	38.1	27.8	30.4	35–40	35.4
Hypolimnion	U dissolv.	31	0.85	0.81	0.89	0.9	0.6	1.0	0.5	13.4	0.9–1.0	35.5
	U total	31	1.18	1.12	1.23	1.2	0.8	1.5	0.7	12.9	1.1–1.2	35.5
	U susp.	31	0.34	0.30	0.37	0.3	0.1	0.5	0.4	25.9	0.3–0.4	54.8
	U dissolv./total (%)	31	72.0	69.5	74.4	72.0	58.3	87.4	29.1	9.3	70–75	38.7
	U susp./total (%)	31	28.4	26.0	30.7	28.0	12.6	41.7	29.1	22.7	25–30	41.9
Fall	U dissolv.	47	0.91	0.88	0.94	0.9	0.8	1.2	0.4	10.3	0.8–0.9	42.3
	U total	47	1.24	1.20	1.27	1.2	1.1	1.6	0.5	8.6	1.2–1.3	27.7
	U susp.	47	0.34	0.31	0.36	0.3	0.2	0.6	0.4	24.0	0.3–0.4	42.6
	U dissolv./total (%)	47	73.5	71.7	75.3	73.8	61.6	87.6	26.0	8.1	70–75	31.9
	U susp./total (%)	47	25.9	24.0	27.8	25.2	7.5	38.4	30.9	25.2	20–25	34.6
Winter	U dissolv.	65	0.89	0.88	0.91	0.9	0.8	1.0	0.2	6.2	0.8–0.9	67.6
	U total	65	1.17	1.15	1.19	1.2	1.0	1.3	0.3	6.8	1.1–1.2	40
	U susp.	65	0.27	0.26	0.29	0.3	0.2	0.5	0.3	22.0	0.2–0.3	67.7
	U dissolv./total (%)	65	76.2	74.9	77.4	75.0	61.5	83.3	21.8	6.7	70–75	41.5
	U susp./total (%)	65	23.7	22.5	24.9	25.0	16.7	33.3	16.6	20.1	24–28	41.5

Measuring unit: μg/l

low variation was found depending on the seasons of the year (Table 4). Because uranium is not an element essential for the functioning of hydrobiocenosis, no correlation was found between the uranium contents and levels of biological production indicators (%O_2 chlorophyll "a" total organic carbon) [Kalff, 2002].

Waters of the Dołgie Lake was richest in dissolved uranium, where the mean concentration was 0.90 μg/l. the results ranged between 0.70 – 1.17 μg/l, and the maximum frequency range was 0.8-0.9 μg/l – at 46.7% frequency. The smallest dissolved uranium concentrations were found in the waters of Swobnica Lake, where the mean dissolved uranium content was of 0.86 μg/l and the range between 0.65 – 1.18 μg/l, with the maximum frequency range of 0.8-0.9 μg/l (frequency 28.8%). Waters of the remaining two reservoirs showed intermediate concentrations of dissolved uranium although the maximum frequency ranges for all samples were the same for all lakes (Table 3 and 4). Similar variations were found for total uranium content, as well as suspended uranium - the average total uranium content varied from 1.23 μg/l in the Dołgie Lake and 1.18μg/l in the Strzeszowskie Lake to 1.13 μg/l in the other two tested aquifers. The total uranium concentration maximum frequency range was of 1.1-1.2 μg/l – for suspended uranium the range was between 0.2-0.3 μg/l (only for the Dołgie Lake, the maximum frequency range was of 0.3-0.4 μg/l) (Tables 3 and 4).

The study revealed that uranium contents in the waters of the studied lakes of the Tywa River drainage basin varied in different seasons with same changes noted for all aquifers. Dissolved uranium levels were highest in the autumn (mean of 0.91μg/l) and winter (0.89 μg/l); they were lower in the spring and summer (0.87μg/l). Highest total uranium contents were also recorded in the autumn (1.24 μg/l); in other seasons concentrations were similar (1.16 μg/l in winter. 1.15 in spring and 1.20 μg/l in the summer. respectively). The largest concentrations of suspended uranium species were found in the autumn as well (average 0.34 μg/l) and in the summer (0.33); they were lower in the spring and winter (mean contents of 0.29 and 0.27 μg/l. respectively) (Tables 3-4).

Seasonal and vertical variation in uranium contents in the samples was the same in all the studied lakes. In all the aquifers throughout the year no uranium concentration variation dependant on sampling depth was found for any uranium species. In the summer epilimnion waters contained 1.23 μg/l of total uranium on average, 0.90 μg/l of dissolved uranium and 0.33 μg/l of its suspended species. In the other layers, i.e. metalimnion and hypolimnion concentrations were of 1.19. 0.85 and 0.33 μg/l for total dissolved and suspended uranium in metalimnion and – respectively – of 1.18. 0.85 and 0.34 μg/l for hypolimnion; the recorded differences were not statistically significant (Tables 3 and 4).

No vertical differentiation in uranium concentrations was due to significant water dynamics in the tested bodies of water mass and the fact that uranium is not a biophile element, thus, vertical changes in the intensity of metabolic processes in lakes did not result in formation of vertical variation of uranium contents [Wetzel 2001, Kalff 2002].

CONCLUSIONS

The conducted studies allow for drawing of the following conclusions regarding the presence of uranium and its various species in lakes waters in the Tywa River drainage area:

1. In the studied aquifers uranium was present in low concentrations similar to those naturally occurring in surface waters.
2. There was no correlation between the concentrations of uranium and intensity of biological production indices.
3. On the basis of studied samples it was found that there is a small variation in uranium concentrations (of the order of hundredth of a μg/l) between different lakes. The most uranium-rich waters were those of the Dołgie Lake, while the lowest levels were recorded in the Swobnica Lake.
4. In the waters of the studied aquifers uranium content varied depending on the season. Dissolved uranium contents were highest in autumn and lowest in the spring and summer. Also total uranium concentrations were found to be greatest in the fall, in other seasons concentrations were lower and similar to each other. Highest suspended uranium levels were recorded in the autumn and summer they were lower in the spring and winter.
5. In the studied lakes no vertical variation in the sample uranium contents was noted.
6. There were no significant statistical differences in uranium concentrations between the waters flowing in and out from the four lakes.

REFERENCES

1. Bernhard. G., Gisel. G. 2007. Bestimmung der Bindungsform des Urans in Mineralwassern. Vom Wasser. 105(3), 7–10.
2. Birke M., Rauch U., Keilert B. 2003. Geochemische Untersuchungen. In: G. Lange, K. Knödel (Eds.). Handbuch zur Erkundung des Untergrundes von Deponien und Altlasten. Band 8: Erkundungspraxis. Springer-Verlag, Berlin, 209–428.
3. Bowen H.J.M. 1979. Environmental chemistry of the elements. Academy, London.
4. Birke M., Rauch U., Lorenz H. 2009. Uranium in stream and mineral water of the Federal Republic of Germany. Environ Geochem Health. 31, 693–706
5. Chappaz A. Gobeil Ch., Tessier A. 2010. Controls on uranium distribution in lake sediments. Geochimica et Cosmochimica Acta 74, 203–214
6. Choiński A.. 2007. Limnologia fizyczna Polski. Wyd. Nauk UAM. Poznań, 1-547.
7. Himiri M., Pastor A., Guardia M. 2000. Determination of uranium in tap water by ICP-MS. Fersenius J. Anal. Chem. 367, 151-156
8. Kabata-Pendias A., Kabata H. 1999. Biogeochemia pierwiastków śladowych. PWN. Warszawa, p. 322.
9. Kalff J. 2002. Limnology. Prentice Hall Ltd. New Jersey, p. 592.
10. Kratz S., Schnug E. 2006. Rock phosphates an P fertilizer as sources of U contamination in agricultural soils. In: B. Merkel & A. Hasche-Berger (Eds.). Uranium in the environment. Springer-Verlag, Berlin, 57–67.
11. Kubiak J., Oszkinis D., Zapert N. 2009. Wstępne dane o warunkach termicznych i tlenowych wód największych jezior zlewni rzeki Tywy. (W): Marszelwski W. (ed.). 2009. Anthropogenic and natural transformations of lakes. Vol. 3. PTLim. Toruń, p. 252.
12. Kubiak J., Machula S., Stepanowska K., Biernaczyk M. 2013. Wpływ zawartości glinu ma biocenozę wód jezior o zlewniach słabo zurbanizowanych. Inż. Ekolog. 35, 65-105.
13. Macioszczyk A., Dobrzyński D. 2007. Hydrogeochemia. PWN, Warszawa, p. 448.
14. Merkel B., Planer-Friedrich B.. Wolkersdorfer C. 2002. Uranium in the aquatic environment. Springer-Verlag, Berlin, p. 1112.
15. Merkel B., Sperling B. 1998. Hydrogeochemische Stoffsysteme. Teil 2: Deutscher Verband fur Wasserwirtschaft und Kulturbau e.V. Wirtschafts- und Verlagsgesellschaft Gas und Wasser, Bonn, p. 397.
16. Mikołajski J. 1964. Geografia województwa Szczecińskiego. STN, Szczecin, p. 221.
17. Pietrzak-Flis Z., Kamińska I., Chrzanowski E. 2004. Uranium isotopes in waters and bottom sediments of rivers and lakes in Poland. Nukleoinka 49 (2), 69–76.
18. Skwarzec B.A., Boryło A.D., Struminńska D., 2002. U isotopes in water and sediments of the southern Baltic. J. Environ. Radioactiv. 61, 345–363.
19. StatSoft Inc. 2009. STATISTICA (data analysis software system) version 10.0. www.statsoft.com.
20. Wetzel R.G. 2001. Limnology. Lake and River Ekosystems. Academic Press, London, p. 1055.
21. WHO 2011. Uranium in Drinking-water. Guidelines for Drinking-water Quality Water.. WHO/SDE/WSH/03.04/118/Rev/12.2.
22. Zielinski R.A., Asher-Bolinder S., Meier A.L., Johnson C.A., Szabo B.J. 1997. Natural or fertilizer-derived uranium in irrigation drainage: A case study in southeastern Colorado, USA. Applied Geochemistry, 12, 9–21.

11

ANALYSIS OF ECOLOGICAL ACTIVITY OF PODLASIE PROVINCE

Agnieszka Kisło[1], Iwona Skoczko[1]

[1] Department of Technology in Engineering and Environmental Protection, Faculty of Civil and Environmental Engineering, University of Technology in Bialystok, 45A Wiejska Str., 15-351 Bialystok, Poland, e-mail: agnieszka.kislo@wp.pl

ABSTRACT

The beginning of the period where the impact of human activity took dangerous size is considered to be 40 years of the nineteenth century. Then the large-scale production of electricity increased production of metallurgical industry and transport development. Significant changes contributed to rapid population growth and urbanization and industrialization processes. Man by intensive exploitation reduces or completely exhausted most natural resources. This has resulted in the availability criterion, which is distinguished by inexhaustible and exhaustible resources. Exhaustible resources may include solar, wind, energy, ocean currents and the energy of the Earth. Exhaustible resources, however, are the elements of nature that as a result of the operation can be completely depleted or destroyed. These resources include: water, soil, plant world, animal world natural resources. The greater the intensity of acquiring human resources, the faster will be the depleted. There is also a very important division of exhaustible resources: non-renewable resources, renewable and very difficult reproducible (living areas). Non-renewable resources are mainly minerals, such as iron ore, crude oil, coal, sulfur, natural gas, useful stones, limestone, salt, and many other minerals). After extraction resources are processed, and the bed from which they were extracted can no longer be used. It is different from renewable resources; they are characterized by the fact that over time they are reconstituted, although, for this to happen balance of the environment through the rational use must be maintained. These resources include, among others, soil, water, flora and fauna, as well as atmospheric air.

Keywords: natural resources, environmental transformation, environmental management, sustainable development.

INTRODUCTION

Anthropopression, that the pressure exerted by humans on natural environment often causes irreparable changes. The basic factors of anthropogenic impact include: waste-that is worn objects and substances(non-sewage) waste-that is, a mixture of waste water and chemical pollution, biological and thermal burden on the environment, pollution of the environment-that is, foreign ingredients in any part of the environment that cause distortion of the characteristics; environmental disturbances-that is, noise, vibration, radiation, malodorous fumes, harmful or nuisance to the environment [4].

Environmental protection is a rational, consistent with the laws of nature and of social shaping of human environment, so that the elements of nature united harmoniously with the products of technology and civilization. The origins of interest in environmental issues is a breakthrough year 60/70 in the twentieth century. Protecting the environment involves activities to enable the restoration of the natural balance. They consist of preventing or counteracting the negative effects on the environment, causing its destruction, damage, pollution, changes in physical, chemical or biological.

Activities under the protection of the environment:

- the use of clean or waste-less technologies,
- the use of technologies that reduce demand for raw materials,
- reducing energy demand,

- reducing greenhouse gas emissions,
- increasing recycling of materials and supplies,
- exploration and implementation of new technologies to reduce emissions,
- developing awareness [2].

Environmental awareness and actions that are taken in the context of its formation are Very important in protecting the environment. The aim of shaping ecological awareness is to sensitize human environmental problems and to develop such aspirations and lifestyles that will enhance the care of the surrounding nature. An important ascertainment extensive surveys of action on the environment is the finding that a prerequisite is to transform the environmental awareness of the Earth's population conditioning attitudes toward ecosystem. As noted by Szymańska [5] "consistent increase of investment for environmental protection is also an important task, as the wake of environmental awareness and education of the next generation in safety and preparation for life in harmony with sustainable development" [7].

Podlasie region belongs to the Polish regions of the purest state of the environment, as evidenced by lower than in other parts of the country the emission of air pollutants, the quantity and quality of discharged wastewater, heavy metals in the soil, the purity of surface and groundwater and other environmental stress factors [4].

Almost entire region of Podlasie is located in the basin of the Vistula River. Only the north-eastern part is in the basin of the Neman River. The main hydrographic form Narew, Biebrza and the Bug, a complement to the network are the lake. The largest of these is Wigry, Hańcza, Szurpiły, Necko, Rajgrodzkie. Podlasie region boasts a beautiful, unspoiled nature and diversity of the landscape. It belongs to the Polish Green Lungs. There are beautiful forests, which have been preserved almost intact: fragments of Knyszyńska Forest, which is one of the largest forests in Poland. Outside the county there are also Narew National Park - with a highly complex system of river channels, which is often called the "Polish Amazon", is a refuge for many animals, especially waterfowl and wildlife reserves [4].

The second distinguishing feature is its multiculturalism - Podlasie and rich history of shaping the influences of various multi-denominational nations. The dominant sector of its economy of the region is agriculture, which is almost entirely associated with the private sector. The level of industrialization in Podlaskie Province is low and based mainly on the production and processing of food, which is dominated by dairy products, meat, fruit and vegetables, brewing, distilling and milling [7].

METHODOLOGY

In order to meet environmental assessment activity two methods were used. First, examination of documents, consisted of the collection, selection and interpretation of documents, making an assessment of the effectiveness of the tasks contained in the strategic management of the environment for the protection of the environment, as well as familiarize themselves with the state of the environment, the organizational structure, and resolutions of local government.. The second method was a method of self-observation (the environment), which is based on the observation of phenomena and drawing conclusions. Strategic documents, in environmental management for the protection of the environment are:

- Environment Programme,
- Strategy Development,
- Waste Management Plan,
- Study of Conditions and Directions of Spatial Development.

DISCUSSION

The first document, which is Environment Programme is aimed at achieving the objectives of the national strategic documents with a particular focus on the National Environmental Policy in Podlaskie. Its essence is coordinated, planned in the Programme activities with government, local government (district offices, offices of cities and municipalities) as well as business and society and their implementation by those entities with effective use of available funds. The program further defines a framework for subsequent projects implemented within the framework of province sectoral programs, and also includes guidelines for environmental programs at the county level. [6].

Development Programme carried out in accordance with current in the range of legislation, as well as on the basis of guidelines to develop environmental programs at the regional and local level prepared by the Ministry of the Environment (Warsaw, December 2002).

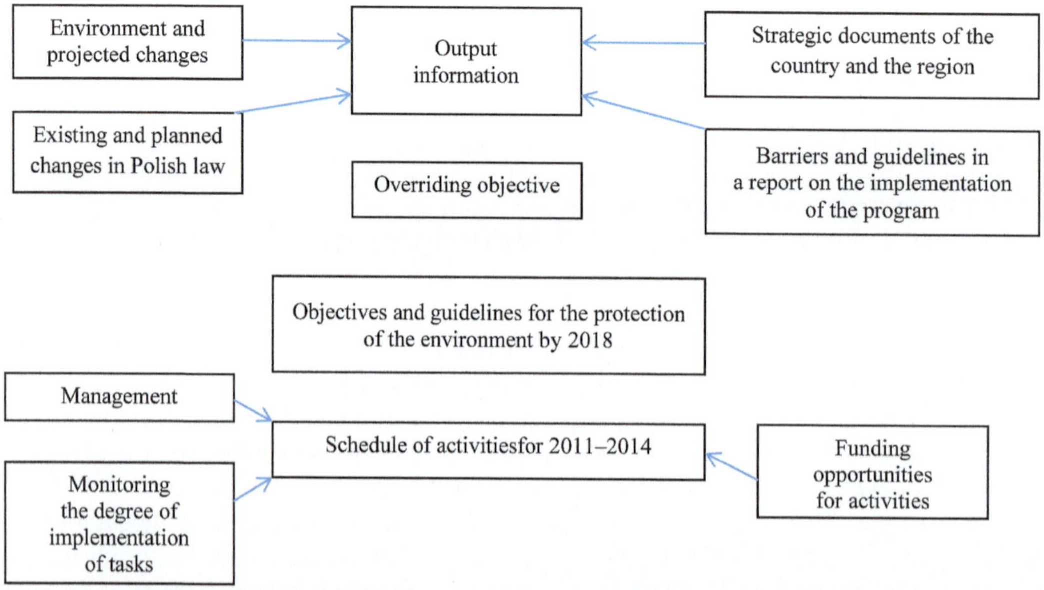

Figure 1. The process of creating Environment Programme [8]

Table 1. Actions taken under the Environment Programme in 2007-2010 [6, 9, 10]

Priority	Purpose	Activities
Priority I: Development of environmental infrastructure	**Short-term:** reduce traffic noise	• development of road infrastructure in order to reduce noise and vibration, • construction of bypasses,
	Short-term: reduce emissions of particulate matter into the air	• keeping air quality standards concerning dust, • the use of environmentally friendly technologies, • boiler modernization and investment in the field of renewable energy sources (excluding those which increase the emission of dust in the zones to excessive dust standard),
	Short-term: limite utrophication (agriculture, cleaning of sewage, wastewater village)	• development of water and sanitation services in the province, • modernization of out dated sewage treatment plants, • restore the proper functioning of the drainage of areas at risk of water deficit,
Priority II: Protecting eco-logical region		• protection of human health and life, • active protection of nature and landscape, • care for protected species increasing the number of forest areas in order to protect the atmosphere, • compliance with the requirements of the regional information system and transport routes storage of hazardous materials, • compliance with the requirements of the regional early warning system and support the emergency services, the strengthening of controls to protect against noise in the manufacturing and service facilities, offsetting losses of Nature 2000 sites, • reducing the risk of major accidents and the efficient removal of their consequences, • the development and implementation of air protection programs.
Priority III: Building en-vironmental awareness		• increasing the environmental awareness of the region's population, • promoting co-style production and consumption, • ecological education of children and youth in schools and kindergartens, • the Environmental Protection Podlaskie Voivodeship for the years 2011-2014, • the creation of local centers of environmental education, shaping information policy aimed at abandonment of conventional energy sources, • mobilizing society to take environmental actions, • supporting institutions and non-governmental organizations dealing with the protection of the environment, • raising the effectiveness of compliance with environmental regulations.

Table 2. The term problems, 2011-2014 [6, 9, 10]

Area environmental	Problem priority problem	Purpose
Air pollution(PA)	**Priority I:** • exceeding the permissible levels of PM10 and levels of objectives for long-term ozone	Long-term objective for the year 2018 Continuation of measures to improve air quality Short-term objectives by 2014: 1. Introduction and implementation guidelines for the protection of Air Programs 2. The fulfillment of legal requirements for air quality by reducing emissions from sources of surface, line and point
Water(W)	**Priority II:** • insufficient degree of channeling, • not enough good water status **Priority II:** • droughts	Long-term objective by 2018 Achieving and maintaining good surface water status and protection of water quality underground and rationalize their use Short-term objective for 2014: 1. Providing good quality drinking water 2. The rationalization of the management of water resources 3. Improving the quality of surface and groundwater 4. Increasing retention basins and preventing the effects of drought and flood flow 5. Restoration and rehabilitation of the ecological continuity of rivers
Protection against noise (H)	**Priority I:** • insufficient environmental monitoring (recognition of areas within zones of exceedance), • oversized noise	Long-term objective by 2018 Reducing noise pollution by reducing its intensity to the level of the existing standards Short-term objective for 2014: 1. Identification and assessment of the exposure of the inhabitants of the region to excessive noise, 2. Eliminating exposure to noise residents.
Renewable energy sources (RES)	**Priority II:** • greater use of renewable energy sources	Long-term objective by 2018 Reducing energy consumption and increase the use of renewable energy sources Short-term objective by 2014: 1. Increasing the use of non-conventional energy sources.
Soils and their impurities (GI)	**Priority III:** • contamination of soils and land with heavy metal sand other substances of anthropogenic origin (wet and dry deposition), • over-acidification of soils.	Long-term objective by 2018 Protecting the earth's surface Short-term objective for 2014: 1. Development of the surface of the earth in accordance with the principles of sustainable development, 2. Indication of contaminated sites and remediation.

The program covers years 2011–2014 with the prospect of 2018. The Period obliges this Programme to be divided into:

- operating period (2011–2014), together with specific short-term objectives and the objectives of their implementation,
- perspective period (years 2015–2018), expressed by one long-term goal for each of the priorities of environmental protection.

The current state of the environment is described in the Programme on the basis of the available data mainly from 2009 in certain cases where it was possible was used more recent data (from 2010) [9].

The guiding principle adopted in the program is the principle of sustainable development, which enables harmonized economic and social development compatible with the protection of environmental values. Therefore, as the overarching goal of the program was adopted: "Sustainable Developmentin the Region Podlaskie Improvements and Promotion of the Environment".

This objective is in line with the vision of development of the region of Podlasie defined in the Strategy for the Development of the Podlaskie province by 2020 the region's economic development must go hand in hand with measures to improve where necessary environment and behavior undoubtedly its exceptional natural values [6].

Another document analyzed is the Podlasie Region Development Strategy. Strategy is the key document defining the principles of software and directions of the long-term concept of development of the region.

The objectives set will be achieved and visible indicators will be achieved when assigned to them. This implies the monitoring of a limited number of indicators strategies for each of the objectives. When select in indicators guided the adequacy, however, the primary constraint

Table 3. Possible sources of funding included in the program, broken down into individual components of the environment [6, 7, 9, 10]

Source of funding	Components of the environment								
	renewable energy sources	water	waste mana-gement	nature and landscape protection	protection against noise	protection against elec-tromagnetic fields	prevention of major industrial accidents	soils and their impuri-ties	environmental education
NFOŚiGW	•	•	•	•	•	•	•	•	•
WFOŚiGW	•	•	•	•	•	•	•	•	•
POZE	•	•	•	•	•	•	•	•	•
Operational Progra-mme Infrastructure and Environment	•	•	•	•	•	•	•	•	•
RPO WP	•	•	•	•	•	•	•	•	•
LIFE+	•	•	•	•	•	•	•	•	•
Priority Environ-mental Education Program	•	•	•	•	•	•	•	•	•
PPOPiK	•	•	•	•	•	•	•	•	•
PPOiZRL	•	•	•	•	•	•	•	•	•
ZRRPB	•	•	•	•	•	•	•	•	•
Program KPOŚK, BPiOOH	•	•	•	•	•	•	•	•	•
PZŚzLIS	•	•	•	•	•	•	•	•	•
PBGKoGZZKiWP	•	•	•	•	•	•	•	•	•
Preferential Loans and borrowings	•	•	•	•	•	•	•	•	•
Loans and advances granted by commer-cial banks	•	•	•	•	•	•	•	•	•
Intelligent Energy Europe Program	•	•	•	•	•	•	•	•	•
Project GDOŚ	•	•	•	•	•	•	•	•	•
POIiŚ	•	•	•	•	•	•	•	•	•

Explanations:

NFOŚiGW – National Fund for Environmental Protection and Water Management.
WFOŚiGW – Regional Fund for Environmental Protection and Water Management.
POZE – program for projects in the field of renewable energy sources and cogeneration.
RPO WP – Regional Operational Programme of Podlasie.
Project GDOŚ – Design of the General Directorate of Environmental Protection.
EFFR – European Regional Development Fund.
WIOŚ – Inspectorate of Environmental Protection in Bialystok.
POiIŚ – Operational Programme Infrastructure and Environment.
PPOPiK – The Priority Protection of Nature and Landscape.
PPOiZRL – The Priority Protection and Sustainable Development of Forests.
ZRRPB – The Priority Protection of areas of high natural - area Forest SustainabilityBialowieza.
BPiOOH – Priority Programme "Construction, reconstruction and restoration of hydro-technical facilities".
PZŚzLiS – The Priority Countering threats to the environment with the liquidation of their consequences.
PBGKoGZZKiWP – Understanding the geological structure of the country and the economy of resources and mineral reserves of groundwater.

was the availability of data. Due to the necessity preserving values of the major source of comparative data to monitor Strategy will be public statistics.

The potential financial resources that can be used by the public sector for the implementation of the Regional Development Strategy for the years 2013–2025 Podlaskie estimated a total of 59.8 billion PLN, which gives about 4.60 billion PLN average [7]. Possible sources of funding of the Development Strategy are presented in Figure 3.

Table 4. Indicators for monitoring purposes [6]

Strategic objective	Indicators objectives	Measure	Source	Base value (year)	Target (2020)
Objective 1: competitive economy	GDP per capita (constant prices of 2005)	PLN	calculations based on data from the Central Statistical Office	22 300 (2010)	30 687
	number of economic entities in the registry REGON per 10 000 inhabitants	pieces	Central Statistical Office	764 (2011)	830
Objective 2: national and international links	the value of exports per capita	EUR	calculations based on data from the Ministry of Finance	930 (2010)	1320
	share of foreign capital invested in the Podlaskie in the total value of the capital in Poland	%	calculations based on data from the Local Data Bank	0.27 (2011)	0.34
Objective 3: quality of life	balances internal and international migration for permanent residence	people per 1000	Central Statistical Office	-1.64 (2011)	-0.4
	the average monthly disposable income for one person in the household	PLN	Central Statistical Office	1188.5 (2011)	1648

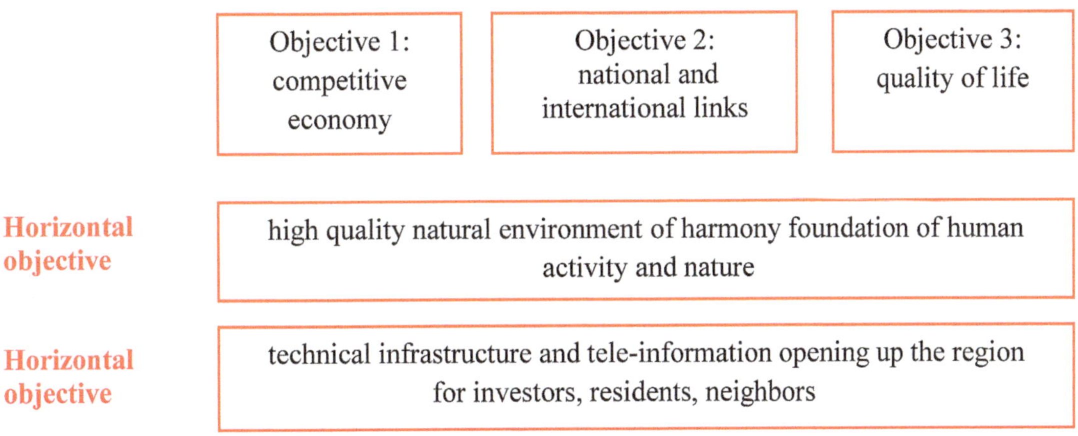

Figure 2. The objectives set out by [7]

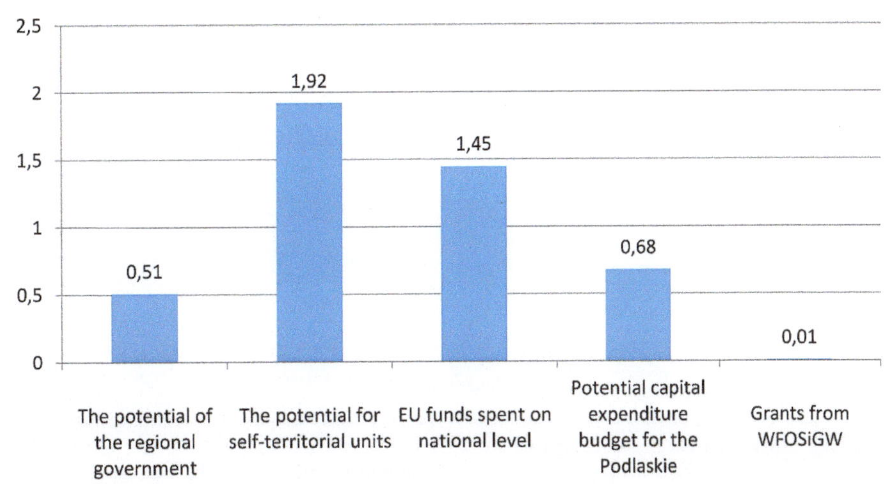

Figure 3. Possible sources of funding Development Strategy province Podlaskie [7]

Waste Management Plan is another document which was analyzed in order to assess ecological activity in Podlaskie. Waste Management Plan of the Podlaskie Voivodeship (hereinafter referred to as WPGO) in accordance with the provisions of the Waste Act (Article 15) including all types of waste generated in the province, and imported into its territory, in particular with regard to municipal waste biodegradable waste, packaging waste, waste the construction, repair and demolition of buildings and road infrastructure, worn out tires and hazardous waste, including used vehicles, electrical and electronic equipment waste, PCBs, asbestos, medical and veterinary waste, waste oil, batteries and accumulators [5].

As part of the project WPGO a series of meetings with the municipalities, which present guidelines for the creation of provincial waste management plan. The proposed variant division of the province of Podlasie region management were assessed municipalities that indicated to the region management Waste municipality will belong. Suggestions municipalities were used to develop a system waste management in this Project WPGO. The Plan aims and objectives relate to the period 2012–2017 [5].

The municipal waste management indicated to achieve the following objectives for 2017:

1. The increase in the share of recovery, in particular with regard to the recycling of glass, metals, plastics, paper and cardboard, as well as energy recovery from waste in accordance with the requirements of environmental protection.
2. Increasing the amount of separately collected hazardous waste occurring in the municipal waste stream.
3. Eliminate the practice of illegal dumping of waste [9].

For the realization of achieving the objectives it is necessary to engage in the following activities:

1. Measures aimed at the prevention of waste, reduce the amount of waste and its negative impact on the environment.
2. Support activities to the proper handling of waste in the collection, transport, recovery and disposal.
3. The implementation of the specified waste management system based on regions waste management (RGO) [9].

It is estimated that the total costs of the tasks in the field of waste management in Podlaskie region by 2017 will amount to 1 912 892 thousand PLN. The costs of the tasks are presented in Figure 4.

CONCLUSIONS

The province of Podlasie includes smaller local authorities with different economic potential and diverse natural conditions and degree of urbanization.

Most units are used to create strategic documents on environmental management. They are additional elements supporting the rational planning of activities and the prevention of risks. They planning activities in the environment, as well as anticipate, prevent or remove the effects of negative changes.

With these documents it is easier to implement the concept of sustainable development. It perfectly reflects the substance of the first sentence of the report of the World Commission on Environment and Development of 1987 – "Our Common Future", i.e.: "At the present level of

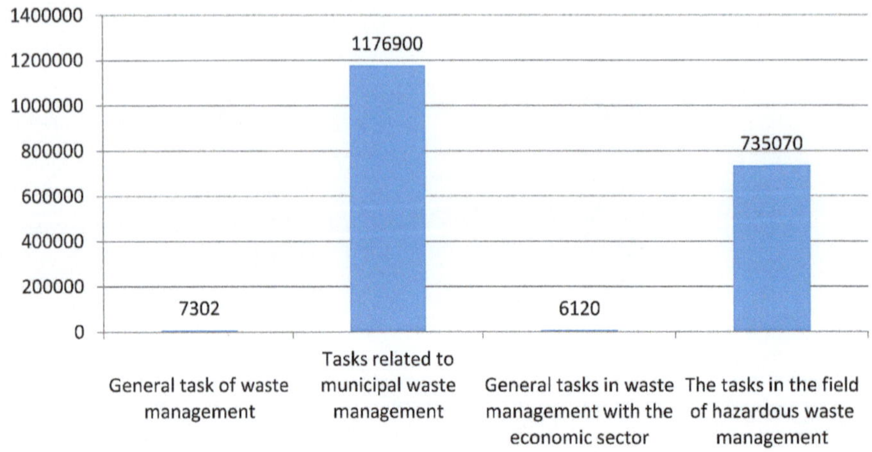

Figure 4. Estimated costs of implementation of the tasks under the Waste Management Program [9]

civilization sustainable development is possible, it is a development that meets the needs of the present generation may be met without compromising the ability of future generations to meet their own needs. It follows all the measures, which aim at striving for the development of social, political or economic aspects andshould not indelibly impoverish natural resources. The development of particular areas should be limited by the capacity of the environment to human activities.

After analyzing those documents it was concluded that individual province regions differ in the level of sustainable development. This is primarily the result of unsystematic development documents, which can make chaos.

Undoubtedly, the document in which their goals were realized in the greatest number is the Environmental Protection Plan. The data presented in the Report show that within almost all components the planned actions have been taken. Short-term objectives were partially met. Long-term objectives were only partially achieved, and therefore, in many cases, the report suggests their continuation. The present state of the environment in Podlaskie is satisfactory, although it is necessary to continue environmental policy in the region, in order to preserve natural values in the region and maintain good environmental status.

Additionally, waste management plan was implemented effectively. The main objective was to reduce the number of waste segregation "at source", and the selection of recycled waste, broken down by means of recovery and disposal of waste generated in the municipal sector. strategies and tools to promote selective waste collection have an important role in achieving this goalThrough self-observation of the environment can be said to developed and undoubtedly improved its environmental activity without affecting significantly the landscape values for the last years. This is due to the fact that the development of the Podlasie province was allocated EUR 2 863 656 282.58 within the framework of the Regional Operational Programme 2007–2013 Podlaskie. This yielded 1,160 completed projects. Nearly half of these projects is to campaign for the wider environment. They concerned the development of environmental infrastructure, i.e., sewage treatment plants, landfills, water treatment stations, etc.

The field that begins to develop in the Podlaskie is renewable energy. Although this is a relatively new area, it has already gained many followers. Incentives to take advantage of renewable energy sources are solar subsidies, which can be as high as 70%. Cost of investment returns after about two years. Relatively few sources of energy use hydropower, biomass and wind energy in the Suwalki region. The impact of renewable energy on the quality of the natural environment is relatively small.

The investments resulted in the improvement of the sanitary condition of surface water due to the construction of new and modernization of existing plant. Expansion of the sewage network also contributed to the improvement of their quality. Requiring treatment was 39.3 hm^3 (97.8%), and purified 39.0 hm^3 (97%), including: mechanically – 0.8 hm^3 (2%), biologically 13.7 hm^3 (34,1%) and biological nutrient removal with increased 24.5 hm^3 (60.9%). From a sanitary sewage system uses 60.2% of the inhabitants of the region [4].

In summary, for the evaluation of ecological conditions in the province of Podlasie two methods were used. The first of these was the analysis of strategic documents on environmental management. On the basis of this method, it was concluded that local authorities in Podlaskie are at different levels of sustainable development. Shortcomings and deficiencies observed in the presence of strategic documents. Analyzing documents for the region as a whole unit it can be concluded that the documents that help in environmental management, are implemented fairly, define specific objectives, which are successively implemented. Phenomena and processes occurring in the proximal and distal environment province, the context of the new EU financial perspective, as well as internal conditions and the structure of the objectives make it necessary in the coming years to involve many institutions and teams of human and financial resources in key areas of structural intervention.

The second method used to assess the activity of the ecological province of Podlasie was a method of self-observation. On the basis of this method it was estimated that environmental activity of the region is at a high level. This fact stems mainly from the fact that local authorities utilize resources for the wider environment very well, causing the improvement of the environment over recent years.

REFERENCES

1. Berkowska E. 2004. Transpozition europian law to the Polish environmental law. Biuro Studiów i Ekspertyz, Wydział Analiz Ekonomicznych i Społecznych, no. 1023/2004.
2. Nowak-Far A. 2000. Environmental Protection in UE. Warszawa, 4(15).
3. Pyłka-Gałkowska E. 2004. Ecology and environmental protection. Wydawnictwo Oświata, Warszawa.
4. Szczepanowski A.E. Qualities and tourist attractions the region of Podlasie.
5. Szymańska U., Zębek E. 2008. Law and environmental protection – legal, economic, ecological and technical aspects of natural environmet protection. Wydawnictwo Uniwersytetu Warmińsko-Mazurskiego w Olsztynie, Olsztyn.
6. Environmental Protection Plan for the Podlasie province 2011–2014.
7. Development Strategy for the Podlasie Province 2020.
8. Environmental Impact Assessment of the the Podlasie Province, 2011–2014.
9. Waste Management Plan 2012–2017 Podlaskie Voivodeship.
10. Act from 8th March 1990 about Municipal Government (Dz. U. z 2001 No. 142).

12

THE REMOVAL INVESTIGATION OF NONYLPHENOL ETOXILAT SURFACTANTS IN ACTIVATED SLUDGE SYSTEMS

Batool Ahansazan[1], Baharan Moazenipour[2]

[1] Departmen of Chemical Industries, Valiasr Technical College of Tehran, Sepah Str., Tehran, Iran, e-mail: ahansazan@tvu.ac.ir

[2] Technical Faculty, University of Tehran, Enghelab Ave., Tehran, Iran, e-mail: baharan_mp@yahoo.com

ABSTRACT

The most significant source of environmental pollution derived from perilous wastes is the circumstantial and intentional emancipation of specific industrial wastes including resistant and/or toxic pollutants to natural environments. Although, biological treatment methods have been commonly found as most effective alternatives in the removal of persistent compounds in industrial wastewaters, they require some increase for obtaining acceptable removal efficiencies, due to the presence of refractory or toxic compounds in the wastewaters. In this study, the use of surfactant of nonylphenol ethoxylates (NPE) in the removal of persistent organic pollutants by biological treatment processes was investigated as an enhancement technique. The application of surfactants can enhance soil and groundwater remediation by increasing contaminant locomotion and solubility to ameliorate the performance of practical conventional remediation technology and by barricading the departure of contaminants to speed the rate of biodegradation of contaminants in environment. The proven effectiveness of surfactants in soil and groundwater remediation has been considered reasonable to expect that surfactants can also enhance the removal of persistent organic pollutants in wastewaters. Different concentrations of nonylphenol ethoxylates (NPE) (1000, 1500, 2000 ppm) were tested to optimize biosurfactant-enhanced degradation of persistent pollutants in wastewaters. The results of this study demonstrate that the biodegradation of persistent organic pollutants in wastewaters is elevated by the use of biosurfactants. The principal mechanism that raises the biodegradation is the augment solvability of poorly soluble compounds in the wastewater. According to the results of this study, it can be anticipated that biosurfactant-enhanced degradation would result in faster and more complete degradation.

Keywords: nonylphenol ethoxylates, surfactants, treatability, activated sludge systems, COD.

INTRODOCTION

APEs have been significant surfactants for more than 50 years. The most joint APE in mercantile use today is the group of nonylphenol ethoxylates (NPEs), comprising about 80 percent of the market, while octylphenol ethoxylates (OPEs) contain most of the remaining 20 percent of the market. NPEs are used primarily for industrial applications, including pulp and paper production, textile structure, and use in the formulation of crop protection chemicals. They are also used in industrial, institutional, and homemaker cleaners and detergents. While NP is used primarily as a raw material in the generation of NPE, some is also used in the production of plastics, resins and stabilizers. In contrast, OP is made in significantly lower volumes and is used primarily as a chemical intermediate in the generation of phenolic resins [APERC, 2006]. Derivative of Alkylohenol ethoxylates is given in Table 1.

Table 1. Acronyms of Alkylphenol Ethoxylates and their derivatives

Compound or Chemistry	Acronym
Alkylphenol ethoxylates	APE
Alkylphenol	AP
Alkylphenol ether carboxylates	APEC
Carboxylated alkylphenol ether carboxylates	CAPEC
Nonylphenol ethoxylates	NPE
Nonylphenol	NP
Octylphenol	OP
Octylphenol ethoxylates	OPE

Environmental issues arising from NP

The US Environmental Protection Agency (US EPA) has recently consummated ambient aquatic life water modality criteria for nonylphenol (NP) [US EPA, 2005]. These criteria were expanded under Section 304(a)(1) of the Clean Water Act and indicate the concentration of NP in water at which aquatic life are safe from acute and chronic adverse effects. The US EPA has counsel an acute freshwater touchstone of 28 µg/L and a chronic freshwater criterion of 6.6 µg/L for NP. For saltwater, the proposed sharp and chronic criteria are 7.0 µg/L and 1.7 µg/L.

Environmental Fate of APEs and NPE

Biodegradation is the overcoming removal process for APEs and NPE in water, settlement and soil. The biodegradation of APEs and NPE has been palmate studied for the past 50 years in a diversity of laboratory studies that have searched both primary and final degradation. Accurate screening tests that measure the ready biodegradability of different APEs and NPE, laboratory modeling tests of potential degradation in specific environmental compartments, and field confirmation studies show that NPE and APEs are extensively biodegraded in surface waters, alluvium, and soil. Collectively, the data from these tests indicate that NPE and APEs, and their degradation moderator are not forestalled to be persistent in the environment. Removal yield for laboratory scale semi-continuous or batch activated sludge and continuous flow activated sludge systems are generally over 90 percent.

Investigations accomplished in this Topic

A collation of the definitive biodegradability (transformation of organic carbon to CO_2) of AE and NPE was manufactured in an improved Sturm test [Sturm, 1973]. Habituate activated sludge cultures from the AE and NPE units were incubated with 50 mg/l surfactant and the rate of CO_2 formation measured over 14 days at 25 °C. AE was quickly and extensively oxidized to 70–75% of the theoretical C02 yield. Organic carbon from the NPE ethoxylate, however, was metabolized to 25–30% of the theoretical CO_2 yield during the same period headings should not be indented).

Studies by Yoshimura [1986], and Stephanou and Giger [1982] have also shown that waste water sewage from act NPE-type surfactants are extremely toxic (e.g., LC50 – 48 hours) to fish and *Daphnia* in the presence of residual ethoxylate (5–10 mg/l), nonylphenol ethoxyacetate metabolites (9–10 mg/l) and nonylphenol (0.15–0.3 mg/l).

MATERIALS AND METHODS

Experimental system

A schematic diagram of the experimental setup is showed in Figure 1. A continuously stirred tank reactor (CSTR) was used in the experimental study. Volume of the aerobic reactor was 4 liter. The influent wastewater was continuously fed through the top of the reactor by a feed pump and the reactor was aerated by an air pump. The effluent from the precipitation tank was collected in an effluent tank.

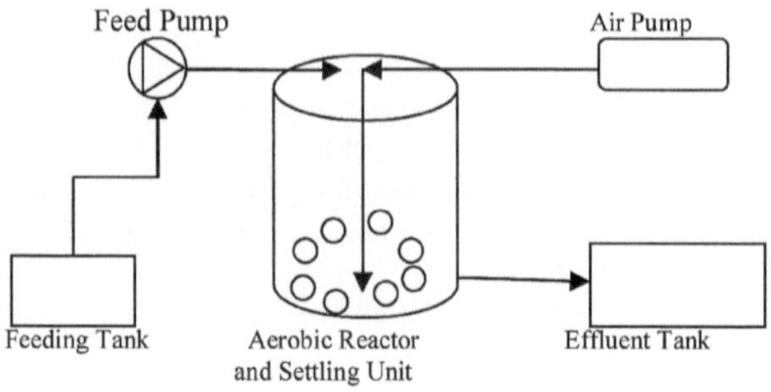

Figure 1. A schematic diagram of the activated sludge bioreactor used in experimental studies

Preparation of activated sludge

A mixed culture was used in the aerobic reactors. The activated sludge culture was received from the wastewater treatment plant of Oil National Company in Tehran, Iran. The aerobic reactor was inoculated with this culture. The constituents of wastewater used throughout the studies are given in Table 2. Whilst the main substrate requirement is for carbon, the growth also depends on the intake of nitrogen and phosphorus. The optimum ratio of C:N:P in the mixed liquor is generally thought to be 100:5:1. The ratio of these nutrients in settled domestic sewage is variously reported as 100:17:5 or as 100:19:6. This indicates that nitrogen and phosphorus will not be limiting for growth. Trace ingredients, which include S, Na, Ca, Mg, K, and Fe are also required, and are available in plenty in domestic sewage. By contrast, wastewater from brewing, pulp and paper, and food-processing industries can be deficient in nitrogen and phosphorus. Nutrients therefore need to be added to the mixed liquor to receive maximum bacterial growth and to optimise carbonaceous treatment. From an operational point of view, lack or insufficiency of a critical nutrient may result in incomplete treatment, because the bacteria are unable to grow optimally.

Phosphor, Nitrogen, Carbon and Oxygen are required that were added every day. In the starting test, first and second day the sludge settling time was two hours then collected the dead cells. On the third day sludge feeding began. The amount of feed was equal to glucose 7.5 g/day, Ammonium phosphate 0.34 g/day and Ammonium nitrate 0.93 g/day. The feeding was done for a week. So the culture was ready for testing.

Table 2. Inorganic constituents of microorganismos

Chemical Compounds	Percent
MgO	8
K_2O	6
Fe_2O_3	1
P_2O_5	50
SO_2	15
Na_2O	11
CaO	9

Experimental procedure

Four reactors with the same structure and volume, as explained above, were used in parallel tests. Experiments were start up batchwise. Activated sludge from wastewater treatment plant was added to the reactors as seed. The similar value of sludge was inoculated in four parallel reactors. The test reactors (R1, R2 and R3) contained biosurfactant. The NPE biosurfactant concentration were added to R1, R2 and R3 respectively: 1000, 1500, 2000 ppm. In the control reactor (R4), the feed water did not contain any surfactant in order to specify the effect of surfactant. Each of the test series dissolve in distilled water and make up 1 litter. All of the reactor was aerated by an air pump. These conditions were Fixed for the entire duration of the experiment. The temperature and pressure were 25 °C and 625 mmHg and pH adjusted to 7. A schematic diagram of the experimental setup is showed in Figure 2.

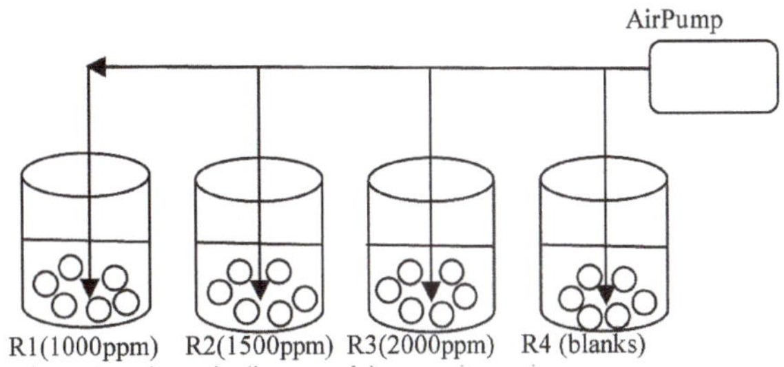

Figure 2. Schematic diagram of the experimental setup

RESULT AND DISCUSSION

A set of experiments were performed in three different surfactant concentration, for two weeks. The method was based on measurements of American Public Health Association (APHA), 1998.

The rate of MLSS and MLVSS of activated sludge in the experimental period is shown in Figure 3. When sludge age was increased for 14 days, MLSS and MLVSS ratio were increased 2605 to 2805 mg/l and 2555 to 2780 mg/l respectively.

COD removal efficiency with the increase of residence time as shown in Table 3. At the 3 days, COD removal efficiencies was from 5.19% to 48.24 in R1, it was from 7.61% to 48.91 in R2 and 8.14 to 34.89% in R3. Reduction was observed in the first week, equal to 97.49% in R1,90.58% in R2 and 84.69% in R3. Although,in this period at R1 and R2 greater amount from COD were eliminated, in the last, the overall results show that the highest COD decrease was in R3.

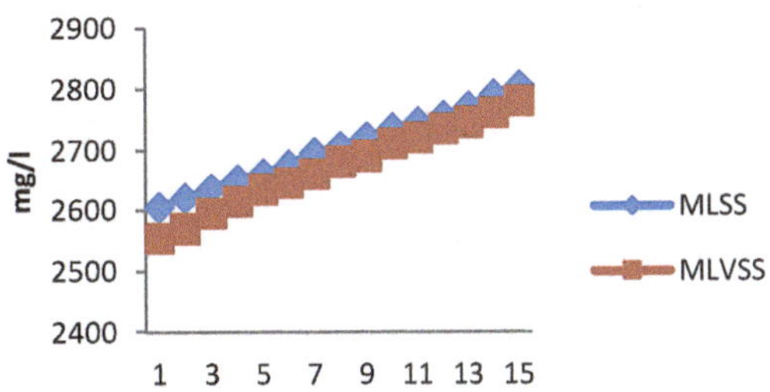

Figure 3. The rate of MLSS and MLVSS of activated sludge in during the test

Table 3. Comparison of operational parameters of the reactors with different surfactant concentration

Reactor	Run	SRT (day)	COD Removal Efficiency (%)
R1	Run 1	1	5.19
	Run 2	3	48.24
	Run 3	7	97.49
	Run 4	14	97.78
R2	Run 1	1	7.61
	Run 2	3	48.91
	Run 3	7	90.58
	Run 4	14	97.42
R3	Run 1	1	8.14
	Run 2	3	34.89
	Run 3	7	84.69
	Run 4	14	98.34

Figure 4 depicts a shift of COD volumetric removal rates with over time. Sharp slope increase in COD volumetric removal rate with the time in R1, R2 and R3 because of high decreases in COD concentrations occurs in the first 7 days. After that, the removal rate is linearly. The results show, decline slope in the R1 and R2 is more than R3. But at the end of the test, the maximum removal occurs at the highest concentration. At the end of two weeks, more than 98% of COD removal was observed in R3 with concentration Equivalent 2000 ppm NPE

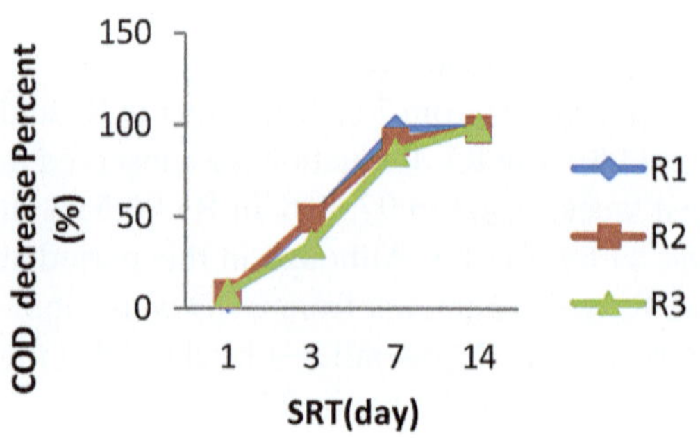

Figure 4. The effect of surfactant concentration on COD volumetric removal rates in the test reactors

CONCLUSIONS

In this paper, the authors investigate the impact of NPE at activated sludge system and COD removal in conventional systems of biological treatment. Increasing the concentration of surfactant would lead to treatment plant efficiency enhance. Also the results showed that increasing the concentration of solids MLSS makes the removal efficiency increase.

According to the results of this study, it can be completely proven by further investigations that biosurfactant-enhanced degradation would result in prompter (reduced treatment times) and modified water quality, and overall cheaper treatment costs.

The interaction between biosurfactant and the pollutants in wastewater is a very complex phenomenon and is not the subject of this study. But it is possible to say that biosurfactants are very effective in the COD removal as demonstrated in this study. So, the effectiveness of different biosurfactants in the removal of more pollutants should be investigated in the future studies in order to distinguish the best combination.

REFEENCES

1. American Public Health Association (APHA), 1998. Standard methods for the examination of water and wastewater, 20th ed.
2. Stephanou E. and Giger W. 1982. Persistent organic chemicals in sewage effluents. 2. Quantitative determinations of nonylphenols and nonylphenol ethoxylates by glass capillary gas chromatographs. Environ. Sci. Technol. 76,
3. Sturm R.N. 1973. Biodegradability of nonionic surfactants: Screening test for predicting rate and ultimate biodegradation. J. Amer. Oil. Chem. SOC. 50, 159-167.
4. Yoshimura K. 1986. Biodegradation and fish toxicity of nonionic surfactants. J. Amer. Oil Chem. SOC. 63, 1590-1596.
5. Ahel M., Giger W., Koch M. 1994. Behaviour of Alkylphenol Polyethoxylate Surfactants in the Aquatic Environment – I. Occurrence and Transformation in Sewage Treatment. Water Res., 28, 5, 1131-1142.
6. Alkylphenols & Ethoxylates Research Council 2006. Alkylphenols and Alkylphenol Ethoxylate Product Uses, Washington, DC.
7. Clara M., Kreuzinger N., Strenn B., Gans O., Kroiss H. 2005. The Solids Retention Time – a Suitable Design Parameter to Evaluate the Capacity of Wastewater Treatment Plants to Remove Micropollutants. Water Res., 39, 97-106.
8. DiCorcia A., Samperi R., Marcomini A. 2000. Occurrence and Abundance of Dicarboxylated Metabolites of Nonylphenol Polyethoxylate Surfactants in Treated Sewages. Environ. Sci. Technol., 34, 3914-3919.
9. Drewes J.E., Hemming J., Ladenburger S.J., Schauer J., Sonzogni W. 2004. An Assessment of

Endocrine Disrupting Activity Changes in Water Reclamation Systems Through the Use of Bioassays and Chemical Measurements. Proceedings of the 77th Annual WEFTEC Conference, New Orleans, LA, 1, 77-88.

10. Fauser P., Vikelsoe J., Sorenson P.B., Carlsen L. 2003. Phthalates, Nonylphenols and LAS in an Alternately Operated Wastewater Treatment Plant – Fate Modeling Based on Measured Concentrations in Wastewater and Sludge. Water Res., 37, 1288-1295.

11. Fujita M., Ike M., Mori K., Kaku H., Sakaguchi Y., Asano M., Maki H., Hishihara T. 2000. Behaviour of Nonylphenol Ethoxylates in Sewage Treatment Plants in Japan Biotransformation and Ecotoxicity. Water Sci. & Technol., 42, 7-8, 23-30.

12. Johnson A.C., Aerni H.-R., Gerritsen A., Gibert M., Giger W., Hylland K., Jurgens M. Nakarni, T., Pickering A., Suter M. J.-F., Svenson A., Wettstein F.E. 2005. Comparing Steroid Estrogen and Nonylphenol Content Across a Range of European Sewage Plants with Different Treatment and Management Practices. Water Res., 39, 47-58.

13. Klecka G., Zabik J., Woodburn K., Naylor C., Staples C., Huntsmann B. 2006. Exposur Analysis of Alkylphenol, Alkylphenol Ethoxylates, and their Metabolites in Surface Water Systems within the United States: Comparison to US EPA Water Quality Criteria. Final Report for Alkylphenols and Ethoxylates Research Council (APERC); The Dow Chemical Company, Midland, MI.

14. Lee H-B., Peart T.E. 1998. Occurrence and Elimination of Nonylphenol Ethoxylates and Metabolites in Municipal Wastewater and Effluents. Water Qual. Res. J. Canada, 33 (3), 389-402.

15. Nasu M., Goto M., Kato H., Oshima Y., Tanaka H. 2001. Study on Endocrine Disrupting Chemicals in Wastewater Treatment Plants. Water Sci. & Technol., 43 (2), 101-108.

16. Naylor C.G. 1992. Environmental Fate of Alkylphenol Ethoxylates. Soap/Cosmetics/Chemical Specialties, 68, 27-32.

17. Shao B., Hu J., Yang M. 2003. Nonylphenol Ethoxylates and Their Biodegradation Intermediates in Water and Sludge of a Sewage Treatment Plant. Bull. Environ. Contam. Toxicol., 70, 527-532.

18. Staples C., Mihaich E., Carbone J., Woodburn K., Klecka G. 2004. A Weight of Evidence Analysis of the Chronic Ecotoxicity of Nonylphenol Ethoxylates, Nonylphenol Ether Carboxylates, and Nonylphenol. Human Ecol. Risk Assess., 10, 999-1017.

19. U.S. Environmental Protection Agency 2005. Aquatic Life Ambient Water Quality Criteria – Nonylphenol. U.S. Environmental Protection Agency Report 822-R-05-005, Washington, DC.

20. Zahn R., Wellens H. OECD Guideline for Testing of Chemicals. Adopted by the Council on 17th July 1992.

21. Cserhati T., Forgaces E., Oros G. 2002. Biological activity and environmental impact of anionic surfactants. Environ. Int., 28, 337-348.

22. Fauser P., Vikelsoe J., Sorensen P.B., Carlsen L. 2003. Phathalates, nonylphenols and LAS in an alternately operated wastewater treatment plant – fate modeling based on measured concentrations in wastewater and sludge. Water Research, 37, 1288-1295.

23. Maguire R.J. 1999. Review of the persistence of nonylphenol and nonylphenol ethoxylates in aquatic environments. Wat. Qual. Res. J. Can., 34(1), 37-78.

24. Bizukojc E.L., Bizukojc M. 2006. Effect of selected anionic surfactants on activated sludge flocs. Enzyme and Microbial Technology, 39, 660-668.

25. Yang K., Zhu L., Xing B. 2007. Sorption of sodium dodecylbenzene sulfonate by montmorillonite. Environmental Pollution, 145, 571-576.

26. Onder E., Koparal A.S., Ogutveren U.B. 2007. An alternative method for the removal of surfactants from water: electrochemical coagulation. Separation and Purification Technology, 52, 527-532.

CONTENT OF COPPER AND NICKEL IN SOILS OF VISTULA RIVER CATCHMENT

Elżbieta Skorbiłowicz[1], Anna Samborska[1]

[1] Department of Technology in Engineering and Environmental Protection, Technical University in Bialystok, Wiejska 45A, 15-351 Bialystok, Poland, e-mail: e.skorbilowicz@pb.edu.pl; anulkasamborska@gmail.com

ABSTRACT

Metal pollution attracts growing interest mainly due to their toxic effects on the entire natural environment. It is therefore necessary to monitor, not to exceed acceptable standards of elements in ecosystems. Heavy metals in soils are derived from natural and anthropogenic sources. The purpose of this article is to present the contents of copper and nickel in agricultural soils within the Vistula river catchment based on research carried out by the Institute of Soil Science and Plant Cultivation in the period since 1995 to 2010. The assessment of a degree of soil contamination by these elements according to standards has been also performed. It has been proven that about 91% of agricultural soils in Vistula river catchment shows natural (0°), while approximately 9% higher (I°) copper concentrations. It was found that in the total agricultural area of the Vistula river catchment, soils not contaminated with nickel (0°), with elevated (I°), and weakly contaminated soils (II°) are respectively 85%, 11%, 4% of the analyzed soils area. There was a significant relationship between the content of copper and nickel in soils and such features as organic matter and pH. Analysis of results confirmed that the highly developed industrial activity affects the increased amount of pollutants in soils; the greatest accumulation occurred in soils of the upper Vistula river catchment, which is caused primarily by interaction of the Upper Silesian Industrial District.

Keywords: soil, copper, nickel, Vistula river catchment.

INTRODUCTION

Metal pollution attracts growing interest mainly due to their toxic effects on the entire natural environment [Hsio et al, 2007, Skorbiłowicz E., 2012]. It is therefore necessary to monitor, not to exceed acceptable standards of elements in ecosystems. Heavy metals in soils are derived from natural and anthropogenic sources. The most important natural source of heavy metals in the soil, which is the bedrock, provides diverse amounts of metals, depending on its type, mineralogical composition and origin [Karczewska, 2002]. Anthropogenic sources of metals are primarily associated with steel industry and metallurgy [Kabata-Pendias and Pendias 1999]. Gruca-Krulikowska and Wacławek [2006] claim that rapidly developing industry and transport, or improper use of pesticides in agriculture, contributes to the acidification of soils and hence the accumulation of heavy metals in the environment. Gray and McLaren [2005]have shown that the activity of metals in the soil solution is dependent on the soil pH, metal concentration, and the total content of potential sorption centers within the soil. Ottosen et al. (2001) in their study showed that metals have a tendency to transfer onto larger areas in contaminated soil, mainly in the finest fractions.

The purpose of this article is to present the contents of copper and nickel in agricultural soils within the Vistula river catchment based on research carried out by the Institute of Soil Science and Plant Cultivation in the period from 1995 to 2010. The assessment of a degree of soil contamination by these elements according to standards has been also performed [Kabata-Pendias et al. 1993].

METHODS

Vistula is the longest and largest Polish river, which flows into the Baltic Sea, and its length is 1047 km. Source of the Vistula river is located in the southern part of Polish part of Silesian Beskid. Tributaries of the Vistula represent almost 54% of the Polish area. Vistula river consists of 3 sections: Upper Vistula, Central Vistula, and Lower Vistula. Table 1 shows dominant soil types in the Vistula river catchment divided into provinces.

Table 1. Types of soils in Vistula river catchment

Province	Predominating soil type
Lesser Poland	Lithosols, Cambisols, Haplic Phaeozems
Subcarpathian	Lithosols, Arenosols, Podzols
Holy Cross	Cambisols, Podzols, Arenosols
Lublin	Cambisols, Podzols, Arenosols, Haplic Phaeozems
Łódź	Podzols, Arenosols, Luvisols, Phaeozems
Masovian	Podzols, Arenosols, Luvisols
Warmian-Masurian	Cambisols, Podzols, Arenosols
Kuyavian-Pomeranian	Eutric Fluvisols, Cambisols, Podzols, Luvisols, Arenosols
Pomeranian	Arenosols, Luvisols, Eutric Phaeozems, Eutric Cambisols

Climate in Poland should be counted to transitional between oceanic in western part of the country and continental temperate in the eastern part. The climate is characterized by frequent variability and different states of the weather. The growing season amounts to about 200 days, the shortest it continues in the southern part of the country, while the longest in western Poland. Average annual rainfall is about 600 mm, while mean annual air temperature is 8 °C. Winds are characterized by high changeability, although western winds predominate. In winter, winds blow mainly from east [Kożuchowski, 2011].

The research presented in this work have been carried out by the Institute of Soil Science and Plant Cultivation, Puławy, on agricultural soils. The study contains the analysis of copper and nickel contents in years: 1995, 2000, 2005, and 2010, in nine provinces (Lesser Poland, Subcarpathian, Holy Cross, Lublin, Łódź, Masovian, Warmian-Masurian, Kuyavian-Pomeranian, and Pomeranian) located in the catchment of the Vistula river. The studied object is shown in Figure 1. The total number of soil samples was 129. Sampling was done at a depth of 0–20 cm and from the surface of approximately 100 m^2 area [http://www.gios.gov.pl/chemizm_gleb/].

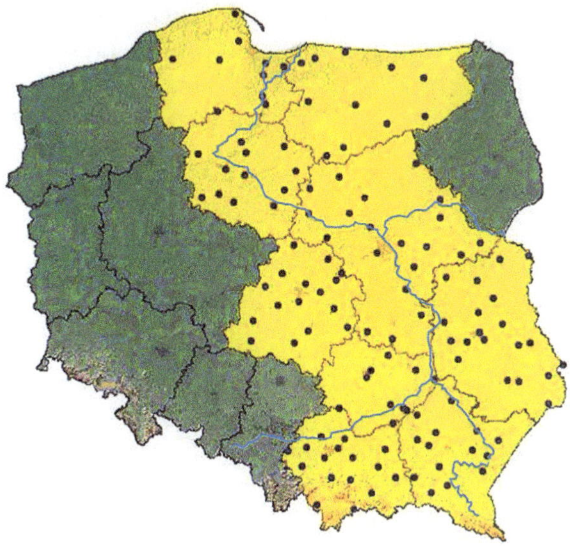

Figure 1. Soil sampling points in Vistula river catchment

Copper and nickel were determined using a spectrophotometer by means of ASA technique. Digestion was performed applying closed CEM Mars-5 system in teflon vials using hydrochloric and nitric acids. The pH of H_2O and KCl suspensions was determined by potentiometry. The organic matter in the soil samples was determined by volumetric modified Tiurin method that involves the wet combustion of organic substances.

RESULTS AND DISCUSSION

The data in Table 2 indicate that copper content varies slightly, and the average level is in the range from 4.7 ± 2.3 mg·kg^{-1} to 15.6 ± 7.6 mg·kg^{-1}. The minimum amount of copper was recorded for 1.2 mg·kg^{-1} in Lublin province, while the highest 32.3 mg·kg^{-1} in Subcarpathian province. Differentiation of copper content in agricultural soils of Vistula river catchment is associated with many factors, such as mineralogical composition and grain size of the soil, bedrock origin, and human activities. Kabata-Pendias and Pendias [1999] reports concentration of copper in Polish soils within the range: 50–140 mg·kg^{-1}. According to the Decree of the Minister of Environment from 9 September 2002 on standards for soil quality and land quality standards, the limiting copper concentration, which is 150 mg·kg^{-1} DM, is not exceeded in soils of Vistula river catchment.

Data in Table 2 indicate that the largest areas of soils with natural (0 °) copper content are present in the lower Vistula river catchment. Soils of the upper Vistula river catchment are characterized by increased copper content in provinces: Lesser Poland, Subcarpathian, and Holy Cross, which exceed 10%. Also increased copper content occurred in Warmian-Masurian region: (0 °) – 84%, (I °) – 16% (Table 2).

Referring to lowlands, the average copper content in soils is 3 mg·kg^{-1}, while in the bedrock, this amount is slightly higher than 7.1 mg·kg^{-1} [Pasieczna, 2003]. According to research by Zawadzki (2002), copper content in soils of central part of the country occurred at the level of 10 mg·kg^{-1}. Studies conducted in the area of Lublin revealed that copper content was 3.1–11.4 mg·kg^{-1} [Zgłobicki and Rodzk 2007].

Spatial distribution of copper in Vistula river catchment (Figure 2) in the period under research shows that the highest accumulation of this element occurred in the upper catchment of the river.

Table 2. Copper content and soil proportions according to the level of contamination with this element

Area	Arithmetic mean and standard deviation [mg·kg^{-1}]				Share of soils in contamination levels (%)					
	1995	2000	2005	2010	0°	I°	II°	III°	IV°	V°
Lesser Poland N = 68	14.0±6.3	14.7±6.5	15.6±7.6	14.3±6.1	90.0	10.0	0.0	0.0	0.0	0
Subcarpathian N = 56	10.7±7.1	11.4±7.1	10.7±6.2	11.9±7.5	86.0	14.0	0.0	0.0	0.0	0.0
Holy Cross N = 36	9.0±7.9	9.0±8.3	8.8±9.0	6.6±5.3	89.0	11.0	0.0	0.0	0.0	0.0
Lublin N = 80	6.7±3.4	6.2±3.0	6.0±2.9	6.2±3.5	90.0	10.0	0.0	0.0	0.0	0.0
Łódź N = 64	4.7±2.3	5.1±2.6	5.1±3.3	6.2±5.9	100.0	0	0.0	0.0	0.0	0.0
Masovian N = 80	5.0±4.3	4.9±3.8	5.3±4.5	5.5±4.6	87.5	12.5	0.0	0.0	0.0	0.0
Warmian-Masurian N = 44	9.1±7.2	8.8±6.8	8.0±6.0	8.0±4.9	84.0	16.0	0.0	0.0	0.0	0.0
Kuyavian-Pomeranian N = 52	5.4±2.7	5.0±2.2	5.6±2.6	5.0±2.5	100.0	0.0	0.0	0.0	0.0	0.0
Pomeranian N = 36	11.5±6.9	12.1±7.7	11.5±7.1	9.7±6.4	100.0	0.0	0.0	0.0	0.0	0.0
Vistula river catchment	8.4±2.2	8.6±2.4	8.5±2.3	8.2±1,5	91.0	9.0	0.0	0.0	0.0	0.0

N – total number of samples in 1995–2010.

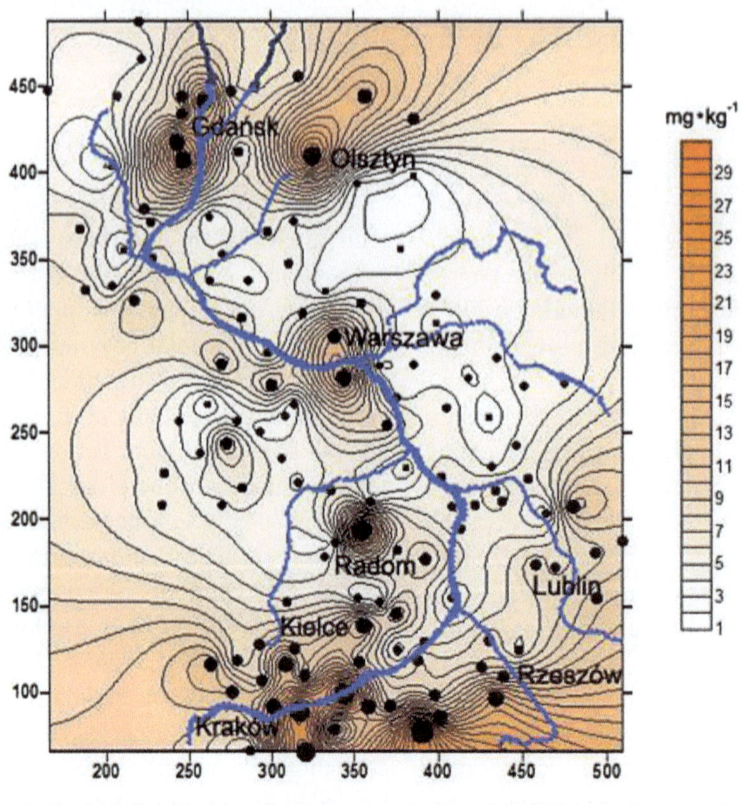

Figure 2. Spatial distribution of copper content in soils of Vistula river catchment

This is due to the impact of the Legnica-Głogów Copper District (LGOM). Annual emissions of dust containing heavy metals, especially Cu and Pb, into the atmosphere from steel works of LGOM progressively increased since the launch of plants (Legnica 1954 as well as Głogów I and II in 1971 and 1977) till the mid-eighties, when each year these plants emitted more than 2600 t of metallurgical dusts, including more than 200 t Cu and 150 t Pb. Controlling the emission in the next few years did not solve the problem of existing soil contamination with heavy metals, mainly Cu and Pb, as well as Zn, Cd, As and others. He contents of Cu and Pb in soils decreases with increasing the distance from steel works. The highest contents of these metals were recorded in the immediate vicinity of these plants (at distances <200 m), which amounted to 9800 mg·kg^{-1} Cu and 45 080 mg·kg^{-1} Pb in the area of Legnica smelter as well as 5000 mg·kg^{-1} Cu and 18 400 mg·kg^{-1} Pb in the region of Głogów steel work [Karczewska, 2002]. Quantity of analyzed element in the south-western part of the country was, according to Kaszubkiewicz and Kawałko [2009], from 6.5 to 1016 mg·kg^{-1}. A much smaller proportion of Cu recorded in the heavily industrialized village. Studies performed by Lu, Bai [2006] indicate that the average copper content in the soils of China amounted to 44.1 mg·kg^{-1} Cu in the eastern part. According to a study conducted in Botswana, South Africa, concentration of this metal ranges from 20 mg·kg^{-1} to 372 mg·kg^{-1} (Cngole and Ekosse 2012). The copper content of soils in New Zealand averaged from 2 to 2116 μg·g^{-1} [Gray and McLaren, 2005]. Excessive copper concentrations in the soil involve a negative impact on the whole environment.

Analysis of middle and lower parts of Vistula river catchment reveals the highest copper concentrations were found near Warsaw, Olsztyn, and Tricity Gdynia-Sopot-Gdańsk (Figure 2).

The arithmetic mean of nickel in soils of Vistula river catchment was recorded at 10.5 ± 2.4 mg·kg^{-1} (Table 3). Nickel content analysis showed that the maximum quantities of the element, like in the case of copper, occurred in the lower and upper catchments of the river. In Lesser Poland province, the average nickel content in 1995 was 21.3 ± 17.3 mg·kg^{-1}. In subsequent years, mean nickel content increased to 23.6 ± 17.3 mg·kg^{-1} (2010). High amounts of nickel were also reported in Subcarpathian province, which ranged from 15.4 ± 6.2 to 17.0 ± 7.1 mg·kg^{-1}. Amounts of nickel in soils of the lower river basin in Pomeranian province appeared to be on a similar level (15.9 ± 11.2 mg·kg^{-1}) in 1995, whereas the concentration of nickel slightly decreased in 2010 (15.4 ± 10.9 mg·kg^{-1}). The city of Gdańsk exerts significant impact on an elevated state of soil contamination in the lower Vistula river catchment; there are major pollution emitters including: "Lotos" refinery, shipbuilding industry, as well as landfill GZHF "Phosphates". Large accumulation of nickel in soils of the southern Poland is associated with continued development of industrial activities. According to Telerak [2005], Polish soils contain small amounts of heavy metals; their higher accumulation occurred in the industrialized parts of the country, such as the southern and the south-west regions. The largest soil surface area of natural nickel content (0°) (Table 3, Figure 3) is located in Lodz and Kuyavian-Pomeranian provinces

Table 3. Nickel content and soil proportions according to the level of contamination with this element

Area	Arithmetic mean and standard deviation [mg·kg^{-1}]				Share of soils in contamination levels [%]					
	1995	2000	2005	2010	0	I	II	III	IV	V
Lesser Poland N=68	21.3±17.3	22.6±6.5	23.3±7.6	23.6±17.3	75.0	18.0	7.0	0.0	0.0	0.0
Subcarpathian N=56	16.1±11.1	17.0±7.1	15.4±6.2	16.3±11.6	61.0	32.0	7.0	0.0	0.0	0.0
Holy Cross N=36	6.4±3.6	5.6±8.3	6.3±9.0	7.2±6.5	97.0	3.0	0.0	0.0	0.0	0.0
Lublin N=80	8.2±4.8	7.4±3.0	8.4±2.9	6.5±3.1	96.0	4.0	0.0	0.0	0.0	0.0
Lodz N=64	5.0±1.9	4.5±2.6	5.5±3.3	4.4±1.5	100.0	0.0	0.0	0.0	0.0	0.0
Masovian N=80	5.0±4.4	5.1±3.8	5.0±4.5	4.9±4.4	95.0	5.0	0.0	0.0	0.0	0.0
Warmian-Masurian N=44	12.1±10.5	10.9±6.8	10.2±6.0	11.2±8.7	81.0	5.0	14.0	0.0	0.0	0.0
Kuyavian-Pomeranian N=52	5.6±3.2	4.8±2.3	5.8±2.6	6.1±3.8	100.0	0.0	0.0	0.0	0.0	0.0
Pomeranian N=36	15.9±11.2	16.2±7.7	14.5±7.1	15.4±10.9	56.0	33.0	11.0	0.0	0.0	0.0
Vistula river catchment	10.6±5.2	10.5±2.4	10.5±2,3	10.6±5.1	85.0	11.0	4.0	0.0	0.0	0.0

N – total number of samples in 1995–2010.

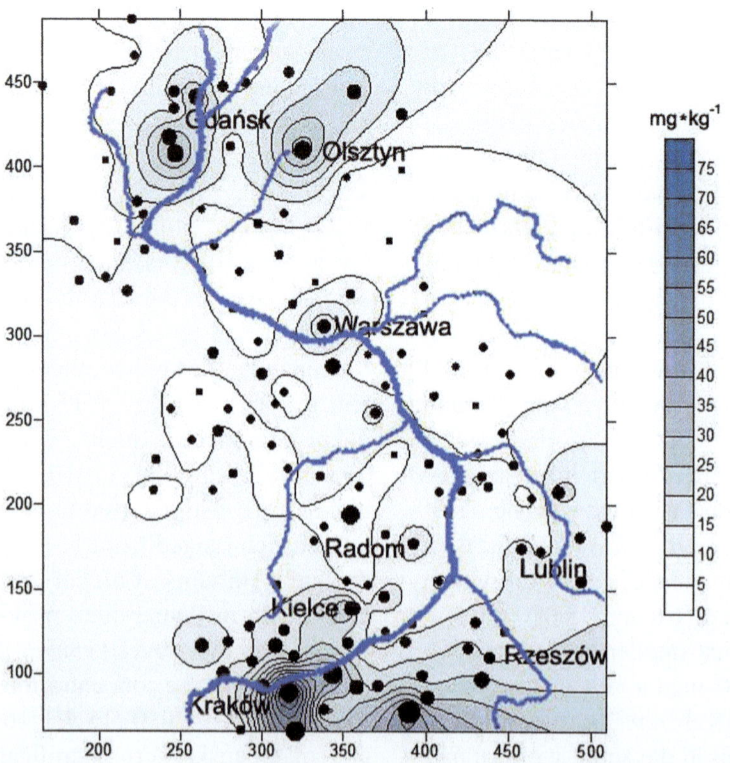

Figure 3. Spatial distribution of nickel content in soils of Vistula river catchment

and it is caused by the lack of anthropogenic contamination by this element.

Zawadzki [2002] showed the amount of nickel in soils of central Vistula river catchment at the level of 1–31 mg·kg^{-1}. The central part of Poland is characterized by a small amount of nickel in soils, which ranges from 4.02 to 31.3 mg·kg^{-1} [Jaremko and Kalembasa, 2011]. Soils of the south-western part of Poland contain average proportions of analyzed element that ranges from 6.2 to 68.3 mg·kg^{-1} [Kaszubkiewicz and Kawałko, 2009]. Areas of mining areas are more contaminated with nickel, which was proved by Cngole and Ekosse [2012], who determined the amount of this element between 22.3 mg·kg^{-1} to 266 mg·kg^{-1}. In Serbian soils, the average nickel content was high amounting up to 114,2 ppm [Jakovlejevic, 1996].

The main factor determining the solubility of heavy metals in soils is pH, which is the well-known relationship and confirmed by numerous authors. Cadmium is considered the most mobile (and most readily soluble) heavy metal, which is subject to mobilization already at pH of about 6.5, while other metals are mobilized at much stronger soil acidification: Zn (pH about 6.0) as well as Cu and Ni [Karczewska, 2002]. The soil organic matter, in particular humus, is also very impor-

Table 4. General soil properties (0–20 cm)

Area	Percentage of samples with pH in H_2O			Percentage of samples with pH in 1N KCl			Percentage of samples with humus content		
	<6.6	6.6–7.2	>7.2	<6.6	6.6–7.2	>7.2	≤1.0	1.0–2.0	>2.0
Lesser Poland	9	4	4	13	3	0	1	3	14
Subcarpathian	7	4	3	13	1	0	0	8	6
Holy Cross	5	4	0	9	0	0	1	7	1
Lublin	8	4	8	12	5	3	1	16	3
Lodz	12	3	1	14	3	0	0	13	4
Masovian	16	2	2	18	2	0	0	15	5
Warmian-Masurian	6	2	3	8	2	1	1	5	5
Kuyavian-Pomeranian	9	1	3	10	0	3	1	9	1
Pomeranian	2	4	3	7	2	0	0	3	6

Table 5. Correlation coefficient

Dependencies	Pearson coefficient	Significance level (p)
Cu – pH_{H2O}	0.347	0.000
Cu – pH_{KCl}	0.321	0.000
Cu – S.O	0.515	0.001
Ni – pH_{H2O}	0.386	0.000
Ni – pH_{KCl}	0.357	0.000
Ni – S.O	0.555	0.000

pH_{H2O} – acidity in water.
pH_{KCl} – acidity in 1N KCl.
S.O – organic matter.

tant in binding heavy metals from the soil solution. Table 4 presents general soil properties: pH and organic matter. The study proved that copper and nickel contents depend on soil characteristics such as pH and organic matter. The correlation coefficient values for the contents of copper and nickel in soils vs. their properties (Table 5) indicate the existence of significant dependence between: Cu content in the soil and the organic substance (R = 0.515) and pH value (R = 0.341, R = 0.321). Similar correlations were observed in case of nickel (Table 5).

CONCLUSIONS

1. The agricultural soils in the Vistula river catchment are characterized by low content of copper, the average value of which was 8.2 ± 1.5 $mg·kg^{-1}$, while the average nickel content was 10.6 ± 5.1 $mg·kg^{-1}$.
2. It has been proved that about 91% of agricultural soils in the Vistula river catchment shows natural (0°), while approximately 9% higher (I°) copper concentrations.
3. It was found that in the total agricultural area of the Vistula river catchment, soils are not contaminated with nickel (0°), with elevated (I°), and weakly contaminated soils (II°) are respectively 85%, 11%, 4% of the analyzed soils area.
4. There was a significant relationship between the content of copper and nickel in soils and such features as organic matter and pH. This indicates that the levels of analyzed metals in soils are also determined by other factors.
5. Analysis of results confirmed that the highly developed industrial activity affects the increased amount of pollutants in soils; the greatest accumulation occurred in soils of the upper Vistula river catchment, which is caused primarily by interaction with the Upper Silesian Industrial District.

REFERENCES

1. Cngole V.M., Ekosse G.I.E., 2012. Copper, nickel and zinc contamination in soils within the precincts of mining and landfilling environments. Int. J. Environ. Sci. Technol., 9: 485-494
2. Gray C.W., Mclaren R.G., 2005. Soil factors affecting heavy etal solubility in some New Zeland. Agriculture and Life Sciences Division, Lincoln University, Canterbury, New Zealand; Agriculture and Environment Division, Rothamsted Research, Harpenden, Herts, AL5 2JQ, UK.
3. Gruca-Krulikowska S., Wacławek W., 2006. Metale w środowisku. Cz. II. Wpływ metali ciężkich na rośliny. Chemia. Dydaktyka. Ekologia. Meteorologia 11, nr 1-2.
4. Hsiao M.C., Wang H.P., Huag C.H., Chang J., Wei Y., 2007. Tracking of copper in contaminated soils. Journal of Electron Spectroscopy and Related Phenomena 156–158, 208-210.
5. Jakovljevic M.D., Kostic N.M., Stevanovic D., Blagojevic S., 1996. Factors infkuencing the distribution of heavy metals in the alluvial soils of the Velika Morava valley, Serbia, Faculty of Agrivulture, Nemanjina 6.
6. Jaremko D., Kalembasa D., 2011. Specjacja niklu w ornych glebach płowych opadowo-glejowych wysoczyzny siedleckiej, Inżynieria Ekologiczna, nr 27, 19-24.
7. Kabata-Pendias A., Pendias H., 1999. Biogeochemia pierwiastków śladowych, PWN, Warszawa.
8. Kabata-Pendias A., Matowicka-Terelak T., Piotrowska M., 1993. Ocena stopnia zanieczyszczenia gleb i roślin metalami ciężkimi i siarką, Bib. Monitoringu Środowiska, Warszawa.
9. Karczewska A., 2002. Metale ciężkie w glebach zanieczyszczonych emisjami hut miedzi – formy i rozpuszczalność, Instytut Gleboznawstwa i Ochrony Środowiska Rolniczego, Wrocław.
10. Kaszubkiewicz J., Kawałko D., 2009, Total content of heavy metals in soils and plant sat the area of Jeleniagóra district, Ochrona Środowiska i Zasobów Naturalnych nr 40, 2009, 177-189,
11. Kożuchowski K., 2011. Klimat Polski, Nowe spojrzenie, wyd. Naukowe PWN, Warszawa.
12. Lu S.G., Bai S.Q., 2006. Study on the correlation of magnetic properties and heavy metals conent in urban soils of Hangzhou City, China. Journal of Applied Geophysics 60, 1-12.

13. Ottosen L.M., Hansen H.K., Ribeiro A.B., Villumsen A., 2001. Removal of Cu, Pb and Zn in applied electric field in calcareous and non-calcareous soils. Journal of Hazardous Materials B85, 291-299.

14. Pasieczna A., 2003. atlas zanieczyszczań gleb miejskich w Polsce. Państwowy Instytut Geologiczny, Warszawa.

15. Pasieczna A., 2011. Naturalne i antropogenicz-ne czynniki zanieczyszczenia środowiska. Państwowy Instytut Geologiczny, Państwowy Instytut Badawczy.

16. Rozporządzenie Ministra Środowiska z dnia 9 września 2002 r. w sprawie standardów jakości gleby oraz standardów jakości ziemi.

17. Siebielec G., Smreczak B., Klimkowicz-Pawlas A., Maliszewska-Kordybach B., Terelak H., Koza P., Hryńczuk B., Łysiak M., Miturski T., Gałązka R., Suszek B., 2012. Monitoring chemizmu gleb ornych w Polsce w latach 2010-2012. IUNG.

18. Terelak H., 2005. Metale ciężkie i siarka w glebach użytków rolnych Polski. Problemy Ekologii, tom 9, nr 5, 259-264.

19. Terelak H., Motowicka-Telerak T., Stuczyński T., Pietruch C., 2000. Pierwiastki śladowe (Cd, Cu, Ni, Pb, Zn) w glebach użytków rolnych Polski. Inspekcja Ochrony Środowiska, Warszawa.

20. Skorbiłowicz E., 2012. Studia nad rozmieszczeniem niektórych metali w środowisku wodnym zlewni górnej Narwi. Oficyna Wydawnicza Politechniki Białostockiej, rozprawy naukowe Nr 222.

21. Zawadzki J., 2002. Badania korelacji przestrzennych zawartości wybranych pierwiastków śladowych w glebach Warszawy i okolic. Ochrona Środowiska nr 4.

22. Zgłobicki W., Rodzik J., 2007. Heavy metals in the slope deposits of loess areas of the Lublin Upland (E Poland). Science Direct, Catena 71, 84-95.

23. http://raportroczny.lotos.pl/zrownowazony-rozwoj/srodowisko/ograniczanie-wplywu-na-srodowisko.

24. Monitoring Chemizmu Gleb Ornych w Polsce. http://www.gios.gov.pl/chemizm_gleb/ (16.12.2013 r.).

14

ASSESSMENT OF GROUND WATER POLLUTION IN PARKING AREAS

Janina Piekutin[1]

[1] Bialystok University of Technology, 45A Wiejska Str., 15-351 Bialystok, Poland, e-mail: j.piekutin@pb.edu.pl

ABSTRACT

Creation of rain sewer is connected with dehydration of roads and coexisting objects. The paper presents a discussion upon the issue of groundwater contamination by petroleum compounds and other pollutants from transport based on studies of groundwater within the parking lots. The study included 9 parking areas, including 7 in Bialystok, 1 in a residential area outside of Bialystok in Ignatki, and one in Kleosin. The tested waters were subject to determination of COD, total suspension, and petroleum substances expressed as a mineral oil index. The studies have shown that the concentrations of determined parameters were in most cases proportional to the larger runoffs and concentration of petroleum compounds increased with the increase of suspension. It has been shown that from part of the parking lots, the meteoric water was discharged directly into watercourses and exceeds the permissible limits regulated by the Decree.

Keywords: meteoric water, parking lots, petroleum substances.

INTRODUCTION

The continuous development of communication infrastructure is inevitable today. It carries many negative effects [Barański, Gworek 2004]. The aquatic environment and groundwater are especially exposed for such a change. There are many adverse effects associated with the construction of more and more new roads and accompanying parking lots. In the studied parking areas there are places for parking vehicles. Each of them should have a parking space that would be tailored to its needs. Depending on the type of vehicles and cargo that can be carried by them, there is a different risk of environmental contamination [Łęgosz, Jasiński 2009]. Within the parking areas, primarily meteoric sewage is created, which is the result of transformation of rainfall into the surface runoff or as a result of runoff resulting from snow melting [Aniszewski 1998]. The mixture, which goes to the sewage system or directly into the ground, may have a different composition [Kłodziejczyk 2009]. The quantity and quality of wastewater discharged from the parking lots is not constant. Such a situation is affected by many factors, which may include mainly [Ociepa 2009]: the surface of the parking areas, traffic capacity, methods of surface maintenance and operation, air pollution, season, characteristics of rainfall.

Surface of a parking lot is often covered with setts or asphalt. Green parking lots are also possible. The runoffs from these areas depend on the tightness of the used surface material [Gierak 1995, Kłodziejczyk 2009, Merkisz, Piekarski, Słowik 2006]. Some of them may be well-drained, while others do not have such good properties. It should also be pointed out that the area should be kept clean. During winter, parking lots are cleared, but also sprinkled with chemicals that accelerate snow melting. These runoffs affect the quality of wastewater [Kłodziejczyk 2009, Merkisz, Piekarski, Słowik 2006].

Traffic capacity and the number of parking lots in a given area are also significant facts. First of all, it should be noted that their number varies and depends mainly on the frequency of rainfalls and their intensity. Retention of snow cover and then snow melting is also a cause of increasing pollution load. The snow left on a parking lot contains many harmful compounds [Merkisz, Piekarski, Słowik 2006]. During rapid melting, these contaminants get along with pollutants into the sewage system or directly into the ground or waters. Referring the quality of wastewater: its composition also changes in some time [Pawluk, Fronczyk, Garbulewski 2011]. If it rains after a long-term dry period, impurities accumulated on the surface of parking lots are the most intensely leached during the first phase of rainfall [Łęgosz, Jasiński 2009, Sadkowski 1990].

The quality of wastewater discharged from parking areas directly to the soil or water is regulated by the law [Regulation of Ministry of Environment on 26 July 2006] [Gierak 1995].

METHODS

The study was conducted in two series included 9 parking lots of various usage types – from maneuvering areas within companies through those in the housing estates. The water samples were collected from points in Bialystok, Kleosin, and Osiedle Ignatki:

- I sampling point – water sample was collected from the ditch near the park at Wierzbowa Street in Białystok near Entertainment Club "Krąg". The area includes the parking lot for clients as well as go-cart track. The parking lot has no drainage system, sewage from the runoff directly gets into the ground and then to groundwater.
- II sampling point - at Wierzbowej Street. The parking lot belongs to the car workshop and vehicle inspection station. The parking lot is paved with setts. Sewage from the parking area is discharged to the wastewater system through inlets and wells. Runoffs from this and previously described point get into the Biała river.
- III sampling point – at Pogodna Street. The secure parking lot is within the sampling point. It is designed for cars and includes about 50 parking spaces. It has no setts. Only the staging places are lined with cobblestones. Polluted water remaining after the rainfall can be found in some places. There are 156 covered garages in the south side. Their surface is covered with gravel, not paved and has no drainage. Wastewater flow down through the concrete surface drainage directly into a ditch, from which water was collected.
- IV sampling point – near the maneuver area that is used to teach applicants for car driving license. This place has asphalt and cobblestones. The parking lot has no drainage system. The wastewater discharge is carried out directly into the sewer drain.
- V sampling point – is located south of the plant "KAN" in Kleosin. In the area of the industrial plant, there is a car park designed primarily for employees and for trucks coming for manufactured products. The drainage is in a form of grooves.
- VI sampling point – is located in a residential housing Ignatki outside the city of Bialystok. The water was collected in the vicinity of the parking lot near the blocks of flats. There are drains in the parking lot. The land slope is in the direction of the test ditch. The test flow is connected to an inlet that supplies the runoff from the parking lot.
- VII sampling point – near the parking of trucks owned by a transportation company. The parking lot is located on the property, through which the ditch flows. The plants and seedlings selling business is carried out within this property. Water from the ditch is used for watering them. At this point, there is also a valve for water stacking.
- VIII sampling point – neighborhood of the property with two parking lots in Hetmańska Street. One is paved with setts, while the second with plates. In the majority, the drainage is in a form of grooves and wells.
- IX sampling point – near the car park of KFC and a petrol station at John Paul II Aleey. Both of these objects have drainage systems. At the petrol station, it is the line drainage with separators for petroleum substances and wells.

Water samples were subject to determination of COD, total suspension, and petroleum compounds expressed as mineral oil index. The analyses were made in accordance with the available methods of investigation. For determination of petroleum compounds, the method using a gas chromatograph was applied. After approximately 4-week intervals, studies were repeated for each of nine sampling points.

RESULTS AND DISCUSSION

The first collection for the first 3 points occurred during a considerable amount of residual layer of snow. Studies were repeated at the same locations after 4 weeks. In the subsequent three points, samples were also collected when the

Figure 1. Residual water at the garages in the vicinity of III sampling point

Figure 2. Wastewater discharge from garages area into the studied receiver (III sampling point)

Figure 3. Wastewater discharge into the sewer drain (IV sampling point)

Figure 4. Parking lot near the VII sampling point

snow cover was still present on the soil layer, while repeating the tests after 4 weeks during the thaw. Samples marked as 7 to 9 were obtained when a layer of snow was already low. Studies were repeated after 4 weeks when there was no snow, and there was no rainfall for a long time.

The highest value of COD (Table 1) was recorded during the collection at the point No 2, where it amounted to 1500 mg O_2/dm^3 and at point No 3 with 732 mg O_2/dm^3. When the measurements were repeated, COD values decreased or increased for some points, which could be due to uneven penetration of additional organic pollutants into the waters. The lowest values occurred during the second sampling at points 7, 8, and 9. Rain did not occur then for a long time, which could indicate a lesser contamination along with runoff. For the points No 5 and 6, the atmospheric conditions were different, which resulted in uneven penetration of contaminants into the groundwater.

At the first sampling point, the suspension at the beginning was 102 mg/dm^3, but when measurements were performed after a certain time, the contents decreased to 80 mg/dm^3. Additionally, it can be seen that the COD content is high as the amount of petroleum substances at point No 1. Along with the reduction of suspension content, the amount of petroleum derivatives also decreased. Similar changes can be observed in two consecutive samples. In these cases, more suspension can be recorded during the snow cover and snowfall. Whereas, it was less during subsequent sampling at the same points, when a thaw occurred. The dependence of COD and content of petroleum substances on a suspension is also similar.

The opposite situation can be observed during the following sample collection for the next two samples. During the sampling at point No 4, the suspension was only 38 mg/dm^3 with a growth to

Table 1. Total suspension and COD_{Cr} in studied samples

Place of taking sample	COD [mg O_2/dm³]	Suspension [mg/dm³]
I	106.5ᵃ/337ᵇ	102/80
II	1500/590	310/260
III	732/365	168/160
IV	417/660	38/209
V	593/396	108/88
VI	349/489	56/62
VII	304/106	10/34
VIII	301/58	23/10
IX	315/62	41/10

Descriptions: a – value for the first sampling, b – value for the second sampling

209 mg/dm³ during the following measurement. The content of COD and petroleum compounds at point No 4 was less, but when studies were repeated after some time, the levels of remaining indicators increased along with the increase of the suspension amount. A similar situation is for point No 6, although the suspension differed by only 6 mg/dm³, which is not such a significant difference (Table 1). As it can be seen, the sum of total suspension is very different for different points. Composition of wastewaters from runoffs is very diverse and it is difficult to determine their typical composition. There is no certain uniformity of wastewater discharge as it is in the case of domestic sewage. Filtering the samples, even though they seemed to be quite transparent, the suspension remained on all filters. This means that a fine and hardly falling suspension made up a large part.

The studies also included determinations of petroleum substances that have been converted into a mineral oil index. These substances are very well sorbed by previously discussed suspension [Rychliński 2006, Surygała 2000, Szymański 1995]. According to some data, this value is as high as 82–99%. Floating colored patches of gasoline prove a very close contamination of the aquatic environment [Pawluk, Fronczyk, Garbulewski 2011]. Many of these, however, is dissolved in water and is able to move a considerable distances, and therefore it is difficult to univocally determine the source of such contamination. According to the Regulation of Minister of Environment of 9 November 2011 concerning the quality of surface waters, water in which the petroleum compounds do not exceed 0.2 mg/l are qualified as class I and II.

The data are presented graphically in charts comparing the first and repeated tests at the same sampling points at different times.

The amount of petroleum substances during the first measurement at point No 1 is much larger and amounts to 60.085 mg/dm³. Studies were conducted when there was a plenty of snow, and the temperature was negative. During the second collection, the value declined to 14.969 mg/dm³ (Figure 5). During the first sampling, considerable snow cover was present at point No 2. During the first sampling, there was a thaw. In this case, during the first collection test, the content of petroleum substances was higher amounting to 49.890 mg/dm³. During the second sampling, this quantity slightly decreased to 42.518 mg/dm³ (Figure 5). The determined quantity of petroleum substances at point No 3 was 60.722 mg/dm³ and it increased to 103.239 mg/dm³ after some time (Figure 5). During the sampling at point No 4, the snow cover still remained, yet it was less than during the sample collection for

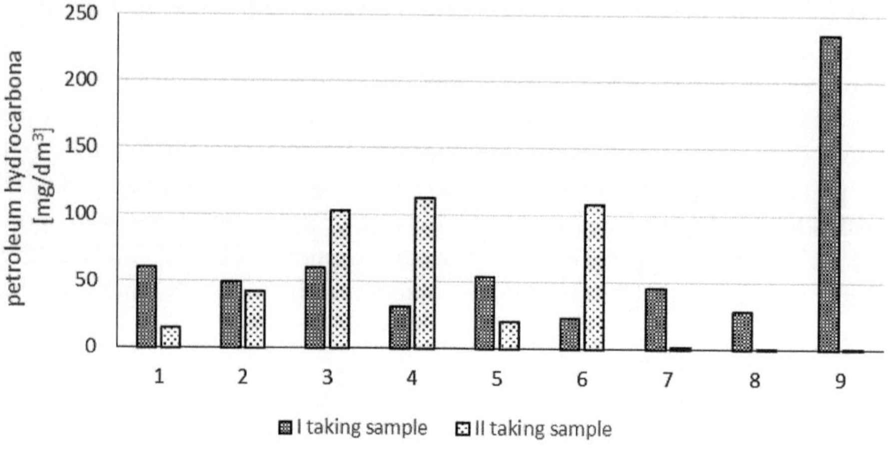

Figure 5. Amounts of petroleum substances at studied sampling points during the first and second sampling

the points No 1-3, and the concentration was 31.206 mg/dm^3. The second sampling at point No 4 was made, when it was raining for a long time; this result was higher more than three and a half times and amounted to 112.506 mg/dm^3 (Figure 5). For the collection at point No 5, values in the first case amounted to 53.664 mg/dm^3, while measurements repeated after a few weeks revealed decreased level to 20.280 mg/dm^3 (Figure 5). Taking into account the melt and rainfall during second sampling at this point, it was to be expected that the results would be significantly higher. The area, on which the test flow is situated, is periodically flooded by sewage from runoff from residential apartment blocks localized behind the railway embankment.

Like in the above two cases, samples were collected during the same collections in similar weather conditions. For sample No 6 (Figure 5), content of petroleum substances was determined for 23.280 mg/dm^3, but after a few weeks, this level increased almost five-fold to a value of up to 108.307 mg/dm^3. Such a situation could be caused by the thaw and rain. In addition, one inlet probably collecting rainwater, is mounted to the test flow. During the sampling at point No 7, only in some places snow occurred. However, during the collection of the second sample, weather conditions significantly changed. It is illustrated in the above plot (Figure 5). There was no rain for a long time, and snow cover completely disappeared. For point No 7, content of petroleum compounds was 45.836 mg/dm^3, while examining the water at the same sampling point, this amount was only 1.627 mg/dm^3. During thaws, the quantity of petroleum substances increased, while during the dry period, it significantly decreased. The conditions, under which the samples were collected, were similar to those discussed for point No 7. The value of petroleum compounds for point No 8 amounted to 29.177 mg/dm^3, but after a few weeks, this level decreased to just 0.444 mg/dm^3. This parking area is equipped with drains, so a considerable amount of wastewater is drained out of it. During the sampling discussed in the plot (Figure 5), weather conditions were similar to those in the two previous cases. For sample No 9, petroleum compounds content was 236.014 mg/dm^3, and during the subsequent collection at the same point, this quantity was determined only for 0.393 mg/dm^3. The sampling point was located near the John Paul II Avenue near KFC and gas station. During thaws, petroleum substances were washed away from the gas station. A sudden increase in the amount of rainwater disabled the separators of petroleum substances. The concentrations determined during the subsequent study decreased significantly, which could be due to a lack of rainfall.

CONCLUSIONS

1. During the surface runoff due to thaw or rainfall, an increase in petroleum substances and suspension contents takes place.
2. Vicinity of high-traffic roads affects the elevation of petroleum substances concentration.
3. Increased amount of suspension causes an increase in petroleum compounds content in the studied flows.

REFERENCES

1. Aniszewski A. 1998. Rozprzestrzenianie i adsorpcja zanieczyszczeń w płynącej wodzie gruntowej. Wydawnictwo Uczelniane Politechniki Szczecińskiej, Szczecin.
2. Barański A., Gworek B. 2004. Ocena ryzyka zdrowotnego i środowiskowego pochodzącego od gruntów zanieczyszczonych produktami naftowymi. Instytut Ochrony Środowiska, Warszawa.
3. Gierak A. 1995. Zagrożenie środowiska produktami ropopochodnymi.
4. Kłodziejczyk U. et al. 2009. Ekologiczne zagadnienia odwodnienia pasa drogowego. Warszawa.
5. Kościelniak S. et al. 1994. Wskazówki metodyczne do oceny stopnia zanieczyszczenia gruntów i wód podziemnych produktami ropopochodnymi i innymi substancjami chemicznymi w procesach rekultywacji. Wydawnictwo TRIO, Warszawa.
6. Krajewski I. 2000. Metoda oceny zagrożenia jakości wód podziemnych przy pomocy systemu Drastic. Inżynieria Środowiska XI nr 385, Zeszyty Naukowe Akademii Rolniczej we Wrocławiu, Wrocław.
7. Łęgosz A., Jasiński W. 2009. Zalecenia projektowania, budowy i utrzymania odwodnienia parkingów i MOP. Warszawa.
8. Merkisz J., Piekarski W., Słowik T. 2006. Motoryzacyjne zanieczyszczenia środowiska. Wydawnictwo Akademii Rolniczej w Lublinie, Lublin.
9. Mizerski W. 2009. Geologia dynamiczna. Wydawnictwo Naukowe PWN, Warszawa.
10. Osmulska-Mróz B., Sadkowski K. 1990. Zasady ochrony środowiska w projektowaniu, budowie i

utrzymaniu dróg. Dział 07 – ochrona wód w otoczeniu dróg. Warszawa.

11. Pieutin J. 2013. Oczyszczanie wód opadowych z substancji ropopochodnych. Ekonomia i Środowisko, 4, 60–66.

12. Pawluk K., Fronczyk J., Garbulewski K. Wpływ wód spływowych i roztopowych z infrastruktury drogowej na jakość środowiska gruntowo – wodnego na terenie aglomeracji warszawskiej, Materiały konferencyjne, Warszawa 2011.

13. Rychliński S. (Ed.) 2006. Metody usuwania zanieczyszczeń węglowodorowych ze środowiska gruntowo-wodnego. AGH, Kraków.

14. Surygała J. 2000. Zanieczyszczenia naftowe w gruncie. Oficyna Wydawn. Politechniki Wrocławskiej.

15. Szymański K. 1995. Ocena zanieczyszczeń wód podziemnych. Wyższa Szkoła Inżynierska w Koszalinie.

LONG-TIME EFFECT OF HARD COAL ASH ON THE CONTENT OF SOME ELEMENTS IN SOIL

Mirosław Wyszkowski[1], Arkadiusz Chełstowski[2], Zdzisław Ciećko[3], Radosław Szostek[4]

[1] Department of Environmental Chemistry, University of Warmia and Mazury in Olsztyn, 4 Łódzki Sq., 10-727 Olsztyn, Poland, e-mail: miroslaw.wyszkowski@uwm.edu.pl

[2] Department of Environmental Chemistry, University of Warmia and Mazury in Olsztyn, 4 Łódzki Sq., 10-727 Olsztyn, Poland, e-mail: arkadiusz.chelstowski@wsse.waw.pl

[3] Department of Environmental Chemistry, University of Warmia and Mazury in Olsztyn, 4 Łódzki Sq., 10-727 Olsztyn, Poland, e-mail: zdzislaw.ciecko@uwm.edu.pl

[4] Department of Environmental Chemistry, University of Warmia and Mazury in Olsztyn, 4 Łódzki Sq., 10-727 Olsztyn, Poland, e-mail: radoslaw.szostek@uwm.edu.pl

ABSTRACT

The aim of the study was to determine the long-time effect of meliorating doses of hard coal ash (0, 100, 200, 400, 600 and 800 t · ha^{-1}) on the content of some elements in soil with the application of different organic substances: manure, straw and bark. Nineteen years after the application of ash there was still an increase in the total content of all elements in soil. Its magnitude depended on soil horizon and it was highest in the humus horizon. The increase in the content of nutrients was, in general, proportional to the applied doses of ash. In the surface layer of soil, the increase was particularly high for sodium and subsequently for calcium, magnesium, sulphur, potassium and phosphorus, arranged in the order of decreasing values. The content of elements in deeper soil layers was significantly lower than in the 0–25 cm zone. A gradual reduction in the content, together with the increase in depth, was recorded only for magnesium and sulphur. The lowest concentration of phosphorus and calcium was detected in the 51–75 cm layer, while in the case of potassium it was in the 26–50 cm zone. The substances introduced into the soil in order to reduce the impact of ash generated by hard coal combustion did not exert any significant impact on the content of elements 19 years following their application.

Keywords: hard coal ash, manure, straw, bark, soil, elements.

INTRODUCTION

The current legal regulations on air protection in Poland have contributed to a significant reduction in dust emissions by the power engineering sector. The total emission estimated at 155 thousand tonnes in 2000 decreased to 61 thousand tonnes in 2010 (this data includes emissions from the occupational power industry, industrial and technological power engineering industry and does not cover local boiler houses, domestic furnaces, workshops and agriculture) [13]. These changes have mainly resulted from the installation of increasingly efficacious dust removal equipment. Reduction in dust emissions generates an increase in the volume of volatile ash disposed on storage yards near power engineering plants. It is widely recognized that this ash poses a serious hazard to the natural environment mainly due to higher content of trace elements [18]. However, the presence of macro- and microelements and compounds with strong adsorptive properties indicate a potential for their environmental utilization [17, 18]. Although the effects of volatile hard coal ash on the properties of soil and plants have been well investigated in the literature, there is a limited source of data on the persistence of changes caused by their introduction into soil.

The objective of the study was to determine a long-time effect of meliorating doses of hard coal ash on the content of elements in soil with the application of different organic substances: manure, straw and bark.

MATERIALS AND METHODS

The studies were based on a field experiment by Prof. dr. hab. Grzegorz Nowak and Prof. dr. hab. Zdzisław Ciećko, the University of Agriculture and Technology in Olsztyn (presently, University of Warmia and Mazury in Olsztyn), in autumn, 1984 in a randomized sub-block model within Łęg Starościński village (Lelis commune) situated in the Ostrołęckie county (Poland) on soil classified as poor rye complex with a granule size composition of loamy sand. The soil was moderately abundant in available phosphorus (55 mg P · kg^{-1} d.m.), with a high content of available potassium (152 mg K · kg^{-1} d.m.), available magnesium (55 mg Mg · kg^{-1} d.m.) and total boron (0.45 mg B · kg^{-1} d.m.), average content of total manganese (20.5 mg Mn · kg^{-1} d.m.) and total zinc (4.16 mg Zn · kg^{-1} d.m.) and with low content of copper (1.75 mg Cu · kg^{-1} d.m.). The sorption capacity was 125 mmol(+) · kg^{-1} d.m. and the pH in water and in 1 mol KCl · dm^{-3} was 6.5 and 5.6, respectively. The experiment was conducted in a randomized block model in four repetitions with the inclusion of two experimental factors using incremental doses of ash (0, 100, 200, 400, 600 and 800 t · ha^{-1}). In order to reduce the impact of ash, different substances were introduced into the soil in amounts equivalent to 10 tonnes d.m. · ha^{-1}. The studies were carried out in four series: without substances and with the application of manure, straw and bark. The hard coal ash and organic compounds were introduced to the soil once in autumn in 1984 (at the time of experiment setting). The content of elements in hard coal ash and organic substances was presented in Table 1.

The ash was collected directly from the electro-filters at the "Ostrołęka" Power Plant Complex. The ash contained highest concentration of magnesium, iron, manganese, copper and cobalt. The reaction of ash measured in 1 mol KCl · dm^{-3} was 9.2. Manure was obtained from a nearby swine operation and bark from the paper and cellulose production plant located in Ostrołęka. Manure contained highest amounts of phosphorus and zinc; oat straw – nitrogen and potassium and bark – calcium. The surface of each experimental plot was 54 m^2 with a 1 m width of protective zones between the plots and 3 m between the sub-blocks. During the first six years of the studies, typical agricultural plants were cultivated (potatoes – 1985; oat for forage + lupine for forage –1986; rye for forage + legumes-grass mixture for forage – 1987; legumes-grass mixture for forage – 1988–1991) and subsequently, starting from 1992, it was used as rough grazing land, non-fertilized with mineral fertilizers. Mineral fertilization with NPK and agricultural engineering procedures was identical on all experimental objects and NPK doses for 19 years equals 1215 kg N, 500 P and 1220 kg K per hectare.

After 19 years, in order to determine the long-time effect of ash on the selected soil parameters, samples from all objects and from 4 horizons (i.e. 0–25, 26–50, 51–75 and 76–100 cm) were collected. Following drying at room temperature and sieving through a 1 mm mesh, the total content of phosphorus with vanadium-molybdenum method was determined and the concentration of other elements (potassium, calcium, magnesium, sulphur and sodium) was measured with an atomic absorption spectrophotometer Unicam 939 Solar. In order to determine concentrations of elements, the soil was wet mineralized in a mixture of perchloric acid and nitric acid.

The results were statistically processed with Statistica 10 software using an ANOVA two-way analysis of variance [14]. The relations between the soil horizons and the content of elements in

Table 1. Content of some elements in hard coal ash, manure, straw and bark at the start of experiment in autumn 1984

Hard coal ash and organic substances	N	P	K	Ca	Mg	Fe	Mn	Cu	Zn	Co
	[g · kg^{-1} of d.m.]						[mg · kg^{-1} of d.m.]			
Hard coal ash	-	1.7	2.9	15.0	7.1	27.9	496.0	61.0	48.0	13.0
Manure	5.3	2.5	3.1	3.6	1.3	2.15	235.4	16.4	151.2	1.33
Straw	5.6	1.5	17.2	3.9	1.0	0.16	80.3	6.1	46.3	0.14
Bark	0.5	0.7	1.3	20.3	0.8	0.87	124.2	7.3	46.8	0.74

the soil expressed as simple Pearson's correlation coefficients and the standard deviation for average content of elements in soil were calculated with Microsoft Excel 2010 [11].

RESULTS AND DISCUSSION

The analysis of the content of elements in the soil demonstrated that, despite the period of 19 years that had passed from the moment of application, the incremental doses of ash generated an increase in the total content of all elements in the soil (Tables 2 and 3). The magnitude of this increase depended on soil horizon – it was highest in the surface layer of the soil. Within this horizon, the increase was particularly high for sodium and, in the case of other elements, it decreased in the following order: calcium, magnesium, sulphur, potassium and phosphorus. However, it

Table 2. Long-time effect of hard coal ash on the total content of phosphorus, potassium, calcium and magnesium in the soil (the average from four series) [g · kg^{-1} of soil]

Hard coal ash dose [t · ha^{-1}]	Soil horizon [cm]				r
	0-25	26-50	51-75	76-100	
Total phosphorus [g · kg^{-1} of soil]					
0	0.56	0.33	0.17	0.20	-0.902 **
100	0.61	0.33	0.17	0.20	-0.894 **
200	0.63	0.34	0.17	0.19	-0.906 **
400	0.65	0.34	0.18	0.20	-0.898 **
600	0.67	0.34	0.19	0.20	-0.899 **
800	0.68	0.35	0.20	0.20	-0.907 **
Average	0.63±0.04	0.34±0.01	0.18±0.01	0.20±0.01	-0.902 **
$LSD_{0.05}$ for: hard coal ash dose - 0.022; soil horizon - 0.018; interaction - n.s.					
Total potassium [g · kg^{-1} of soil]					
0	0.36	0.38	0.40	0.38	0.632 *
100	0.44	0.39	0.41	0.39	-0.710 **
200	0.47	0.40	0.42	0.43	-0.439
400	0.52	0.40	0.41	0.44	-0.546
600	0.53	0.40	0.41	0.45	-0.502
800	0.55	0.39	0.41	0.50	-0.222
Average	0.48±0.07	0.39±0.01	0.41±0.01	0.43±0.04	-0.432
$LSD_{0.05}$ for: hard coal ash dose - 0.024; soil horizon - 0.020; interaction - 0.049					
Total calcium [g · kg^{-1} of soil]					
0	1.11	0.66	0.41	0.43	-0.909 **
100	1.35	0.81	0.48	0.51	-0.912 **
200	1.65	1.03	0.53	0.62	-0.908 **
400	2.38	1.27	0.69	0.73	-0.907 **
600	3.09	1.36	0.70	0.82	-0.874 **
800	3.44	1.44	0.72	0.91	-0.861 **
Average	2.17±0.96	1.10±0.31	0.59±0.13	0.67±0.18	-0.888 **
$LSD_{0.05}$ for: hard coal ash dose - 0.079; soil horizon - 0.064; interaction - 0.160					
Total magnesium [g · kg^{-1} of soil]					
0	0.34	0.31	0.29	0.24	-0.983 **
100	0.40	0.32	0.29	0.29	-0.894 **
200	0.47	0.33	0.30	0.29	-0.882 **
400	0.78	0.42	0.31	0.30	-0.889 **
600	0.87	0.44	0.31	0.31	-0.880 **
800	0.93	0.46	0.31	0.32	-0.877 **
Average	0.63±0.26	0.38±0.07	0.30±0.01	0.29±0.03	-0.894 **
$LSD_{0.05}$ for: hard coal ash dose - 0.025; soil horizon - 0.020; interaction - 0.050					

Explanation: r – correlation coefficient; differences significant for: * – P≤0.05; ** – P≤0.01.

Table 3. The long-time effect of hard coal ash on the total content of sulphur and sodium in the soil (the average from four series) [g · kg^{-1} of soil]

Hard coal ash dose [t · ha^{-1}]	Soil horizon [cm]				Average	r
	0-25	26-50	51-75	76-100		
Total sulphur [g · kg^{-1} of soil]						
0	0.27	0.08	0.06	0.04	0.11	-0.863 **
100	0.30	0.17	0.09	0.04	0.15	-0.979 **
200	0.34	0.14	0.07	0.04	0.15	-0.928 **
400	0.43	0.14	0.07	0.04	0.17	-0.898 **
600	0.46	0.13	0.06	0.04	0.17	-0.878 **
800	0.48	0.14	0.06	0.04	0.18	-0.883 **
Average	0.38±0.09	0.13±0.03	0.07±0.01	0.04±0.00	0.16±0.02	-0.905 **
$LSD_{0.05}$ for: hard coal ash dose - 0.012; soil horizon - 0.010; interaction - 0.024						
Total sodium [g · kg^{-1} of soil]						
0	0.019	0.021	0.301	0.013	0.089	0.239
100	0.147	0.021	0.276	0.015	0.115	-0.147
200	0.443	0.037	0.262	0.001	0.186	-0.687 **
400	1.529	0.039	0.307	0.006	0.470	-0.773 **
600	1.341	0.029	0.105	0.047	0.381	-0.766 **
800	0.562	0.034	0.031	0.004	0.158	-0.802 **
Average	0.674±0.624	0.030±0.008	0.214±0.116	0.014±0.017	0.233±0.155	-0.754 **
$LSD_{0.05}$ for: hard coal ash dose - 0.026; soil horizon - 0.021; interaction - 0.052						

Explanation: r – correlation coefficient; differences significant for: * – P=0.05; ** – P=0.01.

should be noted that the increase in the content of sodium was detected only up to the dose of 400 t · ha^{-1} and was 80-fold in comparison with the control object (without ash).

The increase in the total content of calcium and magnesium in the surface soil layer was linear and amounted to 210% and 174%, respectively, on the object fertilized with 800 t · ha^{-1} in comparison with the control object. As impacted by the application of ash, the increase in the content of sulphur was also directly proportional to their doses. It amounted to max. 78% as compared to the control object. In the case of potassium and phosphorus, the increase in their concentration in the surface layer of the soil was significantly lower: under the dose of 800 tonnes of ash per ha it amounted to, respectively, 53% and 21%.

In the subsequent horizon of soil, an increase of all elements was observed. It was smaller than in surface soil layer. In other horizons of soil, the highest increase in contents was observed in case of magnesium and potassium. Within the zones 26–50 and 76–100 cm, calcium was lower by nearly half than in surface layer, while it amounted to 76% in the 51–75 cm zone.

In other three soil horizons (26–50, 51–75 and 76–100 cm), the increase in the content of total magnesium was directly proportional to the dose of ash reaching, respectively, 48%, 7% and 33%.

The content of analyzed elements was positively correlated with soil pH and negatively with soil hydrolytic acidity [3]. The increase in pH_{KCl} (from 5.2 to 7.2) and decrease in hydrolytic activity (2.5 times) in the surface soil layer was observed in the object fertilized with 800 t ashes · ha^{-1} in comparison with the control object. In 26–50 cm horizon, the changes of pH_{KCl} were the same but hydrolytic activity smaller than in surface soil layer.

The content of elements in deeper soil layers was significantly lower than in the 0–25 cm zone (Tables 2 and 3, Figure 1).

It should be emphasized that a gradual reduction in the content, together with increasing soil depth was recorded only for magnesium (correlation between soil horizons and content of magnesium in soil, r = -0.894) and sulphur (r = -0.905). The lowest concentration of phosphorus and calcium was detected in the 51–75 cm layer while potassium was in the 26–50 cm zone. The biggest differences between the 0–25 and 76–100 cm layers were recorded for the content of sodium and sulphur, whereas the lowest differentiation between the examined zones was detected for

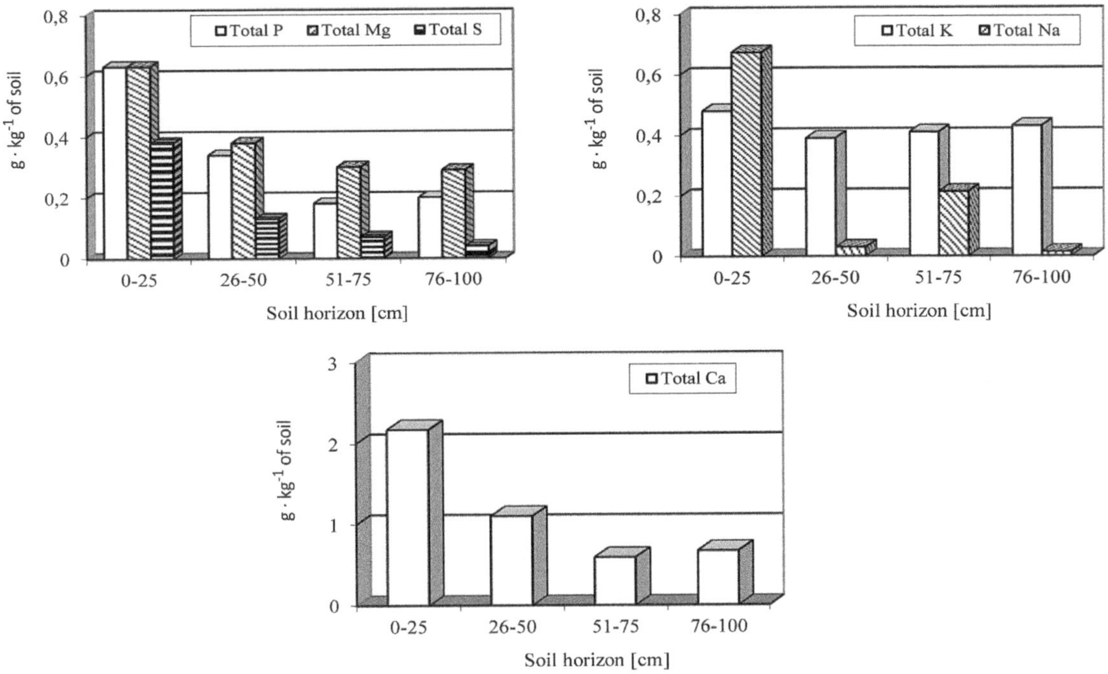

Figure 1. Total content of elements in the soil profile regardless of ash dose (the average for the individual horizons) [g · kg^{-1} of soil]

potassium. The substances applied to the soil, in order to reduce the impact of ash generated by the combustion of hard coal, did not influence significantly the content of elements in the soil 19 years after their application.

The increase in the total content of tested elements recorded in the present study confirms the findings reported by other authors. The changes in the concentration of phosphorus and potassium in the soil are consistent with the results reported by_Właśniewski [19] and Lee et al. [7, 8]. Właśniewski [19] demonstrated that application of the highest dose of ash (134 t · ha^{-1}) caused a marked increase in total content of phosphorus by 361%, magnesium by 633%, and a slight increase in potassium content by 38% (compared to reference). Lee et al. [8] observed that the application of fly ash in the dose of 40, 80 and 120 t · ha^{-1} in two paddy soils increased the available phosphorus significantly. An increase in the total content of potassium with fertilization with hard coal ash was indicated by Lee et al. [7]. Conversely, different results were reported by Matsi and Keramidas [9] who observed that over time following the application of furnace waste materials, the soil surface layer became depleted of potassium.

Mittra et al. [12] demonstrated that, together with ash, a substantial amount of potassium, calcium and magnesium were introduced into soil, for instance, 32 kg of P, 25 kg of K, 33 kg of Ca and 17 kg of Mg were introduced with a dose of 10 t · ha^{-1}. The studies conducted by Kovacik et al. [6] demonstrated that ash-fly ash mixture statistically significant increase of the content of available calcium. An increase in the total content of calcium and sodium with fertilization with hard coal ash in the dose of 50–400 t · ha^{-1} was also indicated by Gourab and Vadakepuram [4].

Hjemar [5] and Theis and Gardner [16] demonstrated that sodium that originated from ash introduced into soil diffused into the soil solution relatively quickly and passed deeper into the soil profile faster than potassium. Tejasvi and Kumar [15] reported a slight increase in the content of phosphorus, potassium, sulphur, calcium and magnesium following the application of ash.

The results of our studies confirm the findings by Meller [10] indicating that nutrients introduced into soil with ash become mobile at different rates depending on the degree of their solubility. The studies conducted by Meller [10] demonstrated that sodium, potassium, calcium and subsequently magnesium were most rapidly transformed into soluble forms as a result of precipitation water activity. Conversely, in an acidic environment calcium, manganese, magnesium and lead are the first quantitatively mobilized elements, followed by sodium and potassium [5, 10, 16]. Higher doses of ash have a large effect on chemical composition of plants [1, 2].

CONCLUSION

1. Nineteen years after the application of ash there was still an increase in the total content of all elements in the soil. Its magnitude depended on the soil horizon and it was highest in the humus horizon. The increase in the content of nutrients was in general proportional to the applied doses of ash.
2. In the surface layer of the soil, the increase was particularly high for sodium and subsequently for calcium, magnesium, sulphur, potassium and phosphorus, arranged in the order of decreasing values.
3. The content of elements in deeper soil layers was significantly lower than in the 0–25 cm zone. A gradual reduction in the content together with the increase in depth was recorded only for magnesium and sulphur. The lowest concentration of phosphorus and calcium was detected in the 51–75 cm layer, while in the case of potassium it was in the 26–50 cm zone.
4. The substances introduced into the soil in order to reduce the impact of ash generated by hard coal combustion did not exert any significant impact on the content of elements 19 years following their application.

REFERENCES

1. Antonkiewicz J., Lošák T. 2007. The effect of hard coal ashes on the quality of maize yield. Part 2. Microelements. Acta Univ. Agric. et Silviculturae Mendelianae Brunensis, LV, 1(5): 9-16.
2. Antonkiewicz J. 2009. Assessment of chemical composition of bushgrass (*Calamagrostis epigejos* L.) occurring on furnace waste and carbide residue lime landfills. Part 1. Contents of N, P, K, Ca, Mg and Na. Ecol. Chem. Eng. A, 16(5-6): 501-513.
3. Ciećko Z., Żołnowski A.C., Kulmaczewska J., Chełstowski A. 2009. Wpływ następczy melioracyjnych dawek popiołów z węgla kamiennego na kwasowość gleby. Zesz. Probl. Post. Nauk Roln., 535: 73-83.
4. Gourab R., Vadakepuram C.J. 2011. Dose-related effect of fly ash on edaphic properties in laterite cropland soil. Exotox. Environ. Safe., 74: 769-775.
5. Hjemar O. 1990. Leachate from land disposal of coal fly ash. Waste Manag. Res., 8: 429-449.
6. Kovacik P., Macak M., Ducsay L., Halcinova M., Jancich M. 2011. Effect of ash-fly ash mixture application on soil fertility. J. Elem., 2: 215-225.
7. Lee C.H., Lee H., Lee Y.B., Chang H.H., Ali M.A., Min W., Kim S., Kim P.J. 2007. Increase of available phosphorus by fly-ash application in paddy soils. Commun. Soil Sci. Plan., 38: 1551-1562.
8. Lee H., Ha H.S., Lee C.H., Lee Y.B., Kim P.J. 2006. Fly ash effect on improving soil properties and rice productivity in Korean paddy soils. Bioresource Technol., 97: 1490-1497.
9. Matsi T., Keramidas V.Z. 1999. Fly ash application on two acid soils and its effect on soil salinity, pH, B, P and on ryegrass growth and composition. Environ. Pollut., 104: 107-112.
10. Meller E. 1999. Zmiany we właściwościach gleby piaszczystej nawożonej zróżnicowanymi dawkami popiołów z elektrowni „Dolna Odra". Zesz. Nauk. AR w Szczecinie, 78: 189-201.
11. Microsoft Excel 2010. www.microsoft.com; 2010.
12. Mittra B.N., Karmakar S., Swain D.K., Ghosh B.C. 2005. Fly ash – a potential source of soil amendment and a component of integrated plant nutrient supply system. Fuel, 84: 1447-1451.
13. Ochrona Środowiska 2012. Informacje i opracowania statystyczne. Główny Urząd Statystyczny, Warszawa; 2012.
14. Statsoft Statistica 10.0 (data analysis software system). www.statsoft.com; 2012.
15. Tejasvi A., Kumar S. 2012. Impact of Fly Ash on Soil Properties. Natl. Acad. Sci. Lett., 35(1): 13-16.
16. Theis T.L., Gardner K.H. 1990. Environmental assessment of ash of ash disposal. Crit. Rev. Env. Contr., 20(1): 21-42.
17. Ukwattage N.L., Ranjith P.G., Bouazza M. 2013. The use of coal combustion fly ash as a soil amendment in agricultural lands (with comments on its potential to improve food security and sequester carbon. Fuel, 109: 400-408.
18. Veeresh H., Tripathy S., Chaudhuri D., Ghosh B.C., Hart B.R., Powell M.A. 2003. Changes in physical and chemical properties of three soil types in India as a result of amendment with fly ash and sewage sludge. Environ. Geol., 43: 513-520.
19. Właśniewski S. 2009. Wpływ nawożenia popiołem lotnym z węgla kamiennego na wybrane właściwości chemiczne gleby piaszczystej i plonowanie owsa. Ochr. Środ. Zas. Nat., 41: 479-488.

16

CHANGES IN SPECIFIC LOADS OF MINERAL COMPONENTS OUTFLOWING FROM CATCHMENT AREA OF RIVER SUPRAŚL IN 2001–2009

Mirosław Skorbiłowicz[1], Piotr Ofman[1]

[1] Białystok University of Technology, Wiejska 45A, 15-351 Białystok, Poland, e-mail: m.skorbilowicz@pb.edu.pl

ABSTRACT

The purpose of this paper was to determine changes in loads of $N\text{-}NH_4^+$, $S\text{-}SO_4^2$, $P\text{-}PO_4^{3-}$, $N\text{-}NO_3^-$, P_{og}, Cl^-, Ca^{2+}, Mg^{2+}, and their origin in waters outflowing from Supraśl catchment. Research was carried out in the period from 2001 to 2009 in three measurement points located around Gródek, Nowodworce and Dzikie. Obtained values of specific load were raging for: $N\text{-}NH_4^+$ 0.051÷2.094 kg $N\text{-}NH_4^+ \cdot ha^{-1} \cdot year^{-1}$, $S\text{-}SO_4^{2-}$ 16.50÷30.00 kg $S\text{-}SO_4^{2-} \cdot ha^{-1} \cdot year^{-1}$, $P\text{-}PO_4^{3-}$ 0.17÷0.45 kg $P\text{-}PO_4^{3} \cdot ha^{-1} \cdot year^{-1}$, P_{og} 0.44÷0.78 kg $P \cdot ha^{-1} \cdot year^{-1}$, Cl^- 14.58÷26.80 kg $Cl^- \cdot ha^{-1} \cdot year^{-1}$, Ca^{2+} 68.2÷77.1 kg $Ca^{2+} \cdot ha^{-1} \cdot year^{-1}$, Mg^{2+} 9.57÷12.29 kg $Mg^{2+} \cdot ha^{-1} \cdot year^{-1}$. In addition, linear Pearson correlation coefficients were calculated between specific loads of individual components and fertilization form in each measuring point. There were obtained statistically significant correlations between NPK fertilization and specific loads of $N\text{-}NH_4^+$, P_{og}, Cl^-, Ca^{2+} i Mg^{2+} and between manure fertilization and $S\text{-}SO_4^{2-}$, Ca^{2+} i Mg^{2+}.

Keywords: catchment, fertilization, mineral components.

INTRODUCTION

Mineral components outflow to catchments surface waters depend on topography, soils type and their conciseness, their ability to buffer and sorption, vegetation, water and climatic conditions and in particular precipitation and anthropogenic factors, which include: land reclamation, agricultural production intensity and presence of settlements and farming facilities in catchment area [Szpakowska and Życzyńska-Bałoniak 1992; Sapek 1996b; Koc et al. 1999; Miller et al. 2001; Spychaj-Fabisiak et al. 2001; Koc et al. 2002a; Koc and Szymczyk 2003; Koc et al. 2003; Szymczyk and Cymes 2004; Kuźniar et al. 2008].

Studies carried out by Skorbiłowicz [2004] in selected river catchments of upper Narew showed that waters flowing through the land used for agriculture purposes had higher concentration of calcium and magnesium ions than waters flowing through forested catchments. On the other hand, water from forested areas is richer in iron and manganese ions in compare to water from agricultural catchments. The reason of increased nutrients' migration into rivers are changes in land use, especially deforestation, afforestation, and drainage of wetlands and peat bogs.

In order to determine agricultural impact on water chemical composition in catchments average values of components concentrations are used in natural waters. Component concentration mean values in water is taken as a measure of chemical composition assessment in water. Specific load seems to be a better factor as allows the assessment of the examined components and specification of their impact in catchment water quality, and in particular their migration the outside catchment. This allows to determine the part of non-point sources of pollution in river or water reservoir [Hansen et al. 2000; Pulikowski et al. 2001; Pulikowski 2004]. Specific loads of substances are a product of average concentration and average flow in micro-catchment [Miler and Murat-Błażejewska 1997].

The aim of this study was to assess trends in specific loads of minerals eluted from Supraśl catchment area and to determinate their origin.

METHODOLOGY

For changes assessment in specific loads of individual components results from Supraśl river (concentration) were used from the research carried out under the State Environmental Monitoring by the Regional Inspectorate of Environmental Protection in Białystok in period from 2001 to 2009. Water samples were collected from three measurement points located on river Supraśl. The first measurement point (PP1) was located near the Gródek, the second (PP2) in Nowodworce and third (PP3) near Dzikie (Figure 1). The concentration of each component was analyzed in accordance with the following Research Procedures and Polish Standards:

- $N\text{-}NH_4^+$ – Research Procedure 051, no. 1, from 07.12.2009
- $S\text{-}SO_4^{2-}$ – PN-ISO 9280:2002
- $P\text{-}PO_4^{3-}$ – PN-EN ISO 6878:2006
- $N\text{-}NO_3^-$ – PN-EN 26777:1999
- Total phosphorus – PN-EN ISO 6878: 2006
- Cl^- – PN-ISO 9297:1994
- Ca^{2+} – PN-ISO 9964-1:1997
- Mg^{2+} – PN-ISO 9964-2:1997

Specific loads of individual components were calculated according to the following equation [Skorbiłowicz 2010]:

$$\phi_{rz} = \frac{c \cdot q}{F}$$

where: ϕ_{rz} – component specific load entered form catchment in $kg \cdot ha^{-1} \cdot year^{-1}$,
c – component average annual concentration in $kg \cdot m^{-3}$,
q – average annual water flow in measurement point in $m^3 \cdot s^{-1}$,
F – catchment area in ha.

Catchment area, average annual water flow and structure of soil usage were adopted in accordance with the data presented by Skorbiłowicz [2010].

Table 1. Catchment area and characteristic water flow [Skorbiłowicz 2010]

Point number	Catchment area [ha]	Average water flow [$m^3 \cdot s^{-1}$]
PP1	97 000	0.67
PP2	152 300	5.36
PP3	193 200	10.13

In this study Pearson correlation coefficients were calculated using a licensed software version of Statistica 10. For statistical analysis, the data from Supraśl catchment (Table 2), and data on the usage of mineral and organic fertilizers was used (Table 3). For specific loads trends calculation the average loads from three measurement points for each year were used. Linear changes trends and mathematical models were performed using Excel 2013.

Supraśl catchment belongs to a group of agricultural catchments. Area used for agricultural purposes is equal to 67%, forest covers 30% and urban area is equal to 3% of total catchment area.

According to GUS data (Table 3), the usage of NPK (mineral) fertilizers in Podlaskie did not change significantly in the period from 2001 to 2009, however, a small upward trend was noted. In the period from 2001 to 2005 average lime fertilizer usage was equal to 58.8 $kg \cdot ha^{-1}$. Beginning

Table 2. Soil usage [Skorbiłowicz 2010]

River	Afforestation [%]	Agricultural area		Urban areas [%]
		Arable areas [%]	Pasture and grasslands [%]	
Supraśl	30	39	28	3

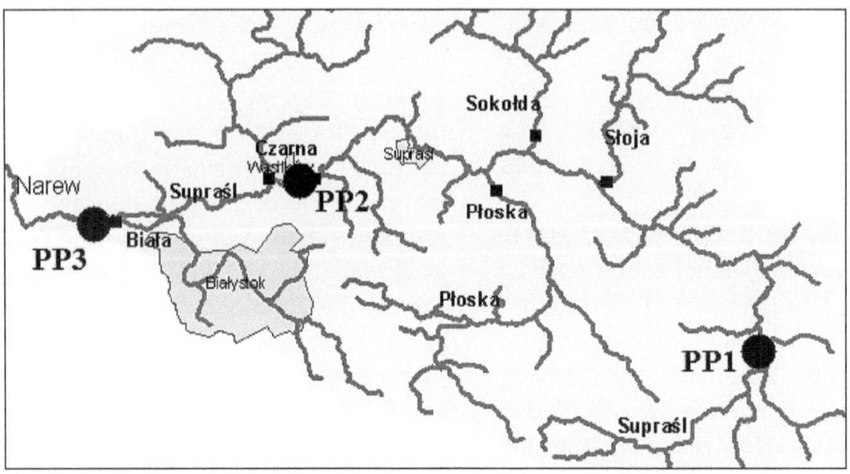

Figure 1. Measurement point localization [Ocena...WIOŚ 2010]

Table 3. Usage of mineral and organic fertilizers in Podlaskie in 2001-2009 [GUS 2010]

Marketing year	Fertilizers		
	NPK	Lime fertilizer	Manure (NPK)
	[kg/1 ha agricultural area]		
2001	78	57	–
2002	80	59	–
2003	81	57	62
2004	86	61	75
2005	87	60	75
2006	91	20	78
2007	90	11	56
2008	94	12	89
2009	93	13	67

from 2006 a significant reduction in de-acidification procedures was observed. Soils organic fertilization (manure) was changing from 56 to 89 kg·ha^{-1} in period from 2003 to 2009. The lowest usage of manure for fertilization purposes was noted in 2007 while the highest was in 2008.

According to IMiGW in Białystok, annual precipitation in Podlaskie was ranging from 514 to 715 mm in period from 2001 to 2009 (Table 4).

Table 4. Annual precipitation in Podlasie [IMiGW Białystok data]

Year	Annual precipitation [mm]
2001	585
2002	514
2003	556
2004	630
2005	560
2006	610
2007	616
2008	613
2009	715

RESULTS AND DISSCUSSION

Analysis of mineral compounds specific loads eluted from catchment is very difficult due to the fact that the amount of leachable components depends on the influence of natural factors, climate, physiographic characteristics, type and structure of land usage and anthropogenic factors such as the amount of used fertilizers and water and sewage management in the catchment [Sojka 2009]. In most cases, anthropogenic factors are increasing specific loads of components outflowing from the catchment area. On the other hand, natural factors can increase or reduce specific loads of each component.

The highest average annual specific loads of examined components in Supraśl catchment are eluted from Dzikie micro-catchment (Table 5), while the smallest form micro-catchment Gródek. This phenomena is primarily dictated by catchment size.

Table 5. Average specific load in measurement points

Component	Unit	Gródek	Nowodworce	Dzikie
$N\text{-}NH_4^+$	kg·ha^{-1}·year^{-1}	0.07	0.11	0.30
$N\text{-}NO_3^-$		0.47	1.30	3.15
$P\text{-}PO_4^{3-}$		0.12	0.24	0.60
P_{og}		0.18	0.49	1.05
$S\text{-}SO_4^{2-}$		6.20	23.67	52.78
Cl^-		2.59	9.24	42.55
Ca^{2+}		15.08	80.43	120.47
Mg^{2+}		1.90	11.73	18.13

$N\text{-}NH_4^+$ specific load values (Figure 2) ranged from 0.11 to 0.33 kg NH_4^+ ha^{-1}·year^{-1}. The lowest load value of this component was observed in 2007, while the highest in 2006. Average annual load values did not differ by more than 0.04 kg $N\text{-}NH_4^+$ ha^{-1}·year^{-1}. The only exception was noted in differences obtained at maximum load, which was observed in 2006. Similar $N\text{-}NH_4^+$ loads were observed by Marchlewska [1991]. $N\text{-}NH_4^+$ load values obtained by Marchlewska ranged from 0.051 to 2.094 kg $N\text{-}NH_4^+$·ha^{-1}·year^{-1}, but it must be noted that the highest $N\text{-}NH_4^+$ load particularly depended on the catchment area. Average annual $P\text{-}PO_4^{3-}$ specific loads (Figure 3) ranged from 0.21 to 0.51 kg $P\text{-}PO_4^{3-}$·ha^{-1}·year^{-1}. The lowest $P\text{-}PO_4^{3-}$ load value was observed in 2009 while the highest in 2001. Similar $P\text{-}PO_4^{3-}$ load values were achieved by Krajewska and Bogdanowicz [2009] in research carried out in Reda, Zagórska Struga, Płutnica and Gizdepka catchments. Load obtained by them were ranging from 0.17 to 0.45 kg $P\text{-}PO_4^{3-}$·ha^{-1}·year^{-1}.

The performed statistical analysis showed a correlation between $N\text{-}NH_4^+$ specific load and mineral fertilization level in Supraśl catchment.

Average annual $N\text{-}NO_3^-$ loads values (Figure 4) were ranging from 1.36 to 2.18 kg $N\text{-}NO_3^-$·ha^{-1}·year^{-1}. The lowest load value of this component was observed in 2009, while the highest in 2002. Average annual load values did not differ by more than 0.50 kg $N\text{-}NO_3^-$ ha^{-1}·year^{-1}. Higher range of

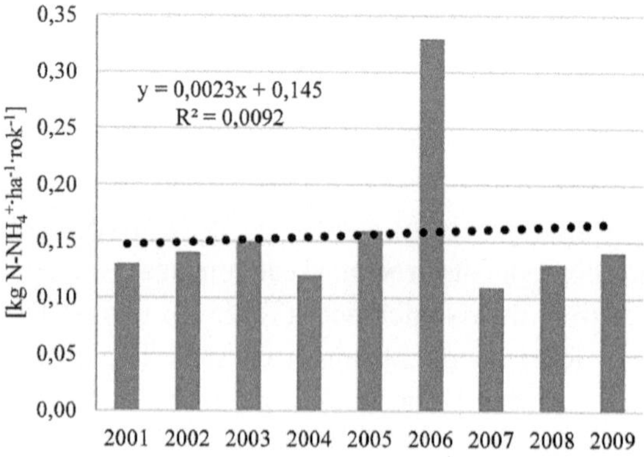

Figure 2. Average annual loads of $N\text{-}NH_4^+$

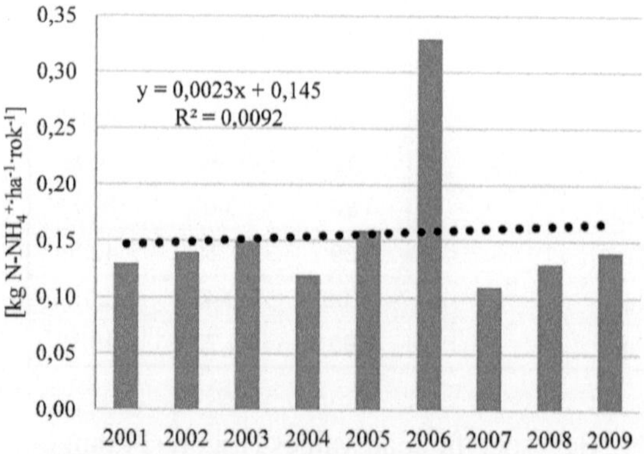

Figure 3. Average annual loads of $P\text{-}PO_4^{3-}$

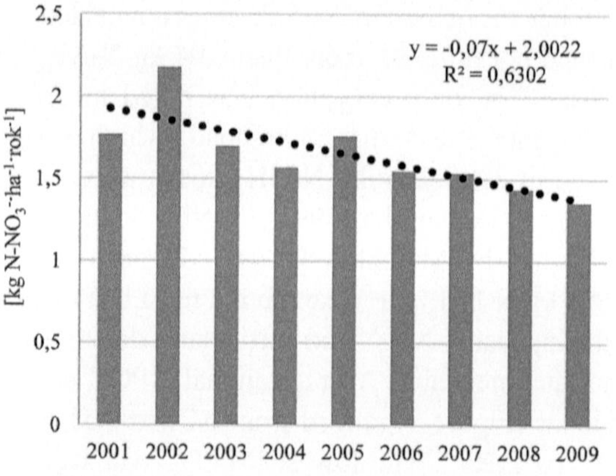

Figure 4. Average annual loads of $N\text{-}NO_3^-$

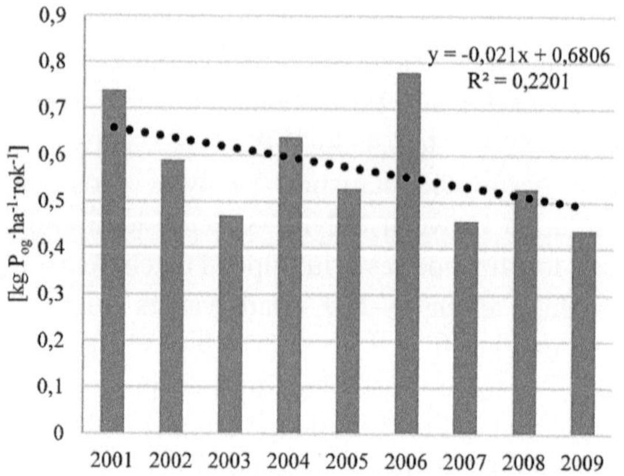

Figure 5. Average annual loads of total phosphorus

$N\text{-}NO_3^-$ load was obtained by Sojka [2009] in Dębina river. The load calculated by the author ranged from 1.74 to 10.25 kg $N\text{-}NO_3^-\cdot ha^{-1}\cdot year^{-1}$.

Average annual total phosphorus loads (Figure 5) were changing in a range from 0.44 to 0.78 kg $P\cdot ha^{-1}\cdot rok^{-1}$. The lowest load of these components was observed in 2009 and the highest in 2006. Total phosphorus load changes trend, in research period, was similar to changes of phosphates load. Similar total phosphorus loads were observed by Grabińska [2007]. Specific load values calculated by the author ranged from 0.17 to 1.27 kg $P\cdot ha^{-1}\cdot year^{-1}$. Those results were obtained for mixed forested and agricultural catchment in the upper Narew.

Statistical analysis performed in this study has shown a correlation between total phosphorus specific load and mineral and organic fertilization level in Supraśl catchment.

Average annual $S\text{-}SO_4^{2-}$ loads (Figure 6), in the research period, ranged from 24.75 to 33.71 kg $S\text{-}SO_4^{2-}\cdot ha^{-1}\cdot year^{-1}$. The lowest specific load was observed in 2007 and the highest in 2004. Similar load range of this component was achieved by Jekatierynczuk-Rudczyk et al. [2006]. $S\text{-}SO_4^{2-}$ load values obtained by the authors changed from 5,44 do 35,03 kg $S\text{-}SO_4^{2-}\cdot ha^{-1}\cdot year^{-1}$ and were observed in Horodnianka catchment. Horodnianka drains agricultural areas near Białystok.

Statistical models obtained in this study point to correlations between $S\text{-}SO_4^{2-}$ specific load and organic fertilization and $S\text{-}SO_4^{2-}$ specific load and annual precipitation.

Average annual chlorides load was ranging from 14,58 to 26,80 kg $Cl^-\cdot ha^{-1}\cdot year^{-1}$. The lowest specific load value of this component was observed in 2001 while the highest in 2002. Obtained chlorides loads values are similar to those that were observed by Ostrowski i Bogdał in agricultural micro-catchment Wronowiec [2006]. Those loads were ranging from 8,41 to 14,74 kg $Cl^-\cdot ha^{-1}\cdot year^{-1}$. Numerical calculations point to a correlation between Cl^- specific load and mineral fertilization level.

Calcium average loads values were changing from 68,2 do 77,1 kg $Ca^{2+}\cdot ha^{-1}\cdot year^{-1}$ during research period (Figure 8). The lowest value of this load was observed in 2008 and the highest in 2001. Similar Ca^{2+} loads values were obtained by Sidoruk and Skwierawski [2006] during studies carried out in waters outflowing from forested areas to Bukwład lake. Average Ca^{2+} load values obtained by the authors was equal to 63,6 kg $Ca^{2+}\cdot ha^{-1}\cdot year^{-1}$.

The highest specific load of Mg^{2+} was observed in 2001 while the lowest in 2008. Those values were equal respectively 12.29 and 9.57 kg $Mg^{2+}\cdot ha^{-1}\cdot year^{-1}$. Alike case of Ca^{2+} loads, Mg^{2+} loads values were decreasing in each year of studies. This phenomenon is probably caused by lime and magnesium fertilizers used in that area. In most cases, lime fertilization is related to simultaneous calcium and magnesium usage. This procedure allows to obtain higher crops. Similar Mg^{2+} loads were obtained by Cymes and Szymczak [2005] in waters flowing through arable areas (15.4 kg $Mg^{2+}\cdot ha^{-1}\cdot year^{-1}$) and grasslands (13.4 kg $Mg^{2+}\cdot ha^{-1}\cdot year^{-1}$) on heavy soils.

Statistical test performed in this study pointed out a correlation between lime and organic fertilization and Ca^{2+} and Mg^{2+} specific loads.

A dependence of $S\text{-}SO_4^{2-}$, Ca^{2+} and Mg^{2+} specific loads from usage of manure fertilizer in micro-catchment area of Gródek catchment was discovered with Pearson correlation coefficients equal respectively r = 0.76, r = 0.83 and r = 0.79. This phenomena could be correlated with the usage of manure fertilization in this area but not with soil types in that catchment. Many studies shown that soils in Gródek micro-catchment are of post bog origin, which are not rich in calcium and magnesium.

From micro-catchment Nowodworce outflows $N\text{-}NH_4^+$, P_{og}, Cl^-, Ca^{2+} i Mg^{2+} specific load positively correlated with the usage of NPK fertilization with coefficients equal respectively r = 0.87, r = 0.79, r = 0.74, r = 0.51 and r = 0.66. The loads of total phosphorus and $P\text{-}PO_4^{3-}$ are positively correlated with organic fertilization value with coefficients equal respectively r = 0.54 and r = 0.61.

Lime fertilization value was positively correlated with Ca^{2+} and Mg^{2+} (r = 0.79 and r = 0.85) load from surface run-offs in Dzikie micro-catchment. In this area, as shown by Skorbiłowicz [2010a], intensively fertilized, acidic light soils are present which are probably the cause of high macroelements' concentration in waters from drainages that flow to many ditches located in this area. This phenomenon was also indicated by Terelaka et al. [1999], Koc et al. [1997] and Sapek [1995] in their research carried out in other catchments. Recent studies carried out by Skorbiłowicz [2010b] also confirm those observation in Supraśl catchment.

No positive Pearson and Spearman correlation coefficients were noted between precipitation (Table 4) and specific load of each com-

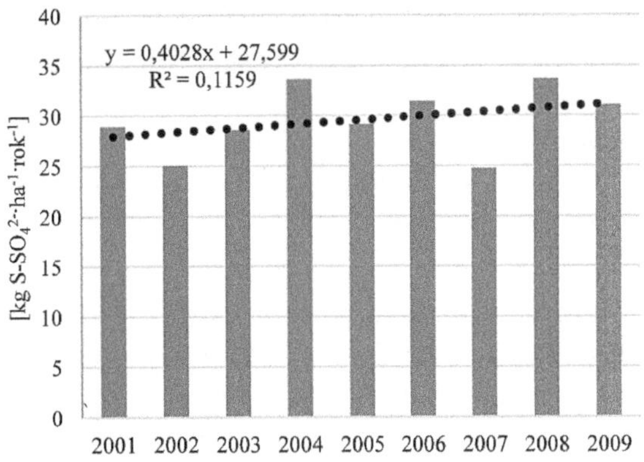

Figure 6. Average annual loads of $S\text{-}SO_4^{2-}$

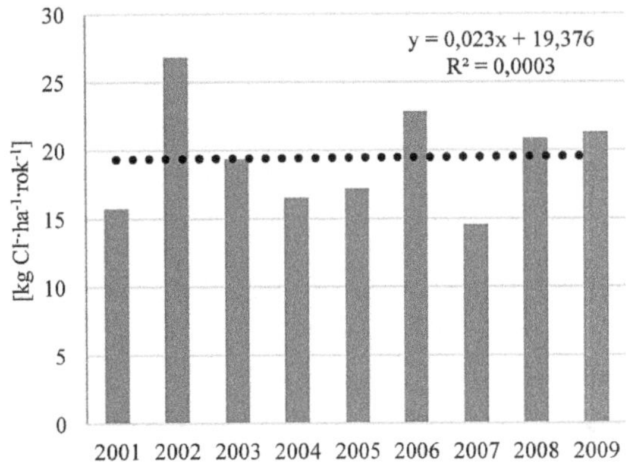

Figure 7. Average annual loads of Cl^-

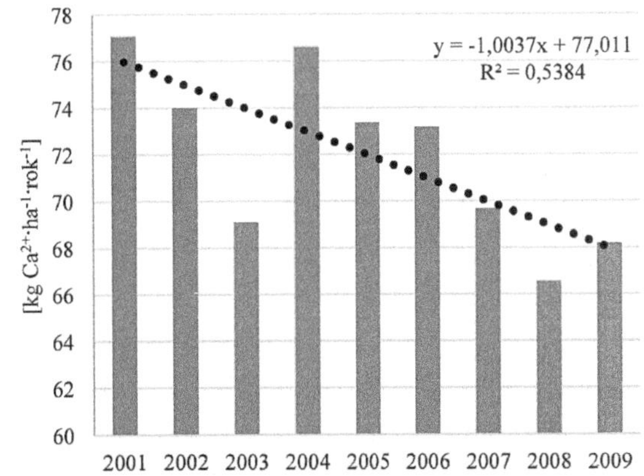

Figure 8. Average annual loads of Ca^{2+}

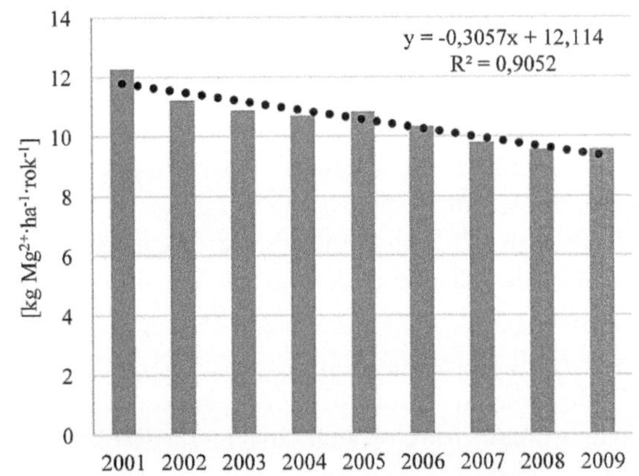

Figure 9. Average annual loads of Mg^{2+}

ponent, except sulfates (r = 0.50). Supraśl catchment is characterized by 30% afforestation (Table 2) and forests have strong a retention impact on natural waters, hence there is probably a lack of clear relationships between the amount of precipitation and specific loads in the studied catchment. Artificial neural network sensitivity analysis performed by Skorbiłowicz [2010b] showed that network was most sensitive to location of a measuring point at Supraśl, and less sensitive to the amount of precipitation in the catchment. Neither did Skorbiłowicz obtain a reliable relationship between component concentration in waters and their average specific loads outflowing from catchment.

CONCLUSIONS

1. Specific loads of N-NO_3^-, P-PO_4^{3-}, P_{og}, Ca^{2+} and Mg^{2+} outflowing form Supraśl catchment were characterized by decreasing trend and specific loads of N-NH_4^+, S-SO_4^{2-} and Cl^- were characterized by increasing trend in the period from 2001 to 2009.
2. Statistical analysis showed dependence of S-SO_4^{2-}, Ca^{2+} and Mg^{2+} specific loads on manure fertilization value in Gródek micro-catchment.
3. From Nowodworce micro-catchment outflows positively correlated specific loads of NH_4^+, P_{og}, Cl-, Ca^{2+} and Mg^{2+} with NPK fertilization value. Total phosphorus and P-PO_4^{3-} loads outflowing form micro-catchment were correlated with organic fertilization value.
4. Ca^{2+} and Mg^{2+} load were positively correlated with lime fertilization value in micro-catchment Dzikie.
5. One statistically significant Pearson correlation was observed between precipitation and sulfates load.

REFERENCES

1. Cymes I., Szymczyk S. 2005. Wpływ sposobu użytkowania terenu, melioracji i czynników naturalnych na stężenie sodu, wapnia i magnezu w wodach gruntowych i ich odpływ siecią drenarską z gleb ciężkich. Inżynieria Ekologiczna, 13, 44–49.
2. Grabińska B., Koc J., Skwierawski A., Sobczyńska-Wójcik K., Rafałowska M. 2005. Stężenia i odpływ fosforu ogólnego z wodami rzecznymi ze zlewni zróżnicowanym użytkowaniu. Inżynieria Ekologiczna, 13, 87–92.
3. Jekatierynczuk-Rudczyk E., Zieliński P., Górniak A. 2006. Stopień degradacji rzeki wiejskiej w bezpośrednim sąsiedztwie Białegostoku. Woda-Środowisko-Obszary Wiejskie, 2(18), 143–153.
4. Koc J., Procyk Z., Szymczyk S. 1997. Czynniki kształtujące jakość wód powierzchniowych obszarów wiejskich. Materiały Seminaryjne IMUZ 39, 222–229.
5. Krajewska Z., Bogdanowicz R. 2009. Zróżnicowanie wielkości eksportu substancji biogennych w zlewisku Zatoki Puckiej. [In:] Janowski A.T., Absalon D., Machowski R., Ruman M. (Eds.) Przeobrażenia stosunków wodnych w warunkach zmieniającego się środowiska. Wydział Nauk o Ziemi Uniwersytetu Śląskiego, Sosnowiec, 177–186.
6. Marchlewska B. 1991. Obciążenie związkami biogennymi zlewni rzek w województwie legnickim, Ochrona Środowiska, 2(43), 19–21.
7. Ostrowski K. Bogdał A. 2006. Loads of selected chemical components delivered by percipitation and flowing away from Wronowiec microcatchment. Acta Scientiarum Polonorum, 5(2), 37–46.
8. Sapek B. 1995. Wymywanie azotanów oraz zakwaszanie gleby i wód gruntowych w aspekcie działalności rolniczej. Materiały Informacyjne IMUZ, Falenty, 30, 1–31.
9. Sidoruk M., Skwierawski A. 2006. Wpływ użytkowania zlewni na ładunek wapnia, sodu, potasu i magnezu w wodach dopływających do jeziora Bukwład. Ecological Chemistry and Engineering, S2(13), 337–343.
10. Skorbiłowicz M. 2010(a). The concentrations of macroelements, zinc and iron ions in waters of the upper Narew basin, NE Poland. Polish Journal of Environmental Studies, 2(19), 397–405.
11. Skorbiłowicz M. 2010(b). Czynniki i procesy kształtujące obieg składników mineralnych w wodach rzecznych zlewni górnej Narwi. Oficyna Wydawnicza Politechniki Białostockiej, Białystok.
12. Sojka M. 2009. Ocena ładunków związków biogennych wymywanych ze zlewni cieku Dębina. Środkowo-Pomorskie Towarzystwo Naukowe Ochrony Środowiska, 11, 1225–1234.
13. Terelak H., Motowicka-Terelak T., Sadurski W. 1999. Wymywanie składników chemicznych z gleb gruntów ornych do wód drenarskich. [In:] Ochrona zasobów i jakości wód powierzchniowych i podziemnych. Wyd. Ekonomia i Środowisko, Białystok.
14. Urząd Statystyczny 2010. Ludność, ruch naturalny, migracje w województwie podlaskim w 2010 r., Białystok.
15. Wojewódzki Inspektorat Ochrony Środowiska w Białymstoku 2011. Ocena stanu czystości wód w zlewni rzeki Supraśl w 2010 roku.

17

ASSESSMENT OF GROUNDWATER CONTAMINATION AROUND RECLAIMED MUNICIPAL LANDFILL – OTWOCK AREA, POLAND

Dorota Porowska[1]

[1] Institute of Hydrogeology and Engineering Geology, Faculty of Geology, University of Warsaw, Żwirki i Wigury 93, 02-089 Warsaw, Poland, e-mail: dorotap@uw.edu.pl

ABSTRACT

The research was conducted around reclaimed landfill, located on the suburb of Otwock, around 25 km south-west of Warsaw. The objective of this study was to identify the chemical composition of groundwater and to determine the landfill impact on the chemical composition of groundwater downgrading from the landfill. Otwock landfill is located in very permeable area, where leachate quickly seeps into groundwater and plays a key role in controlling redox condition (and chemical composition of groundwater) of the downgradient area. High concentrations of HCO_3^-, Cl^-, Ca^{2+}, Mg^{2+}, Na^+, K^+, $Fe_{tot.}$ as well as DOC in groundwater downgradient from the landfill (in comparison to background water) likely indicate that groundwater quality is being significantly affected by leachate percolation. Currently, the load of contamination is released from landfill periodically and slowly moves (70 m/y) in the aquifer along the flow direction. The effect of distance of the piezometer from the pollution source was also investigated. As expected, water from the nearest piezometer to the landfill showed the highest values of contaminant (water temperature, specific electrical conductivity, sodium, iron, chlorides (except for summer and autumn analysis) and calcium (except for winter analysis)). Chemical status of groundwater downgradient from the landfill is poor.

Keywords: landfill, anaerobic conditions, leachate-contaminated water, background water, groundwater quality.

INTRODUCTION

Landfills were constructed without taking too much care of the surrounding environment in early 1960's. Many of them were constructed without appropriate engineering structure to control and minimize the impact of degrading refuse on soil and groundwater quality. Moreover, many of them could contain illegally dumped non-segregated wastes. Landfills were generally not capped after closure to prevent the formation of leachate by infiltration of rainwater, nor were precautions taken to prohibit spreading of leachate to underlying aquifers. Therefore, release of hazardous chemicals to the environment continues, although landfills are closed now. It is commonly known that the quantity of leachate is a function of geological structure (and underlying soil), hydrogeological and climatic conditions as well as the characteristics of the refuse. The quality of groundwater impacted by landfill leachate is highly dependent on the stage of the refuse transformation (the landfill age) and the effectiveness after landfill reclamation. It is commonly observed that despite the reclamation, groundwater quality is being significantly affected by leachate percolation. As a result, leachate plumes have been formed downstream from many landfill sites and have been moved away along the groundwater flow direction.

Similar situation had been suspected around the reclaimed Otwock landfill. The study of hydrogeological conditions and chemical composition of groundwater in the vicinity of this landfill was carried out to asses the impact of the leachate on groundwater quality and to distinguish natural water-rock interactions and anthropogenic influences.

In the interior of the Otwock landfill, typical anaerobic conditions are observed. It is generally expected that the geochemical condition of landfill will evolve toward an aerobic, but currently measured gases indicated that biochemical processes are still active and landfill gas (biogas) is generated [Porowska, Gruszczyński 2006, 2013]. Very permeable and thin layer of compost receives a large amount of rainwater which infiltrates into the subsurface of the landfill and leachate is generated. Otwock landfill is located in a very permeable area, hence leachate quickly seeps into groundwater and plays a key role in controlling redox condition of the downgradient area, where a distinct contamination was detected. The second factor in the formation of leachate is a shallow position of the water table, reaches as far as the base of the landfill when temporarily part of the landfill body resides below groundwater table. Research indicates that leachate affects on groundwater composition at the downgradient piezometers. This landfill, and other old landfills without liners or leachate collecting systems constitutes a threat to the downgradient groundwater quality due to infiltration of leachate into the aquifer. Leachate plume may be attenuated by natural processes such as dilution, dispersion, diffusion, sorption and biodegradation in the groundwater aquifer system along the groundwater flow direction [Cherry (eds.) 1983, Christensen, Kjeldsen 1989, Christensen et al. 1994]. However, the load of contamination is released from landfill periodically and causes almost continuous (in variable concentrations) contamination downgradient from the landfill. Variations in concentrations of chemical species downgradient of the landfill are controlled by change in leachate composition and mentioned above physical attenuation and chemical reaction [Baedecker, Apgar 1984]. Chemical reactions and processes in contaminated water do not remain constant with time.

The objective of this study was to identify the chemical composition of groundwater and determine landfill impact on chemical composition and quality of groundwater downgradient from the landfill.

This paper presents preliminary results from field and laboratory research of groundwater samples collected from aquifer in Otwock landfill. The physical and chemical constituents discussed here in detail include: water temperature, pH, specific electrical conductivity (SEC), oxidation-reduction potential (ORP), dissolved oxygen (DO), HCO_3^-, SO_4^{2-}, Cl^-, Ca^{2+}, Mg^{2+}, Na^+, K^+ and $Fe_{tot.}$, as well as dissolved organic carbon (DOC).

MATERIALS AND METHODS

Research site description

Research was conducted around old municipal landfill (after reclamation) located on the suburb of Otwock, around 25 km south-west of Warsaw (Figure 1).

Otwock landfill covers an area of 2,8 ha and received municipal waste in the period 1961–1991 [Koda et al. 1999]. The amount and type of refuse in Otwock landfill is not known (is not well documented), because landfilling was uncontrolled. The landfill has no bottom lined and the waste is placed directly on the ground. After closing, the landfill was covered with a soil layer consisting of approximately 30–50 cm compost and was planted with grass. Landfill leachate is formed from the infiltration and passage of water through solid waste which results in a combination of physical, chemical and microbial processes that transfer pollutants from waste material to the water. The site history and landfill gases composition as well as hydrogeochemical conditions were studied in a previous investigation [Porowska, Gruszczyński 2006, 2013].

The landfill is located on the Vistula river valley, close to the Świder stream. The Świder stream to the north, and Vistula river to the west drain groundwater from the landfill area. According to the groundwater table contour shown on Figure 1, the general flow direction is northwest.

The bedrock of the landfill is represented by river deposit with high permeability – gravels and varigrained sands, locally contain organic matter distribute within aquifer or as a peat layer. The full extent of these peat deposits is not known, but partly include the northern area where piezometer no 2 is located. The hydraulic conductivity of aquifer ranged between 6.5×10^{-4} m/s and 1.6×10^{-3} m/s, depending on the clay content. Groundwater flow velocity, determined by Małecki eds. [2006], is expected to be about 70 m per year. During the periods of high recharge and high river levels (flood peaks in the early spring), the largest part of the landfill body resides below groundwater table, which is less than 2.7 m below ground surface, generally. In the research period, depending

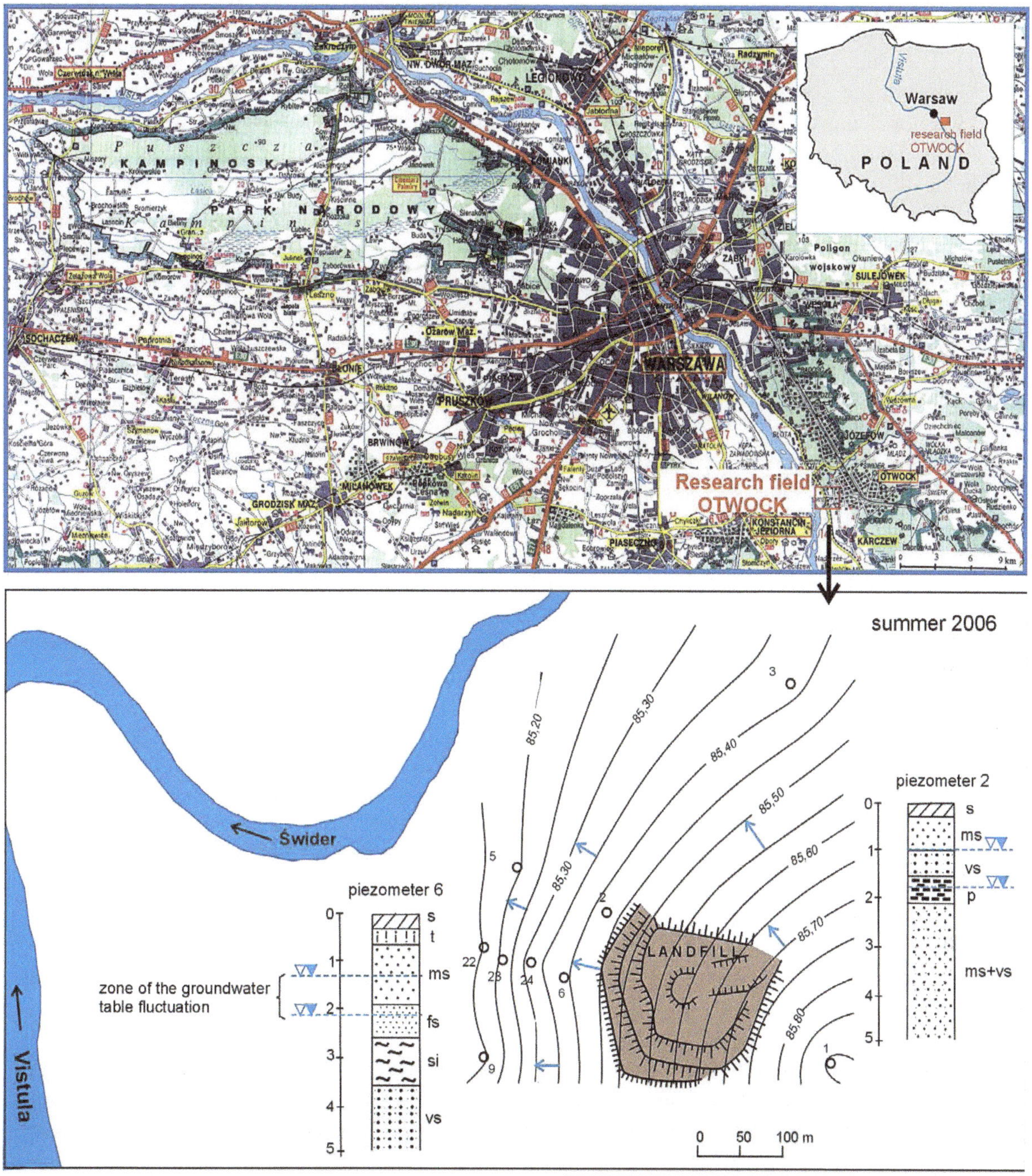

EXPLANATION

2 piezometer number

85,20 groundwater table contour (m asl)

groundwater flow direction

groundwater table

s - soil
vs - varigrained sand
ms - medium-grained sand
fs - fine-grained sand
si - silt
t - till
p - peat

No piez.	Depth to the groundwater table (m)				Amplitude (m)
	25.05.2006	23.08.2006	8.12.2006	6.03.2007	
22	2,10	2,16	2,31	1,51	**0,80**
6	1,98	2,01	2,16	1,36	**0,80**
23	1,79	1,90	nm	nm	-
24	1,91	2,00	2,17	1,33	**0,84**
2	1,53	1,61	1,77	1,00	**0,77**
1	1,80	2,36	2,66	1,67	**0,99**
3	1,77	1,85	1,96	1,32	**0,64**
5	2,40	2,53	1,81	nm	-
9	1,93	2,04	2,20	nm	-

nm - no measured

Figure 1. Location and hydrogeological setting of the study area

on the study site, groundwater table ranged from 1.5 m bgl to 2.7 m bgl (from May to December) and from 1.0 m bgl to 1.7 m bgl (in March) (Figure 1). The annual water level fluctuation is 0.64–0.99 m, but the general groundwater flow direction has not changed.

As it was mentioned above, the water table around the landfill corresponds well to the rainfall and river levels. The thickness of sand between the bottom of the refuse and the water table is relatively thin and additionally is decreased owing to capillary rise (which is greater than 0.3 m in this sediment), thus the aeration zone below the refuse may be thin to non-existent. This short travel distance through the unsaturated zone is unlikely to significantly change the quality of the percolating water. This suggest that spatial differences in chemical compositions of groundwater below the landfill and along the flow direction are due to different chemical processes occurring in the aquifer.

METHODS

To understand a general variation in groundwater chemistry within the study area, the field and laboratory research were conducted quarterly from May 2006 to February 2007 (25.05.2006, 23.08.2006, 8.12.2006 and 6.03.2007). Water samples for chemical analysis were collected within the same aquifer (from similar depth – 3.8 to 4.8 m bgl.) using the PCV piezometers, which were installed in a series along the main direction of groundwater flow as well as above of the landfill (Figure 1). Piezometers localized above of the landfill refereed to background water. Leachate were not analyzed (no water in the ditch on the leachate).

The samples from the piezometers were first taken with an insertion pump with a large pumping capacity, which was pumped long enough to remove the standing water column and/or until constant values were measured for the water temperature, oxidation-reduction potential, content of dissolved oxygen, pH and specific electrical conductivity of the water. Prior to sampling, parameter listed above were measured. A flow-though cell (Eijkelkamp Corp.) was used to improve the comfort and precision of on-line measurements. Water quality characteristics were measured using a portable instruments for analysis in the field: water temperature and pH – pH-meter 330i (WTW); oxidation-reduction potential (ORP) – pH-meter 330i (WTW) with redox electrode PolyPlast ORP; specific electrical conductivity (SEC) – conductivity-meter CC-401 (Elmetron Corp.); dissolved oxygen (DO) – oxygen meter of the ProfiLine Oxi 197 (WTW). Field measured ORP was corrected to the standard hydrogen electrode [Weight, Sonderegger 2000]. Field sampling was conducted with an emphasis on eliminating potential sources of error associated with water collection and transport to the laboratory. Water samples were collected for laboratory tests according to updated sampling procedures described in: Hermanowicz et al. [1999], Weight and Sonderegger [2000)], Witczak et al. [2013] and Polish Standards. All samples were stored at 4°C until the analysis. Laboratory tests included determination of major ions (HCO_3^-, SO_4^{2-}, Cl^-, Ca^{2+}, Mg^{2+}, Na^+, K^+) and NO_3^-, $Fe_{tot.}$ as well as the content of dissolved organic carbon. The analyses were performed at the Central Chemical Laboratory of Polish Geological Institute - National Research Institute in Warsaw and at the Laboratory of Institute of Hydrogeology and Engineering Geology, Faculty of Geology in Warsaw. Water table contour map were prepared using a computer program Surfer 8.0.

RESULTS AND DISCUSSION

A simple assessment of the impact of the leachate on groundwater quality can be made comparing background water chemistry upstream of the landfill (uncontaminated groundwater) and the worst-case measured chemistry in groundwater below the landfill (leachate-contaminated water) [Basberg et al. 1998]. In order to make such a comparison, research were performed in a series piezometers located along the main groundwater flow direction, downstream of the landfill (inside plume where leachate-contaminated water occurs) and upstream of the landfill (uncontaminated groundwater which can be regarded as background water) (Figure 1). Often the range of the landfill impact on groundwater is difficult to designation due to the lack of clear differences between adjacent samples (e.g. gradual increase in the content of each ions). In this case the division between the two zones (uncontaminated groundwater and leachate-contaminated water) was based on the previously performed analysis of saturation indices (with respect to calcite, dolomite and siderite) [Porowska 2010] and based on the relationship between the concentrations of bicarbonates and calcium (Figure 2).

Figure 2 illustrates the relationship between calcium concentrations (mval/dm^3) and bicarbonates (mval/dm^3) of background water and leachate-contaminated water. This plot clearly indicates which piezometers are influenced by landfill contamination. The samples representing natural conditions are located along 1:1 line and suggest that chemical composition of groundwater (main ions: HCO_3^-, Ca^{2+}) is principally controlled by natural biogeochemical processes (CO_2 generation in the soil and aquifer through the root respiration and the decay of natural organic matter). Groundwater samples collected from contaminated area located below 1:1 line manifest high HCO_3^- concentration in comparison to Ca^{2+} concentration caused by anthropogenic processes (activity of the landfill).

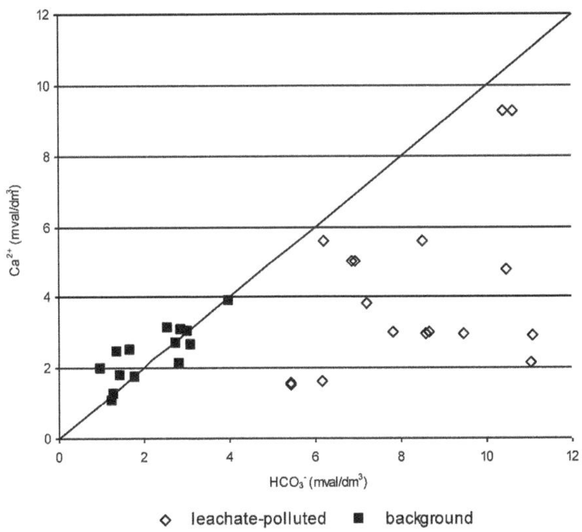

Figure 2. Relationship between calcium and bicarbonates concentrations of groundwater

According to this division, the extreme values of selected parameters and concentrations (as mg/dm^3) of major constituents of groundwater around Otwock landfill are shown in Figure 3.

The field measurements of: groundwater temperature, pH, specific electrical conductivity (SEC) and oxidation-reduction potential (ORP) as well as selected ions: HCO_3^-, Cl^-, NO_3^-, Ca^{2+}, Mg^{2+}, Na^+, K^+ and $Fe_{tot.}$ indicates differences between background water and leachate-contaminated water.

Generally, temperatures of the shallow groundwater are dependent on ambient air temperatures, hence the highest values are expected in summer. It is noteworthy that, groundwater temperature was close to landfill body, what depends on not only high ambient air temperatures in summer but much more on the fermentation heat during decomposition of organic wastes (an exothermic reaction). Evidently differences occurred in winter, the minimum temperatures of background water were 1.4 °C while the leachate-contaminated water reached 6.3 °C (Figure 3). The highest groundwater temperatures are observed inside the leachate plume and tend to decrease downstream. Regardless of the research season, the highest water temperature values was measured in the piezometer localized close to the landfill (Figure 4).

The pH of the background water in Otwock was near neutral (slightly acid or slightly alkaline), except winter season, when the pH was higher and its value ranged from 7.30 to 7.68 (Figure 3). This pH values are typical for mature landfills [Tchobanoglous et al. 1993]. The highest pH values was found in groundwater within contaminated zone, regardless of the research season. However, any significant trend of changes in the pH value was not detected with distance from the landfill (Figure 4). A considerable range was found for pH in groundwater (from 7.32 to 7.73) at the landfill border (Figure 3), caused by a number of physical, chemical and microbial processes in the interior of the Otwock landfill.

According to the literature, the major controls on pH in anaerobic leachate are: (1) degradation of organic matter producing large amounts of CO_2, which forms carbonic acid and lowers pH and (2) production of NH_4^+ and CH_4 and reduction of Fe and Mn oxides, all of which consume hydrogen and raise the pH. High pH values within leachate-contaminated water and in background water during winter season indicates the second process is prevailing in Otwock, generally. But the lowest pH values (in comparison to other contaminated water) at the distance between 60 and 90 m from the landfill, indicating influence of organic matter degradation in the interior of the landfill. This situation is possible, because the load of contamination is released from landfill periodically and slowly moves downgradient from the landfill.

Specific electrical conductivity (SEC) is one of the simplest indicators for groundwater contamination by landfill leachate. The SEC of the

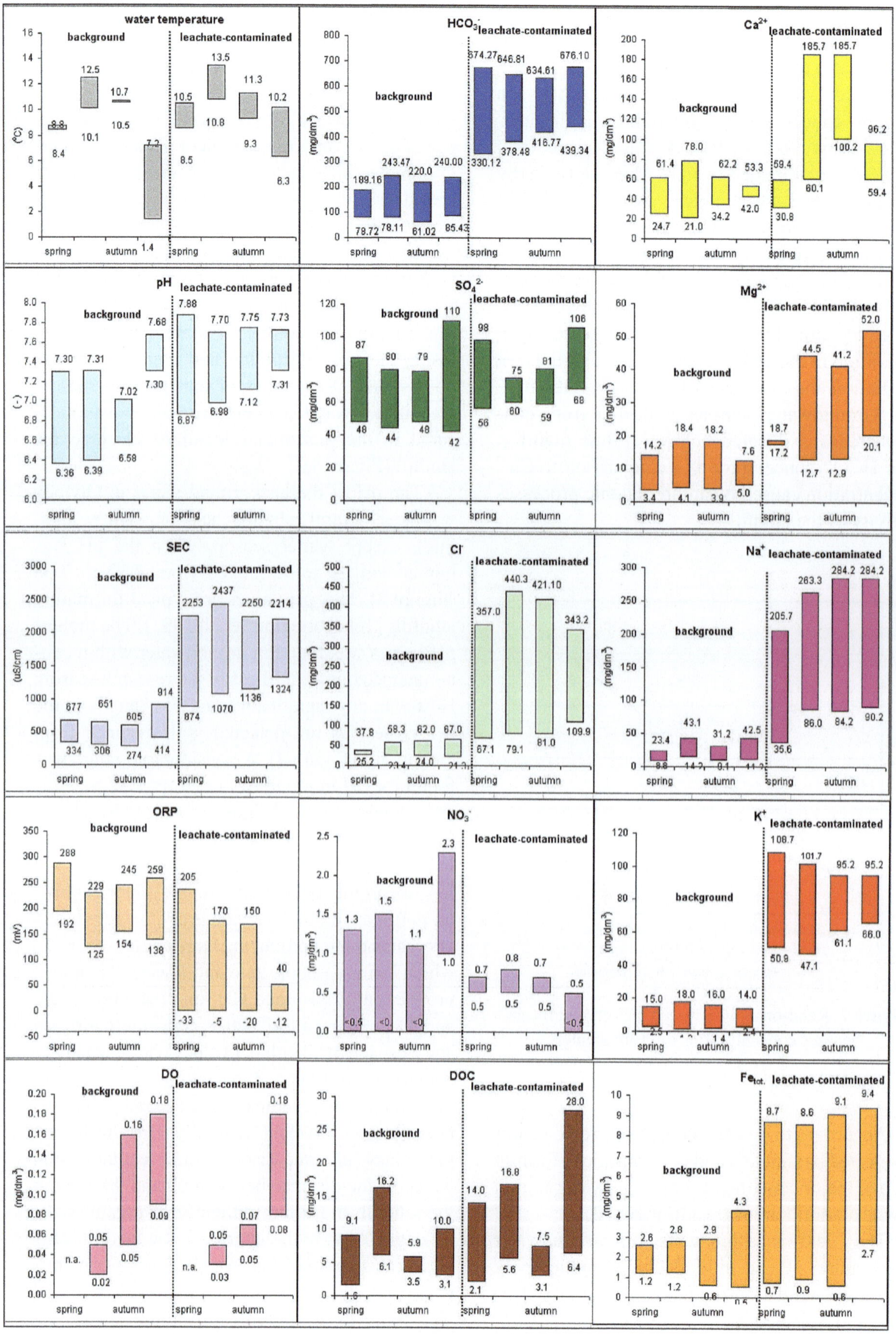

Figure 3. Differences in the values of selected parameters and ions between the background water and leachate-contaminated water

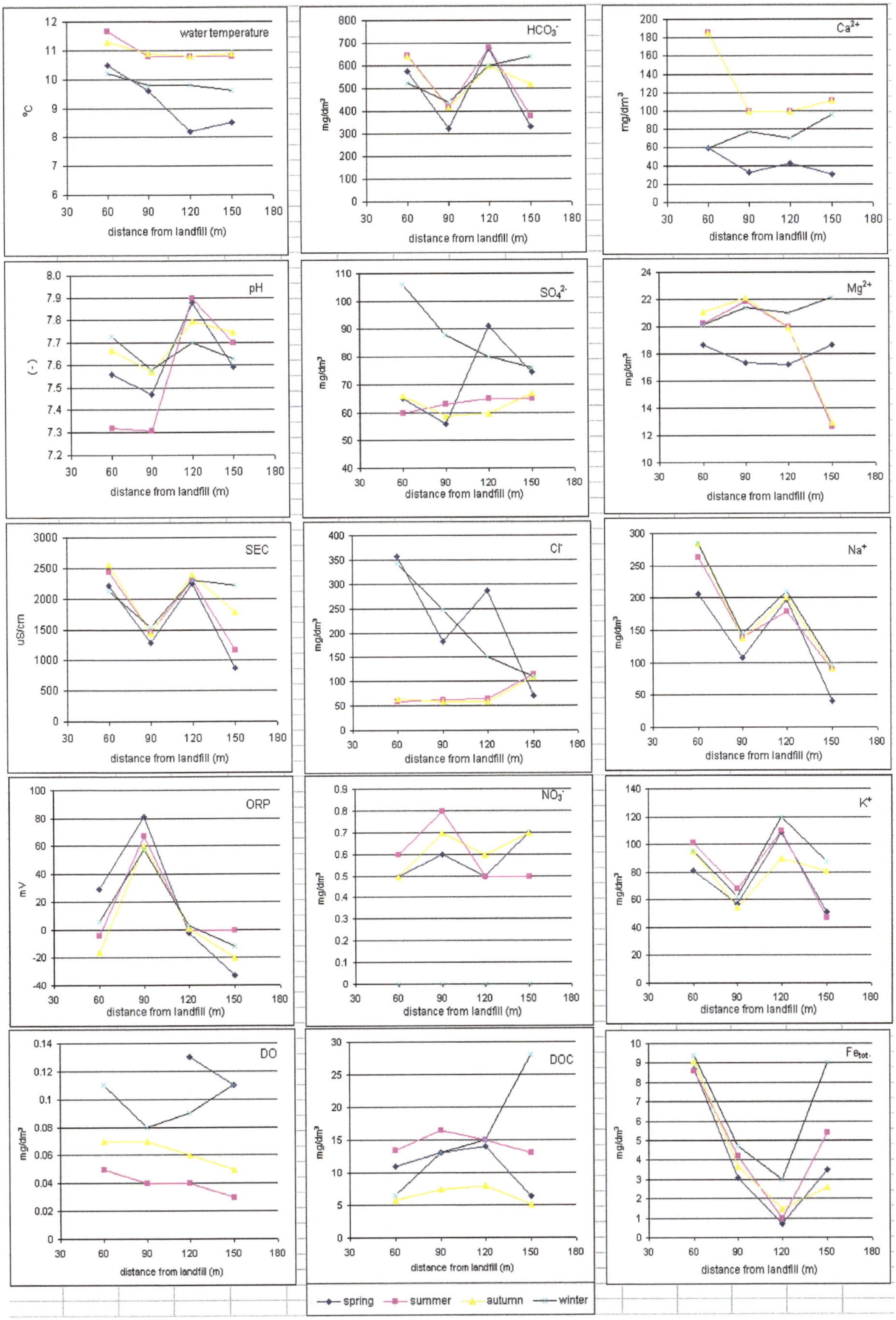

Figure 4. Variations of selected parameters and ions along the groundwater flow direction

background water ranged from 274 to 914 μS/cm and leachate-contaminated – from 874 to 2437 μS/cm (Figure 3). As expected, the groundwater showed very high SEC values within the nearest area of landfill (Figure 4). Similar value was found at the distance between 60 m and 120 m from landfill, indicating two major plumes of leachate moving north-west of the landfill (along the groundwater flow direction). Between two contaminating plumes there is a less contaminated water area. This trend was characteristic for all the measurements, regardless of the research season. Similar trend was observed for the concentration of other ions discussed below (eg. HCO_3^-, Na^+, K^+) (Figure 4). This data indicates that the load of contamination is released from the landfill periodically and slowly moves in the aquifer along the groundwater flow direction. This hypothesis is consistent with other data determined by Małecki eds. [2006] in this area.

Oxidation-reduction potential (ORP) is a useful indicator representing redox condition of groundwater. Redox processes are important in both the production of landfill leachate and the fate of many leachate constituents in the groundwater zone. Redox environments can vary greatly due to contaminant load, groundwater chemistry, geochemistry and microbiology inside the plume and along the flow direction [Christensen at al. 2000a,b, van Breukelen et al. 2003]. A considerable range was found for ORP in groundwater (from -33 mV to 288 mV) in Otwock (Figure 3). The negatives ORP values revealed an anaerobic conditions and were characteristic only for leachate-contaminated water. Outside of the leachate plume, in background water, the minimum ORP values drastically increased up to 125 mV.

ORP does not correspond well to the concentration of the dissolved oxygen (Figure 3), but such a situation is given in the literature [Rose S., Long A. 1988]. Redox potentials cannot adequately predict dissolved oxygen concentrations nor can DO concentrations be used to calculate redox potentials.

Regardless of the research area and season, concentrations of dissolved oxygen are low and not exceed 0,18 mg/dm³ in Otwock. If low oxygen concentration was observed in groundwater above the plume in natural conditions, it can be assumed that there was absent in aquifer before landfilling started. This assumption is very likely in such geological conditions, where organic matter is distributed within the aquifer. According to this, in Otwock large amounts of DO are removed from the percolation water by the oxidation of organic matter occurred in aquifer (river deposit). After all molecular DO is utilized, organic matter is both oxidized and reduced by anaerobic fermentative processes to produce CO_2 and reduced organic species, such as CH_4. In Otwock methane is produced in the interior of the landfill [Porowska, Gruszczyński 2006] and probably subsequently transported in aquifer and/or partly is oxidized. According to Abichou et. al. [2006] in a landfill intermediate cover, part of methane is oxidized to carbon dioxide (even significantly grater than 25%). This hypothesis needs incontrovertible evidences using carbon isotope analysis in Otwock landfill.

As it was mentioned above, groundwater quality is being significantly affected by leachate percolation downgradient from the landfill. The variation occurs due to spatial and temporal changes in the migration of contaminants from landfill leachate indicated by different values of selected parameters and ions in the same piezometer but in other research period.

Relatively high bicarbonates and chlorides concentrations in groundwater proved that they are all taken from an aquifer zone strongly contaminated by landfill leachate (Figure 3). Bicarbonates concentrations in the plume ranged from 330.12 mg/dm³ to 676.10 mg/dm³, whereas within background water HCO_3^- did not exceed 244 mg/dm³ (Figure 3). Initially, bicarbonates in the plume decreases along the groundwater flow direction from about 600 mg/dm³ at the landfill border to about 400 mg/dm³ at further downstream. At the distance of 120 m HCO_3^- values sharply increased and exceeded even 680 mg/dm³ to subsequently decrease to 300–500 mg/dm³ (except for winter season). High HCO_3^- values at the landfill border are likely to be derived from the biodegradation of organic matter in the interior of the landfill. Organic matter in municipal solid waste landfills typically undergoes methanogenic decomposition, producing carbon dioxide and methane as end products, according to reaction (1) [Hayes 2001]:

$$6CH_2O \rightarrow 3CO_2 + 3CH_4 \qquad (1)$$

Oxidation of methane in migrating landfill gas to carbon dioxide can occur if the gas travels through a zone with appropriate conditions (access of oxygen or other potential oxidants, such as nitrates or sulfates). This CO_2 combines with water forming carbonic acid, according to reaction (2) [Fetter 1994]:

$$CO_2 + H_2O \rightarrow H_2CO_3 \quad (2)$$

Carbonic acid dissociates into H^+ and HCO_3^- and may be represented as reaction (3) (depending on pH) [Fetter 1994]:

$$H_2CO_3 \rightarrow HCO_3^- + H^+ \quad (3)$$

High HCO_3^- concentrations at the distance 120 m from the landfill can be explained that the load of contamination is release from landfill periodically and slowly moves downgradient from the landfill, depending on the climatic and hydrogeological conditions (precipitation, river levels and groundwater table fluctuations).

Sulfates concentrations were independent on the sampling site and similar values (ranged from 42 to 110 mg/dm^3) were found in both the leachate-contaminated groundwater and natural water (Figure 3). Regardless of the sampling area it is observed that the sulfates distribution differs from that of other ions and is different in each sampling season (Figure 4). This variability is dependent on climatic and hydrogeological conditions (precipitation and groundwater table elevation). Relatively high sulfates content in this environment indicates a weak reduction of the ion into a form of lower oxidation state (eg sulfides), both inside plume and in natural conditions.

Based on the comparison of hydrogen sulfide in the biogas and sulfates in groundwater [Porowska in press] was found that the sulfates content in the analyzed groundwater, not only due to the activity of the landfill but also of hydrogeochemical processes occurring in the aquifer.

Chlorides are considered a conservative parameter and they are not sensitive to oxidation-reduction reaction. An excess of chlorides concentrations in water is usually taken as an index of pollution and considered as a tracer of groundwater contamination [Macioszczyk, Jeż 1995]. Chlorides can also be used as an ideal indicator of the position of an overall leachate plume. According to this, such concentrations of chlorides confirm the above-mentioned principle that pollution loads are released from the landfill at a certain time (pulse) and their concentrations, are variable in time and in different places. (Figure 4). The highest Cl^- concentrations in groundwater occurred close to the landfill (350 mg/dm^3), and decreased to about 100 mg/dm^3 at 150 m downgradient from the landfill. This suggests a significant dilution of leachate in the aquifer. Such a significant reduction of Cl^- concentration in a short distance, allows the assumption that other processes (e.g. dispersion, diffusion) are of negligible importance in comparison to dilution process. This trend was found in winter while in other seasons it was different (Figure 4). In summer and in autumn Cl^- value only slightly exceeded 50 mg/dm^3 at the distance of 60–120 m at the landfill border. It is noteworthy that, regardless of sampling season and different values along the groundwater flow direction, at the distance of 150 m from the landfill, chlorides fluctuate around 100 mg/dm^3. It can be seen, the hydrogeochemistry changed little in time for the last piezometer. Chlorides values 100–400 mg/dm^3 are typical for mature landfills [Tchobanoglous at al. 1993].

Nitrates concentration only was above 2 mg/dm^3 within the aquifer upstream of the landfill (natural conditions). Usually, higher nitrates concentrations was observed in background water in comparison to the leachate plume. The presence of elevates levels of nitrates in background water is considered to be due to the excess oxidation-reduction potential values (more oxygenated conditions). Nitrates reduction probably causes disappearance of nitrates downstream of the landfill. Processes causing this decrease are discussed below (reaction 5). A single measurement of the concentration of NH_4^+ showed higher values in the contaminated zone (0.74–5.4 mg/dm^3) compared with background water (0.16–0.92 mg/dm^3). Similar trend was observed by Cozarelli et al. [2011] in groundwater at the Norman landfill. Hence, it can be assumed that the dominant species of nitrogen in this contaminated environment is an ammonium ion.

Leachate-contaminated water contain high concentrations of the following ions: Ca^{2+}, Mg^{2+}, Na^+, K^+ and Fe_{tot} in comparison to background water (Figure 2). They show variable concentrations in the plume, but no identical tendency along the groundwater flow direction (except Na^+ and K^+) (Figure 4). Sodium and potassium concentrations are change along groundwater flow direction similar to SEC and bicarbonates, indicating the highest values at the landfill border and at the distance 120 m from the landfill. These parameters confirm the mentioned above principle, that the load of contamination is release from landfill periodically and slowly moves downgradient from the landfill.

Calcium concentrations are low in the plume in comparison to sodium and potassium concentrations. In natural conditions calcium concentrations prevail sodium and potassium concentrations. It can be assumed that low calcium concentrations is caused by precipitation carbonate minerals ($CaCO_3$) in leachate-contaminated aquifer [Porowska 2010]. It is commonly known that supersaturation for $CaCO_3$ as well as other carbonates is observed at other landfill leachate plumes [eg. Van Breukelen 2003, Manning 2001].

Chemical analysis of the groundwater indicated a slight difference in the concentrations of the dissolved organic carbon (DOC) between the leachate-contaminated water and background water (Figure 3). Similar values are likely to be derived from the biodegradation of organic matter in the interior of the landfill as well as organic matter, which occur within aquifer sediments. DOC concentrations differs more with time and at the distance, especially 150 m from the landfill (Figure 4). In this place, DOC concentrations increase from 5 mg/dm^3 in autumn to the highest value of 28 mg/dm^3 in winter. Low DOC concentrations suggest biodegradation within aquifer sediments [Christensen at al. 1998].

Biodegradation of organic matter is a very important process in this aquifer, because it plays a key role in shaping hydrogeochemical processes of the selected ion concentrations in groundwater, such as nitrates or sulfates. Thermodynamically, the oxidation of dissolved organic carbon (DOC) is coupled with the reduction of DO, NO_3^-, Mn(IV), Fe(III), SO_4^{2-}, and CO_2, which is predicted to occur in that order [Champ et al. *1979*]. In general, as the DOC is oxidized, DO is preferentially consumed. After the depletion of DO, the next available electron acceptor will be consumed following the sequence described above (aerobic respiration, denitrification, Mn(IV) reduction, Fe(III) reduction, sulfates reduction, and methanogenesis). Accordingly, the process of denitrification [e.g. Pedersen et al. 1991] and/or dissimilatory nitrates reduction to ammonium (DNRA) [eg. Tiedje et al. 1982] are possible by slight reduction of the conditions in the plume (reaction 4, 5).

$$\text{denitrification: } 5CH_2O + 4NO_3^- + 4H^+ \rightarrow \rightarrow 2N_2 + 5CO_2 + 7H_2O \quad (4)$$

$$\text{DNRA: } 2CH_2O + NO_3^- + 2H^+ \rightarrow \rightarrow NH_4^+ + 2CO_2 + H_2O \quad (5)$$

As it was mentioned above, nitrates concentrations were lower in the leachate plume in comparison to background water in Otwock landfill (Figure 3), caused by nitrates reduction, according to the above reaction: denitrification (4) and/or dissimilatory nitrates reduction to ammonium (5).

The importance of denitrification, as a major pathway of nitrates removal from aquifers, has been reported in many studies. Primary redox processes involve the oxidation of organic carbon and release of reduced redox species (CH_4, H_2S, Fe(II), Mn(II)). These reductants become oxidized in secondary redox processes. At the fringe of a leachate plume in particular, secondary redox processes happen and compete with organic carbon degradation processes for available electron-acceptors [Van Breukelen 2003].

On the base of the research in Otwock it can be assumed that CO_2 released (according to reaction 4 and 5), is dissolved (reaction 2) and dissociates (reaction 3). In the pH range near neutral, dissolution of CO_2 generates bicarbonates ion, which prevail other species of CO_2. It also needs to be considered that high HCO_3^- concentrations in the plume downstream from the landfill can be caused by biodegradation of organic contaminants proceed from the interior of the landfill (reaction 1).

One of the several methods for assessing groundwater quality is to compare the physical and chemical parameters with the limits for groundwater listed in Polish Standards [Journal of Laws No. 143, item 896] (Table 1).

According to this classification of good quality water belongs to classes I–III, and the poor quality water – to classes IV–V. Comparison of groundwater quality along the groundwater flow direction with the highest admissible values in class I–V shows that the: pH, water temperature, NO_3^-, Mg, SO_4^{2-}, Ca lie within the normative range to the class I–III, whereas the content of other parameters: DOC, SEC, DO, Cl^-, K^+, Na^+, HCO_3^-, Fe_{tot} is exceeded and water belong to the classes IV–V (Table 1). Summarizing, chemical status of groundwater along the flow direction is poor.

CONCLUSIONS

Field and laboratory researches of groundwater around Otwock landfill were performed in order to identify the chemical composition of groundwater and assess the landfill impact on the quality of groundwater downgradient from the landfill. The following conclusions can be drawn from the physical and chemical data derived from this study:

1. Despite the storage of waste was finished in 1991, the site has been reclaimed, groundwater quality is being significantly affected by leachate percolation downgradient from the

Table 1. Assessment of groundwater quality along the groundwater flow direction according to Polish Standards [Journal of Laws No. 143, item 896]

Parameter	Unit	Limit values in class I–V					Distance from the landfill [m]			
		I	II	III	IV	V	60 (n=4)	90 (n=4)	120 (n=4)	150 (n=4)
pH	–	6.5–9.5			<6.5 lub >9.5		7.32–7.73	7.31–7.58	7.70–7.78	7.59–7.75
Dissolved organic carbon (DOC)	mgC/l	5	10	10	20	>20	5.8–13.5	7.5–16.5	8.0–15.0	5.1–28.0
Specific electrical conductivity (SEC)	uS/cm	700	2500	2500	3000	>3000	2121–2550	1288–1536	2253–2400	879–2214
Temperature	°C	<10	12	16	25	>25	10.2–11.7	9.6–10.9	8.2–10.8	8.5–10.9
Dissolved oxygen (DO)	mg/l	>1	0.5–1	<0.5	<0.5	<0.5	0.05–0.11	0.04–0.08	0.04–0.13	0.03–0.13
NO_3^-	$mgNO_3/l$	10	25	50	100	>100	0–0.6	0–0.8	0–0.6	0–0.7
Cl^-	mgCl/l	60	150	250	500	>500	60.0–356.98	59.0–247.57	60.0–286.42	70.01–114.50
Mg^{2+}	mgMg/l	30	50	100	150	>150	18.65–21.10	17.30–22.10	17.20–21.00	12.65–22.10
K^+	mgK/l	10	10	15	20	>20	81.35–101.70	55.10–68.15	90.00–120.00	47.05–87.50
SO_4^{2-}	$mgSO_4/l$	60	250	250	500	>500	60–106	56–88	60–91.2	65–76
Na^+	mg Na/l	60	200	200	300	>300	205.7–284.2	108.6–144.6	180.0–210.0	40.75–96.5
Ca^{2+}	mgCa/l	50	200	200	300	>300	59.4–185.7	32.3–100.2	42.8–100.0	30.8–111.6
HCO_3^-	$mgHCO_3/l$	200	350	500	800	>800	523.55–646.81	325.27–439.54	600.00–680.00	330.12–638.27
Fe_{tot}	mgFe/l	0.2	1	5	10	>10	8.6–9.4	3.1–4.7	0.72–3.0	2.6–9.0

landfill. Hydrogeological and climatic conditions play a key role in groundwater chemistry in landfill and surrounding area. Most of the water reaches to the landfill and subsequently to the aquifer by precipitation. A highly permeable, sandy aquifer provides favorable conditions for the transport of contamination into the aquifer. In addition during periods of high groundwater table large part of the landfill body resides below groundwater table. In this situations the load of contamination is release from landfill and moves in the aquifer along the flow direction.

2. Chemical composition of groundwater and hydrogeochemical processes downgradient from the landfill do not remain constant with time. Plume of leachate migrates from the landfill along the groundwater flow direction (70 m/y). In comparison to landfills reported in the literature, the contaminated zone is restricted to relatively small area, owing to location close to Vistula and Świder rivers.
3. Contaminations of groundwater below the landfill have been identified by comparing the chemistry of the water above and below the landfill and by comparing the chemistry of water with quality standards for groundwater [Dz. U. Nr 143, poz. 896]. The values of: groundwater temperature, pH, specific electrical conductivity (SEC) and oxidation-reduction potential (ORP) and concentrations of selected ions: HCO_3^-, Cl^-, NO_3^-, Ca^{2+}, Mg^{2+}, Na^+, K^+ and Fe_{tot} indicates differences between background water and leachate-contaminated water and evidence of groundwater contamination. As expected, the nearest piezometer from the landfill showed the highest values of water temperature, specific electrical conductivity, sodium, iron, chlorides (except for summer and autumn analysis) and calcium (except for winter analysis). A comparison of groundwater quality along the groundwater flow direction with the highest admissible values [Journal of Laws No. 143, item 896] shows that the chemical status of groundwater along the flow direction is poor.
4. The load of contamination and concentration in groundwater depends on the following factors: 1) dilution of leachate with uncontaminated groundwater, 2) mixing of anaerobic leachate with oxygenated water that facilitates the decomposition of organic matter and precipitation of metals, and 3) interaction of contaminants with aquifer materials and organic matter.

5. The result of these investigations indicated that leachate contaminate the environment, even after the reclamation of landfill. For this reason, the site developed for future landfilling should have a planned engineering structure to control and minimize the impact of landfill leachate on groundwater quality around the landfill.

REFERENCES

1. Abichou T., Chanton J., Powelson D., Fleiger J., Escoriaza S., Lei Y., Stern J., 2006. Methane flux and oxidation at two types of intermediate landfill covers, Waste Management, 26 (11), 1305–1312.
2. Baedecker M.J., Apgar M.A., 1984. Hydrogeological studies at a landfill Delaware, Ground Water Contamination, National Academy Press, Washington D.C, 127–138.
3. Basberg L., Banks D., Sæther O. M., 1998. Redox processes in groundwater impacted by landfill leachate, Aquatic Geochemistry, 4, 253–272.
4. Champ D.R, Gulens J., Jackson R.E., 1979. Oxidation–reduction sequences in groundwater flow systems, Can J Earth Sci., 16, 12–23.
5. Cherry J.A. (eds.), 1983. Migration of contaminants in groundwater at a landfill: A case study, Journal of Hydrology, Special Issue, 63, 1/2, 1–197.
6. Christensen T.H., Kjeldsen P., 1989. Basic Biochemical Processes in Landfills, Sanitary Landfilling: Process, Technology and Environmental Impact, Academic Press, San Diego, CA, 29–49.
7. Christensen T.H., Kjeldsen P., Albrechtsen H.J., Heron G., Nielsen P.H., Bjerg P.L., Holm P.E., 1994. Attenuation of landfill leachate pollutants in aquifers. Crit. Revs. Eniron. Sci. Tech, 24(2), 119–202.
8. Christensen J.B., Jensen D.L., Grøn C., Filip Z., Christensen T.H., 1998. Characterisation of the dissolved organic carbon in landfill leachate polluted groundwater, Water Res., 32 (1), 125–135.
9. Christensen T.H., Bjerg P.L., Kjeldsen P., 2000a. Natural attenuation: A feasible approach to remediation of groundwater pollution at landfills?, Ground Water Monitoring and Remediation, Vol. 20, (1), 69–77.
10. Christensen T.H., Bjerg P.L., Banwart S.A., Jakobsen R., Heron G., Albrechtsen H.J., 2000b. Characterization of redox conditions in groundwater contaminant plumes, Journal of Contaminant Hydrology, 45, 165–241.
11. Cozzarelli I.M., Böhlke J.K., Masoner J., Breit G.N., Lorah M.M., Tuttle M.L.W., Jaeschke J.B., 2011. Biogeochemical Evolution of a Landfill Leachate Plume, Norman, Oklahoma, Ground Water, Vol. 49, No. 5, 663–687.
12. Dz. U. Nr 143, poz. 896. Rozporządzenia Ministra Środowiska z dnia 23 lipca 2008 r. w sprawie kryteriów i sposobu oceny stanu wód podziemnych.
13. Fetter C.W., 1994. Applied hydrogeology, Prentice Hall, Inc A. Simon & Schuster Company Englewood Clifs, New Jersey USA.
14. Hayes J.M., 2001. Fractionation of carbon and hydrogen isotopes in biosynthetic processes, 225-277, W: Valley J.W., Cole D. (red.), Stable isotope geochemistry. Reviews in mineralogy and geochemistry, vol. 43. Mineralogical Society of America, Washington.
15. Hermanowicz W., Dojlido J., Dożańska W., Koziorowski B., Zerbe J. 1999. Fizyczno+chemiczne badanie wody i ścieków, Wydawnictwo Arkady, Warszawa (in Polish).
16. Koda E., Augustyniak E., Pachuta K., Paprocki P., 1999. Ocena oddziaływania na środowisko nieczynnego wysypiska odpadów stałych w Otwocku, woj. mazowieckie, Arch. MPO, Warszawa (in Polish).
17. Macioszczyk A., Jeż L., 1995. Chlorki czułym wskaźnikiem zanieczyszczeń antropogenicznych wód podziemnych, Mat. VI Symp. „Współczesne problemy hydrogeologii", Kraków-Krynica, 259–267 (in Polish).
18. Małecki J.J. (eds.), 2006. Wyznaczanie parametrów migracji zanieczyszczeń w ośrodku porowatym dla potrzeb badań hydrogeologicznych i ochrony środowiska – poradnik metodyczny, WG UW, MŚ, Warszawa (in Polish).
19. Manning D., 2001. Calcite precipitation in landfills: an essential product of waste stabilization, Mineralogical Magazine, 65, 5, 603–610.
20. Pedersen J.K., Bjerg P.L., Christensen T.H., 1991. Correlation of nitrate profiles with groundwater and sediment characteristics in a shallow sandy aquifer, J. Hydrol., 124, 263–277.
21. Porowska D., Gruszczyński T., 2006. Zmienność składu fazy gazowej składowiska odpadów komunalnych w Otwocku, Przegl. Geol., vol. 54, nr 11, 996–1001 (in Polish).
22. Porowska D., Gruszczyński T., 2013. Czynniki warunkujące zmienność składu fazy gazowej składowiska odpadów komunalnych w Otwocku, Biuletyn PIG, z. XVI/2, nr 456, 457–464 (in Polish).
23. Porowska D., 2010. Zastosowanie wskaźników nasycenia SI do wyznaczenia strefy zanieczyszczenia wód podziemnych wokół składowiska odpadów, W: Stan i antropogeniczne zmiany jakości wód w Polsce, Wyd. Uniwersytetu Łódzkiego, vol. 6, 109–117 (in Polish).
24. Porowska D., Związki siarki w fazie gazowej oraz w wodach podziemnych wokół zrekultywowanego składowiska odpadów komunalnych, Przegl. Geol., (in Polish) in press.

25. Rose S., Long A., 1988. Monitoring Dissolved Oxygen in Ground Water: Some Basic Considerations, Groundwater Monitoring & Remediation, Vol. 8, Issue 1, 93–97.

26. Tchobanoglous G., Theisen H., Vigil S.A., 1993. Integrated solid waste management, New York, McGraw-Hill International.

27. Tiedje J.M, Sexstone A.J., Myrold D.D, Robinson J.A., 1982. Denitrification: ecological niches, competition and survival, Antonie Van Leeuwenhoek, 48 (6), 569–583.

28. Van Breukelen B.M, 2003. Natural Attenuation of Landfill Leachate: a Combined Biogeochemical Processes Analysis and Microbial Ecology Approach, Academisch Proefschrift, Amsterdam.

29. Van Breukelen B.M., Röling W.F.M., Groen J., Griffioen J., Van Verseveld H.W., 2003. Biogeochemistry and isotope geochemistry of a landfill leachate plume, Journal of Contaminant Hydrology, 65, 245–268.

30. Weight W.D., Sonderegger J.L. 2000. Manual of Applied Field Hydrogeology, McGraw-Hill, Printed and bound by RR Donnelly & Sons Company.

31. Witczak S., Kania J., Kmiecik E., 2013. Katalog wybranych fizycznych i chemicznych wskaźników zanieczyszczeń wód podziemnych i metod ich oznaczania, IOŚ, Warszawa (in Polish).

18

CONTENT OF ZINC, LEAD AND CADMIUM IN SELECTED AGRICULTURAL SOILS IN THE AREA OF THE ŚLĄSKIE AND CIĘŻKOWICKIE FOOTHILLS

Agnieszka Józefowska[1], Anna Miechówka[2], Michał Gąsiorek[3], Paweł Zadrożny[4]

[1] Department of Soil Science and Soil Protection, University of Agriculture in Krakow, 21 Mickiewicza Ave., 31-120 Kraków, Poland, e-mail: agnieszka.jozefowska@ur.krakow.pl

[2] Department of Soil Science and Soil Protection, University of Agriculture in Krakow, 21 Mickiewicza Ave., 31-120 Kraków, Poland, e-mail: rrmiecho@cyf-kr.edu.pl

[3] Department of Soil Science and Soil Protection, University of Agriculture in Krakow, 21 Mickiewicza Ave., 31-120 Kraków, Poland, e-mail: rrgasior@cyf-kr.edu.pl

[4] Department of Soil Science and Soil Protection, University of Agriculture in Krakow, 21 Mickiewicza Ave., 31-120 Kraków, Poland, e-mail: rrzadroz@cyf-kr.edu.pl

ABSTRACT

The purpose of this study was to evaluate the state of contamination with zinc, lead, and cadmium in selected soils of the Śląskie and Ciężkowickie Foothills and to determine the impact of the type of agricultural use and selected physico-chemical properties of soils on heavy metal content. The test soils were characterized by natural content of zinc, lead, and cadmium in most cases. Only one type of soil located on Śląskie Foothills developed increased levels of Cd (1.1 mg · kg^{-1}). The content of zinc, lead, and cadmium in the surface layer (0–30 cm) was higher in the soils of Śląskie Foothills than in soils of Ciężkowickie Foothills. The bedrocks from which the soils of these two mesoregions are formed differed significantly only in the content of zinc (it was higher in the soils of Śląskie Foothills). The content of Zn, Pb, and Cd in the surface layer of soil depends on its texture and organic carbon and total nitrogen content. There was also a positive correlation between the content of Pb and Cd and hydrolytic acidity and between the content of Zn and Ca and CEC. Different types of land uses did not influence the content of the metals.

Keywords: Śląskie and Ciężkowickie Foothills, heavy metal, land use.

INTRODUCTION

Carpathian soils are relatively well documented for their content of heavy metals [1, 15, 21]. The most well known soils in this regard are those of national parks [19, 22, 31] and the least recognised are the foothill soils used for agricultural purposes [4, 30]. Among the latter ones are the Ciężkowickie and Śląskie foothills soils. Śląskie Foothills is part of West Beskid Foothills [12], in a short distance from Upper Silesian Industrial Region (Polish) and Ostrava-Karwiński Industrial District (Czech Republic). These foothills are characterized by high rainfall, *via* which dust pollution in the atmosphere can get into the soil. Average annual rainfall for the meteorological station in Wisła Centrum (Śląski Beskid) in 2000-2010 was 1286.7 mm (data from the Institute of Meteorology and Water Management). Climatic conditions in this area favour the accumulation of organic matter in soils [16]. Organic matter stops the metals through the creation of specific connections with them [3]. Therefore, in the soils of this mesoregion a higher content of heavy metals can be expected than in the soils of the Ciężkowickie Foothills, located in the Central Beskid foothills at a distance from industrial centres. Soils of the Ciężkowickie Foothills are characterized by lower organic carbon stocks [18] and the occurrence of lower average annual rainfall (801.7 mm for Biecz-Grudna post in 2000–2010) than those of the Śląskie Foothills.

The purpose of this paper was to evaluate the state of contamination with zinc, lead and cadmium in brown soils and lessive soil, which are characteristic for arable land and fresh grasslands occurring in the Śląskie and Ciężkowickie foothills. The next aim was to determine the effect of the different types of land use soils on heavy metal content and to attempt to identify the relationships between the selected physico-chemical properties and concentrations of zinc, lead and cadmium in the soils.

MATERIALS AND METHODS

The study was conducted using soil material collected from 16 profiles located in the Śląskie (in the villages of Ustroń, Górki Wielkie, and Porąbka) and Ciężkowickie (Swoszowa, Joniny, Czermna, and Dobrocin) foothills. In each location, soil pits were made in the neighbouring arable lands (A) and grasslands (G). The exception was the village of Ustroń, where four profiles located in two different positions were examined. The tested soils are representative of the habitat of fresh agricultural use. They are classified into brown soils and lessive soils [27], and most were formed from weathering or landslide material of the Silesian unit of the Carpathian Flysch. The location of these soils and their detailed characteristics are given in previous papers [12, 17].

For the purposes of this study, soil was sampled from layers of 0–30 cm depth (topsoil) and the bedrock horizons. These samples were dried and sieved (2 mm diameter). In the fine earth parts of the soils the following properties were determined:

- texture, using Prószynski's modification of the aerometric Casagrande method [24],
- pH in 1 mol · dm^{-3} KCl solution, using the potentiometric method [5],
- the total exchangeable base (TEB) cation contents, by means of determining individual cations (Ca^{2+}, Mg^{2+}, K^+ and Na^+) after their extraction with 1 mol · dm^3 CH_3COOHN_4 solution [13]; in extracts Ca^{+2}, K^+ and Na^+ were determined by flame photometry, and Mg^{2+} by atomic absorption spectrophotometry,
- hydrolytic acidity (Hh) with the Kappen method using 1 mol · dm^{-3} CH_3COONa solution [13],
- the content of organic carbon (OC) using the Oleksynowa modification of the Tiurin method [23],
- the content of total nitrogen (TN) with the Kjeldahl method [13] using Kjeltec apparatus,
- the total content of zinc, lead, cadmium and calcium after digestion of soil samples in a mixture of nitric and perchloric acids, by atomic emission spectrometry with inductively coupled plasma (ICP-OES).

Cation exchange capacity (CEC) and base saturation (BS) percentage were then calculated as follows:

$$CEC = TEB + Hh;$$

$$BS\ [\%] = \frac{TEB \cdot 100}{CEC}$$

The assessment of heavy metal contamination in soil was carried out by comparing the results obtained with the limit for heavy metals contents specified in the Regulation of the Minister of the Environment of 9 September 2002 regarding the quality standards for soil and ground [25] and based on the limit for the content of heavy metals recommended by the Institute of Soil Science and Plant Cultivation (IUNG) [8]. To assess the degree of heavy metals accumulation in soil, two indicators were used: the accumulation index (AI) [6] and the enrichment factor (EF) [14]. To calculate EF, calcium (Ca) was used as a reference element [14].

The statistical analyses were carried out using the program Statistica 10PL. The two-way analysis of variance (two-way ANOVA) and coefficients of linear correlation were tested and standard deviations and the average values were calculated.

RESULTS AND DISCUSSION

The analysed soils were characterized by diverse physicochemical, physical and chemical properties (Table 1). According to "Particle Size Distribution and Textural Classes of Soils and Mineral Materials" [10], the soils investigated were silty loam, clay loam, loam, and sandy loam.

The soils from sites located in the Ciężkowice Foothills were characterized by lighter texture than soil from samples collected in the Śląskie Foothills. The topsoil from these two regions contained significantly different amounts of sand and silt fractions (Table 1). The content of the fraction < 0.02 mm allow, classified the investigated soil as medium-heavy and heavy. It was necessary to evaluate the contamination of soils on the basis

Table 1. Selected properties of the soil surface layer (0–30 cm)

Soil properties	Śląskie Foothills		Ciężkowickie Foothills		Region	The way of land use
	A[a]	G[b]	A[a]	G[b]	F[c] P[d]	F[c] P[d]
pH KCl	4.2-4.9[1] 4.4[2] (0.3)[3]	3.8-4.7 4.2 (0.3)	3.8-5.9 4.4 (0.8)	4.0-7.2 5.6 (1.5)	2.08 0.174	0.86 0.372
Hh [mmol(+) · kg⁻¹]	36.7-70.4 53.1 (11.9)	44.6-106.2 72.7 (21.9)	22.2-65.0 48.5 (15.9)	12.0-56.1 36.3 (16.1)	4.46 0.056	0.15 0.709
TEB [mmol(+) · kg⁻¹]	41.8-98.3 73.3 (20.2)	11.6-111.3 64.2 (36.7)	13.9-102.5 57.5 (38.7)	19.0-172.1 90.4 (54.7)	0.05 0.824	0.27 0.612
CEC [mmol(+) · kg⁻¹]	112.2-135.0 126.4 (8.6)	117.8-155.8 137.0 (16.7)	76.4-144.0 106.0 (29.2)	62.5-184.1 126.7 (45.3)	0.87 0.369	0.90 0.362
BS [%]	38.3-72.8 57.3 (12.3)	9.6-71.8 44.5 (22.6)	17.9-82.3 48.7 (25.0)	26.4-92.8 67.1 (26.1)	0.30 0.596	0.05 0.832
Ø 2-0.05 mm	17.1-27.7 24.2 (4.2)	17.4-26.5 23.9 (3.8)	24.7-58.2 34.4 (13.8)	31.2-75.6 52.6 (18.2)	8.16 0.014*	1.73 0.213
Ø 0,05-0,002 mm	55.7-69.1 61.6 (5.3)	58.7-73.2 63.7 (5.7)	34.3-64.0 50.3 (11.1)	29.0-54.4 45.0 (9.7)	9.16 0.011*	0.05 0.831
Ø <0,002 mm	12.3-18.9 15.2 (2.5)	9.4-15.0 12.3 (2.1)	7.5-25.0 15.3 (6.6)	6.3-19.0 10.7 (5.1)	0.08 0.782	2.09 0.174
Ø <0.02 mm	47.3-52.0 49.7 (1.7)	42.2-50.0 46.7 (3.1)	28.2-55.0 42.1 (9.5)	22.3-49.8 35.6 (9.8)	5.19 0.042*	1.36 0.266
OC [g · kg⁻¹]	9.6-14.6 12.4 (1.8)	16.2-20.0 17.5 (1.5)	3.1-9.5 6.9 (2.4)	10.3-14.2 12.0 (1.6)	26.82 0.000*	22.86 0.000*
TN [g · kg⁻¹]	0.9-1.7 1.4 (0.3)	1.6-2.3 1.9 (0.3)	0.8-1.2 1.0 (0.2)	0.9-1.8 1.4 (0.3)	10.70 0.007*	9.32 0.010*
Ca [mg · kg⁻¹]	1325.4-2107.5 1602.2 (315.1)	806.9-2464.8 1506.1 (603.4)	38.7-544.0 253.7 (191.4)	57.6-2737.3 911.7 (1065.7)	6.92 0.022*	0.58 0.461

Explanation: [1] the range; [2] mean; [3] standard deviation; [a] arable land; [b] grassland; [c] value of two-way ANOVA F-test; [d] probability p; * indicated value significant at $p < 0.05$.

of the limit specified in Regulation [25] and Institute of Soil Science and Plant Cultivation (IUNG) recommendations [8]. The content of organic carbon and total nitrogen in the investigated soils differed both between the regions and depending on the type of use (Table 1). Śląskie Foothills soils were characterized by higher total Ca content than Ciężkowickie Foothills soils. This was mostly conditioned by the content of this element in the bedrock.

The zinc content in the Śląskie Foothills soils was higher than in the soils of the Ciężkowickie Foothills. A similar relation was also observed in the concentration of this element in the bedrock from these two regions (Table 2). In the topsoil at all sites in this mesoregion, zinc content (86.3 $mg \cdot kg^{-1}$) was higher than the average content of this component in the soils of the Earth, which is 64 $mg \cdot kg^{-1}$ [26] and depends on the natural content of this element in the bedrock (Table 3). Several authors [2, 20, 29] drew attention to the relation between the content of Zn in soils and bedrock. Based on the calculated accumulation index (AI) it is difficult to determine from which region soils have a higher amount of Zn of anthropogenic origin. Average values of WA for soils of the two regions were similar (1.7 for Śląskie Foothills and 1.9 for Ciężkowickie Foothills) and were higher than the value of this ratio given for the soil of the Earth, which is 1.075 [26]. Based on the average values of the zinc enrichment factor (EF_{Zn}), it can be concluded that the accumulated Zn in the Śląskie Foothills soils is of anthropogenic origin (Table 2). Mean values of EF_{Zn} for soils from those regions were higher than 1.5, which, according to Zhang and Liu [32], indicates anthropogenic accumulation of the metal.

Despite the zinc enrichment of the soil surface layers, its content does not exceed the allowable concentrations [25] and the maximum Zn content found in the soils (94.5 $mg \cdot kg^{-1}$) indicates that they can be classified as soil with a natural content of this element (zero degrees of contamination) [8].

As well as the zinc content, the lead content in the Śląskie Foothills soils was higher than in the Ciężkowickie Foothills soils. However, such relationship has not been reported for the concentration of this element in the bedrock of these two regions (Table 2). In the topsoil of Śląskie Foothills the lead content was higher (31.5 $mg \cdot kg^{-1}$)

Table 2. The content of zinc, lead, and cadmium in the surface layer (0–30 cm) soils and bedrock, accumulation index (AI), and enrichment factor (EF)

Metal	Śląskie Foothills		Ciężkowickie Foothills		Region[a]	The way of land use[a]
	A[a]	G[b]	A[a]	G[b]	F[c] P[d]	F[c] P[d]
0-30 cm [mg · kg^{-1}]						
Zn	75.5-94.5[1] 83.4[2] (7.7)[3]	73.4-113.7 89.1 (15.4)	33-7-67.1 48.0 (12.1)	25.4-57.9 44.8 (12.7)	31.57 0.000*	0.032 0.862
Pb	25.4-32.9 28.3 (2.8)	29.6-48.4 34.7 (7.9)	10.3-20.7 14.5 (3.8)	8.6-20.3 14.6 (4.5)	32.78 0.000*	1.20 0.295
Cd	0.2-1.1 0.5 (0.4)	0.2-0.7 0.4 (0.2)	0.0-0.1 0.0 (0.0)	0.0-0.2 0.1 (0.1)	10.54 0.007*	0.03 0.874
in the bedrock [mg · kg^{-1}]						
Zn	41.9-79.6 60.5 (17.0)	37.7-59.0 48.6 (8.5)	12.2-59.6 43.0 (18.9)	10.3-53.0 25.1 (16.5)	5.07 0.044*	2.69 0.127
Pb	6.6-14.6 10.5 (3.2)	7.7-10.2 8.7 (1.1)	4.0-20.4 11.4 (6.1)	3.1-13.5 6.0 (4.3)	0.14 0.713	2.30 0.155
Cd	0.0-0.3 0.1 (0.1)	0.0-0.2 0.1 (0.1)	0.0-0.0 0.0 (0.0)	0.0-0.0 0.0 (0.0)	1.96 0.187	0.04 0.844
AI						
Zn	1.2-1.8 1.5 (0.3)	1.5-2.1 1.9 (0.3)	0.8-2.8 1.4 (0.8)	1.1-4.0 2.3 (1.1)	0.30 0.593	2.43 0.145
Pb	2.2-3.8 2.9 (0.7)	3.2-4.7 3.9 (0.5)	1.0-3.3 1.6 (0.9)	1.5-4.1 3.0 (1.1)	5.08 0.044*	6.35 0.027*
Cd	3.5-381.0 225.7 (138.6)	2.9-380.0 237.5 (151.1)	1.0-97.0 45.0 (44.4)	11.7-240.8 99.5 (90.9)	5.83 0.033*	0.25 0.625
EF						
Zn	0.8-12.6 4.0 (5.0)	1.1-2.8 1.9 (0.6)	0.4-1.8 1.1 (0.6)	0.1-1.0 0.5 (0.4)	2.13 0.170	0.91 0.358
Pb	10.7-14.9 12.2 (1.7)	9.6-14.9 12.5 (2.1)	4.6-6.8 5.9 (0.9)	6.2-9.4 7.9 (1.3)	36.93 0.000*	1.64 0.224
Cd	3.4-120.8 53.5 (43.3)	2.9-59.4 28.7 (20.7)	0.3-30.0 10.8 (12.2)	2.0-27.6 15.1 (12.5)	3.65 0.080	0.49 0.498

Explanation: see Table 1.

than the average content in the soil of the Earth, which is 22.6 mg · kg^{-1} [26].

The content of this component in the upper layer of soil was not related to the amount of this element in the bedrock (Table 3), which could indicate an anthropogenic origin. The calculated value of the accumulation index (AI) also confirms the anthropogenic origin of this element. Mean values of AI_{Pb} for both regions are higher (in soil of Śląskie Foothills 3.4 and Ciężkowickie Foothills 2.3) than the value of this ratio specified for soil of the Earth, which is 1.364 [26]. It is also worth noting that the accumulation index of lead determined for grassland soils was higher than for arable land soils, regardless of the region (Table 2).

The unnatural lead enrichment in foothills soil is also reflected by the high values of the enrichment factor (EF_{Pb}), which are usually in the range from 5 to 20. Despite the significant lead enrichment of the soil surface layers, the content of these elements, like zinc, does not exceed the permissible levels of this metal specified in the Regulation of the Minister of the Environment on the quality standards for soil and ground [25]. Even the highest content of Pb, 48.4 mg · kg^{-1}, determined in soil collected from grassland in the village of Porąbka, does not exceed the maximum allowable level for the uncontaminated Pb soils [8]. At the same time, all tested soils contain more than 25 mg · kg^{-1} of lead, which is the amount that suggests a human impact on soil contamination by this element according to Kabata-Pendias and Pendias [7].

In both the topsoil and the bedrock, the cadmium concentration in Śląskie Foothills soils was higher than in Ciężkowickie Foothills soils (Table 2). The average content of Cd in the topsoil from this mesoregion was higher (0.5 mg · kg^{-1}) than the average content in the surface layer of soil on Earth, which was 0.145 mg · kg^{-1} [26], whereas in an analogous layer of soil from the Ciężkowickie Foothills the Cd content was significantly lower (0.1 mg · kg^{-1}). The high accumulation index of

Table 3. Linear correlation coefficients defining the relationship between the heavy metals in the topsoil contents and their contents in the bedrock

Layer		in the bedrock		
		Zn	Pb	Cd
0-30 cm	Zn	0.699*	0.558*	0.216
	Pb	0.414	0.319	-0.088
	Cd	0.145	0.166	0.882**

Explanation: Coefficient significant at * $p < 0.05$; ** $p < 0.01$.

cadmium (AI_{Cd}) calculated for the investigated soils significantly exceeded the value specified for soils of the Earth, which is 1.477 [26]. High values of EF_{Cd} could indicate the anthropogenic origin of this metal (Table 2). The cadmium content in the surface layer of the investigated soils in both foothills does not exceed the limit values specified in the Regulation of the Minister of the Environment [25]. Taking into account the limit values given by Kabata-Pendias et al. [8], most of the investigated soils were characterized by a natural content of cadmium (0° of soil pollution). An increased content of this element (greater than 1 $mg \cdot kg^{-1}$) occurred only in the soil of arable land in the village of Porąbka (I° of soil pollution).

The results confirmed the relationship between the concentration of heavy metals and texture, organic carbon content, and sorption properties of the soil already described by several authors [6, 11, 20]. The content of Zn and Pb in soils depended on their texture. This is confirmed by a statistically significant correlation coefficient between Zn and Pb content and the amount of the fraction lower than 0.02 mm (Table 4). Similar results were obtained by Kołodziej et al. [11], who found that the heavy metal content in clayey flysch soils is higher than in loamy flysch soils.

The relation between the content in the soil fraction below 0.02 mm and the concentration of zinc is not confirmed by studies by Kaniuczak et al. [9]. The Zn and Pb content in soils was significantly dependent on the content of organic carbon and total nitrogen. The accumulation of lead in the topsoil is connected with this metal's particularly strong capability to form interconnections with organic matter [7, 22, 28]. These dependencies are not shown in the studies of Kaniuczak et al. [9], who found no relation between the humus content and concentration of heavy metals in agricultural soils derived from the Carpathian flysch. Moreover, the total content of lead and cadmium is positively correlated with hydrolytic acidity, and the zinc content is strictly positively correlated with the content of calcium and the total volume capacity of the sorption complex (Table 4).

The soils from the Śląskie Foothills were characterized by a higher content of heavy metals than the soils from the Ciężkowickie Foothills. On the basis of these studies and available literature, it can be assumed that the Śląskie Foothills soils are more exposed to the possibility of contamination from the air. In the bedrock of the Śląskie Foothills there is a higher concentration

Table 4. The correlation coefficients between the heavy metals content in the topsoil and selected soil properties at $p < 0.05$

Soil properties	Zn	Pb	Cd
pH KCl	-0.227	-0.358	-0.230
Hh [mmol(+) · kg^{-1}]	0.346	0.523*	0.510*
TEB [mmol(+) · kg^{-1}]	0.192	0.037	-0.133
CEC [mmol(+) · kg^{-1}]	0.500*	0.415	0.176
BS [%]	0.136	-0.065	-0.261
Ø <0.002 mm	0.422	0.230	-0.014
Ø <0.02 mm	0.767***	0.653**	0.277
OC [g · kg^{-1}]	0.700**	0.809***	0.314
TN [g · kg^{-1}]	0.756***	0.780***	0.207
Ca [mg · kg^{-1}]	0.546*	0.437	0.405

Explanation: Coefficient significant at * $p < 0.05$; ** $p < 0.01$; *** $p < 0.001$.

of metal, and the properties such as the soil parts below 0.02 mm and humus content are conducive to the accumulation of zinc, lead, and cadmium.

CONCLUSION

1. The investigated soil of the Śląskie and Ciężkowickie Foothills, in most cases, is characterized by natural content of zinc and lead as well as cadmium.
2. The Śląskie Foothills soils had significantly higher accumulation index of Pb and Cd and enrichment factor of Pb than the Ciężkowickie Foothills soils.
3. In the analysed soils of the Śląskie Foothills, zinc, lead, and cadmium content in the 0–30 cm layer was higher than in the soils of the Ciężkowickie Foothills.
4. Different types of land use did not affect the content of heavy metals.
5. Only the accumulation index of lead was higher in grassland soils than in arable land soils.

Acknowledgements

The studies were financed by the Ministry of Science and Higher Education, Poland, under the grant No. N N310 312434.

REFERENCES

1. Barančoková M., Barančok P., Mišovičová D. 2009. Heavy Metal loading of the Belianske Tatry Mts. Ekológia (Bratislava), 28(3): 255-268.
2. Czarnowska K. 1983. Wpływ skały macierzystej na zawartość metali ciężkich w glebach. Zesz. Probl. Post. Nauk Roln., 242: 51-61.
3. Dziadowiec H. 1993. Ekologiczna rola próchnicy glebowej. Zesz. Probl. Post. Nauk Roln., 411: 269-282.
4. Hajduk E., Kaniuczak J., Waśniewski S. 2007. Wpływ przemysłu na zawartość metali ciężkich w glebach Pogórza Strzyżowskiego i Dołów Jasielsko-Sanockich. Zesz. Probl. Post. Nauk Roln., 520: 55-63.
5. ISO 10390. 1998. Jakość gleby. Oznaczanie pH.
6. Kabata-Pendias A. 1991. Wyznaczanie „tła" zawartości metali śladowych w glebach. Krajowa konferencja „Ekologia, aspekty ochrony środowiska", AGH Kraków, 25-29.
7. Kabata-Pendias A., Pendias H. 2001. Trace Elements in Soils and Plants.
8. Kabata-Pendias A., Piotrowska M., Witek T. 1993. Ocena stopnia zanieczyszczenia gleb i roślin metalami ciężkimi i siarką. Ramowe wytyczne dla rolnictwa. IUNG, Puławy.
9. Kaniuczak J., Kołodziej M., Gąsior J. 1998. Ogólna zawartość Mn, Cu i Zn w glebach terenów górzystych południowo-wschodniej Polski. Zesz. Probl. Post. Nauk Roln., 464: 321-327.
10. Klasyfikacja uziarnienia gleb i utworów mineralnych. 2009. Rocz. Glebozn., 60(2): 7-16.
11. Kołodziej M., Pęcek J., Zych U. 1998. Całkowite zawartości metali ciężkich w glebach terenów górzystych południowo-wschodniej Polski. Zesz. Probl. Post. Nauk Roln., 464: 329-336.
12. Kondracki J. 2009. Geografia regionalna Polski. Wyd. Nauk. PWN, Warszawa.
13. Lityński T., Jurkowska H., Gorlach E. 1976. Analiza chemiczno-rolnicza, PWN, Warszawa.
14. Loska K., Wiechuła D., Korus I. 2003. Metal contamination of farming soils affected by industry. Environ. Int., 30: 159-165.
15. Melke J., Chodorowski J., Dębicki R. 2005. The content of total and DTPA-TEA extracted forms of Zn, Cu, Mn, Pb, Co and Ni in Chornohora (Ukraine) mountain soils. Pol. J. Soil Sci., XXXVIII(1): 51-60.
16. Miechówka A., Gąsiorek M., Józefowska A. 2009. Wpływ sposobu użytkowania na zasoby węgla w glebach Pogórza Śląskiego. Roczn. Glebozn., 60(2): 41-46.
17. Miechówka A., Gąsiorek M., Józefowska A., Zadrożny P. 2011. Content of microbial biomass nitrogen in differently used soils of the Carpathian Foothills. Ecol. Chem. Eng. A, 18(4): 578-584.
18. Miechówka A., Józefowska A., Gąsiorek M., Zadrożny P. 2011. Organic carbon stocks in differently used agricultural soils of Ciężkowickie Foothills. Polish J. Soil Sc., 44: 11-17.
19. Miechówka A., Niemyska-Łukaszuk J., Ciarkowska K. 2002. Heavy metals in selected non-forest soils from the Tatra National Park. Chem. Inż. Ekol., 9(11): 1433-1438.
20. Niemyska-Łukaszuk J. 1993. Formy cynku, ołowiu i kadmu w glebach wybranych regionów Karpat Zachodnich. Zesz. Nauk. AR w Krakowie, Rozp. hab. 187: 60 pp.
21. Niemyska-Łukaszuk J., Miechówka A., Zadrożny P., Gąsiorek M. 2005. Metale ciężkie w glebach Polskich Karpat. Probl. Zag. Ziem Gór., 52: 71-77.
22. Niemyska-Łukaszuk J., Miechówka A., Zadrożny P., Mazurek R. 1998. Metale ciężkie (Cd, Cr, Cu, Mn, Ni, Pb, Zn) w wybranych glebach Babiogórskiego Parku Narodowego. Zesz. Probl. Post. Nauk Roln., 464: 311-319.
23. Oleksynowa K., Tokaj J., Jakubiec J. 1987. Przewodnik do ćwiczeń z gleboznawstwa i geologii.

Część II. Metody laboratoryjne analizy gleby. Wyd. AR, Kraków.

24. PN-R-04032. 1998. Gleby i utwory mineralne. Pobieranie próbek i oznaczanie składu granulometrycznego, PKN.

25. Rozporządzenia Ministra Środowiska z dnia 9 września 2002 r. w sprawie standardów jakości gleb oraz standardów jakości ziemi. Dz.U. 2002, Nr 165, poz. 1359.

26. Salminen R., Batista M.J., Bidovec M., Demetriades A., De Vivo B., De Vos W., Duris M., Gilucis A., Gregorauskiene V., Halamic J., Heitzmann P., Lima A., Jordan G., Klaver G., Klein P., Lis J., Locutura J., Marsina K., Mazreku A., O'Connor P.J., Olsson S.Å., Ottesen R.-T., Petersell V., Plant J.A., Reeder S., Salpeteur I., Sandström H., Siewers U., Steenfelt A. and Tarvainen T. (Eds). 2005. Geochemical Atlas of Europe. Part 1 - Background Information, Methodology and Maps.

27. Systematyka gleb Polski wyd. 5. Rocz. Glebozn., 62(3): 5-142.

28. Shaheen S.M. 2009. Sorption and lability of cadmium and lead in different soils from Egypt and Greece. Geoderma, 153(1-2): 61-68.

29. Szopka K., Karczewska A., Jezierski P., Kabała C. 2013. Spatial distribution of lead in the surface layers of mountain forest soils, an example from the Karkonosze National Park, Poland. Geoderma, 192: 259-268.

30. Terelak H., Tujaka A. 2003. Występowanie pierwiastków śladowych w glebach użytków rolnych województwa podkarpackiego. Zesz. Probl. Post. Nauk Rol., 493: 245-252.

31. Woźniak L. 1996. Biogenne pierwiastki metaliczne i niektóre toksyczne metale ciężkie w glebach i roślinach Bieszczadów. Zesz. Nauk. AR w Krakowie, Rozpr., 216.

32. Zhan J., Liu C.L. 2002. Riverine Composition and Estuarine Geochemistry of Particulate Metals in China - Weathering Features, Anthropogenic Impact and Chemical Fluxes Estuar. Coast. Shelf Sci., 54: 1051-1070.

19

ENZYMATIC ACTIVITY OF SOILS EXPOSED TO TRANSPORTATION POLLUTANTS, LOCATED ALONG ROAD NO. 957

Barbara Filipek-Mazur[1], Monika Tabak[1], Olga Gorczyca[1]

[1] Department of Agricultural and Environmental Chemistry, University of Agriculture in Kraków, A. Mickiewicza 21, 31-120 Kraków, Poland, e-mail: rrfilipe@cyf-kr.edu.pl; Monika.Tabak@ur.krakow.pl; o.gorczyca@ur.krakow.pl

ABSTRACT

The research was conducted in order to determine the catalase, dehydrogenase, and arylsulfatase activities of soils exposed to transportation pollutants. The research material consisted of soil samples collected from points located along road no. 957 at a section passing through Zawoja (the Malopolska Region), from places at a distance of 5 and 200 m from the road edge. The samples were collected from a 0–10 cm layer, from areas covered with grasses. No considerable diversification in the enzymatic activity of the soils, depending on their distance from the road edge, was found. The mean activity of catalase and dehydrogenases in the soils located 5 m from the road edge was, respectively, 4 and 7% greater than the activity of the soils located 200 m from the road edge. The mean arylsulfatase activity in the soils located 5 m from the road edge was 3% lower than in the soils located at a distance of 200 m. A positive correlation was found between the catalase and arylsulfatase activities, and the dehydrogenase activity in the soils.

Keywords: soil, transportation, dehydrogenase, catalase, arylsulfatase.

INTRODUCTION

Biological activity of soil is, next to their physical and chemical properties, an important element which decides its fertility. It is characterized by enzymatic activity, micro-organism biomass, the composition and number of microorganisms, and also by an activity specific for particular soil conditions [Brzezińska 2006]. A lot of researchers [Gostkowska et al. 1998, Koper et al. 2008] believe that enzymes can be a sensitive index of changes in soil. The number of micro-organisms in soil as well as the activity of enzymes depend on a lot of factors (e.g. soil pH values, water-air relationships, the content of organic compounds) which are formed by management and state of the natural environment [Barabasz, Vorisek 2002]. Environmental pollution from anthropogenic sources, including transportation, causes changes in chemical properties of soils, leading to changes in the enzymatic activity of these soils [Bielińska et al. 2010, Szymczak et al. 2011, Kuziemska 2012, Niemeyer et al. 2012]. Car transportation (combustion of fuels, abrasion of vehicle parts, wear of fluids and operating lubricants) is a source of environmental pollution with, among other things, heavy metals and organic compounds [Wang, Zhao 2008, Dao et al. 2010].

The research was conducted in order to determine the catalase, dehydrogenase, and arylsulfatase activities of soils exposed to transportation pollutants.

MATERIAL AND METHODS

The research material consisted of soil samples collected from points located along road no. 957 at a section passing through Zawoja (the Malopolska Region), from places at a distance of 5 and 200 m from the road edge (Figure 1). The samples were collected from 13 points, from a 0–10 cm layer, from areas covered with grasses. The precise characteristics of the research area, the intensity of car traffic as well as basic physico-chemical properties of the studied soils were presented in a previous publication [Filipek-Mazur et al. 2013].

The determination of the dehydrogenase and arylsulfatase activities in the soils was conducted by colorimetric method on a Beckman UV/VIS DU 640 Spectrophotometer. The dehydrogenase activity was determined using 2,3,5-triphenyltetrazolium chloride as a substrate [Brzezińska and Włodarczyk 2006]. The soil was incubated with the substrate for 24 hours at a temperature of 37 °C. The content of the created 1,3,5-triphenylformazan (TPF), extracted with alcohol, was determined at a wavelength of 485 nm. The catalase activity was determined by manganometric method consisting in shaking the soil with 0.3% H2O2 0.3% H_2O_2 (30 rot.·min^{-1}, 20 min) at room temperature. The arylsulfatase activity was determined after incubating the soil with p-nitrophenyl sulfate for 1 hour at a temperature of 37 °C, determining the content of the formed p-nitrophenol (pNP) at a wavelength of 400 nm [Allef, Nannipieri 1995, Brzezińska, Włodarczyk 2006].

All the analyses were carried out in 4 replications. The obtained results were elaborated statistically – minimum and maximum values were stated, and arithmetic mean and standard deviation were computed. All statistical analyzes were performed using the STATISTICA data analysis software system, version 10 (StatSoft, Inc.).

RESULTS AND DISCUSSION

The results of determining the enzymatic activity of the studied soils are presented in Table 1 and Figure 2. The dehydrogenase activity was slightly higher in the soils located closer to the road (5 m) and was, on average, 0.200 μmol TPF·g^{-1} d.m.·24h^{-1}. For the distance of 200 m from the road, that value amounted to 0.187 μmol TPF·g^{-1} d.m.·24h^{-1} and was 7% lower than the mean value determined in the distance of 5 m from the road. Determination of dehydrogenase activity in soils is used as an indicator of the intensity of respiratory metabolism of all populations of soil micro-organisms, which is used to determine the total microbiological activity of soils. The dehydrogenase activity is commonly used to evaluate the factors which have a negative impact on soil micro-organisms [Brzezińska 2006].

Similar relationships were found in respect to the catalase activity which at the distance of 5 and 200 m from the edge of the road was, on average, 6.36 and 6.11 μmol H_2O_2 · g^{-1} d.m.· min^{-1}, respectively. Closer to the road the value was 4% higher. The earlier tests on pollution of the analyzed soils with heavy metals, including cadmium, showed that the content of this element in the soils located 200 m from the road was slightly higher than at the distance of 5 m [Filipek-Mazur et al. 2013]. This higher cadmium content might have been the cause of the decrease in the catalase activity.

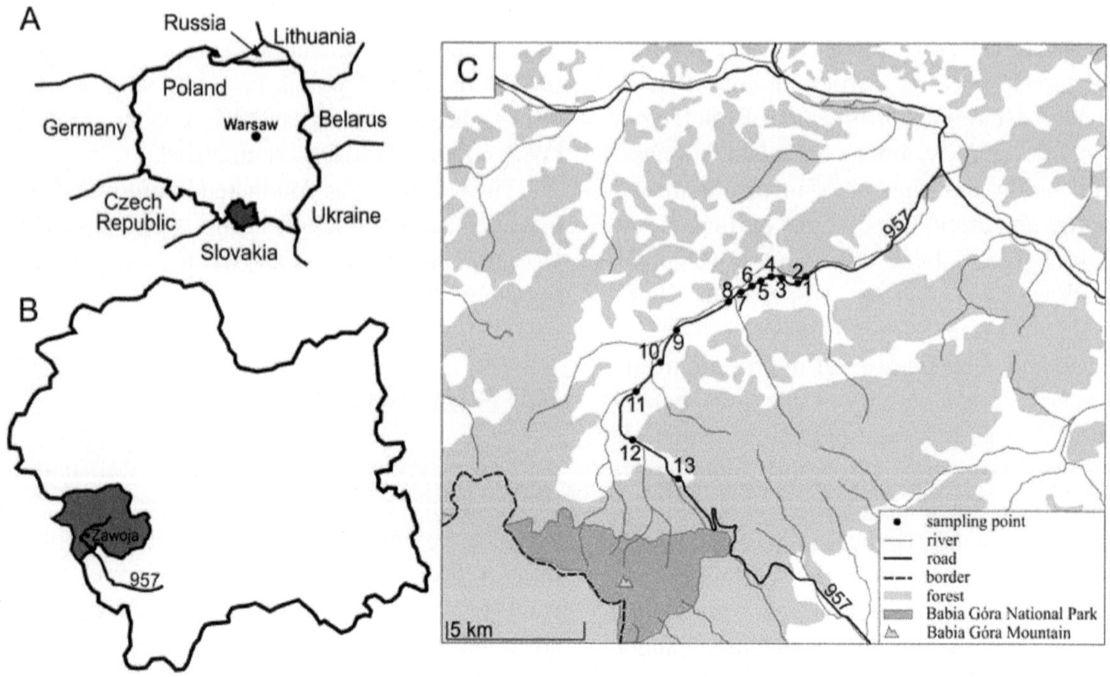

Figure 1. Location of sampling points: A – Poland with marked Małopolska province, B – Małopolska province with marked Suski district, road No. 957 and Zawoja, C – sampling points along road No. 957

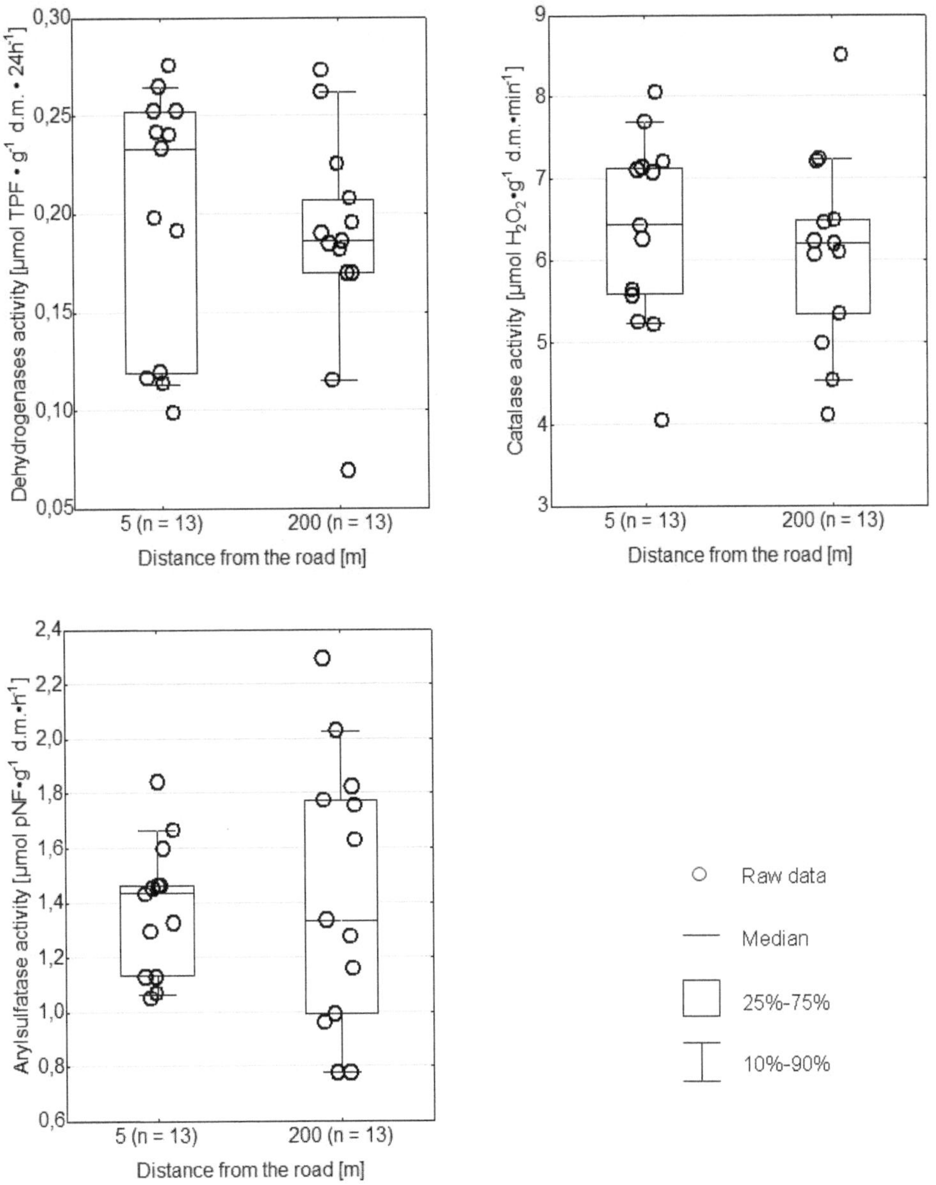

Figure 2. Dehydrogenase, catalase, and arylsulfatase activities in soils from Zawoja region

Szymczak et al. [2011] obtained such an effect in their research, whereas Wyszkowska et al. [2009] showed a slight decrease in the activity of soil catalase under the influence of increased copper content in the soil.

Relationships between the values of arylsulfatase activity were different, depending on the distance of soils from the road. The mean arylsulfatase activity in the soils located 5 m from the road edge was 3% lower than in the soils located at a distance of 200 m (1.38 and 1.43 µmol pNF·g^{-1} d.m.·h^{-1}, respectively). The determined values were lower than the ones obtained in the research of Siwik-Ziomek [2005], but the author analyzed soils from the fertilization experiment. Table 2 shows the values of correlation coefficients between selected soil properties and the values of their enzymatic activity. A positive correlation was found between the catalase and arylsulfatase activities, and the dehydrogenase activity in the soils (the correlation coefficient values were 0.405 and 0.429, respectively). The arylsulfatase activity depended on concentration of hydrogen ions in the soil as well as on the contents of organic carbon and total nitrogen in the soil (the values of correlation coefficients between 0.398 and 0.565). Organic compounds which constitute a source of energy for soil micro-organisms influence an increase in enzymatic activity of soils. Siwik-Ziomek and Koper [2013] noticed a beneficial effect of fertilization with manure on arylsulfatase activity.

Table 1. Dehydrogenase, catalase, and arylsulfatase activities in soils from Zawoja region – statistical parameters

Distance from the road [m]	Parameter	Dehydrogenase activity [μmol TPF·g^{-1} d.m.·24 h^{-1}]	Catalase activity [μmol H_2O_2·g^{-1} d.m.·min^{-1}]	Arylsulfatase activity [μmol pNF·g^{-1} d.m.·h^{-1}]
5 (n = 13)	Mean	0.200	6.36	1.38
	Minimum	0.099	4.03	1.05
	Maximum	0.275	8.05	1.84
	Standard deviation	0.065	1.16	0.24
200 (n = 13)	Mean	0.187	6.11	1.43
	Minimum	0.070	4.11	0.77
	Maximum	0.273	8.50	2.29
	Standard deviation	0.054	1.18	0.49

Table 2. Values of correlation coefficients between selected soil properties and their enzymatic activity

	Dehydrogenase activity	Catalase activity	Arylsulfatase activity
Fraction content $\phi < 0.02$ mm	-0.234	0.188	0.220
H$^+$	-0.191	-0.048	0.454 *
Organic C	-0.002	0.135	0.398 *
Total N	-0.060	0.174	0.565 **
Total S	0.001	0.050	0.375
Organic S	0.006	0.047	0.386
Sulfatic S	-0.132	0.090	-0.170
Arylsulfatase activity	0.429 *	0.353	–
Catalase activity	0.405 *	–	–

* significant at $p < 0.05$; ** significant at $p < 0.01$

Such a dependence was not observed with respect to catalase and dehydrogenases. Scientific literature shows a positive correlation between the activity of dehydrogenases and catalase, and the content of organic carbon in soils [Brzezińska 2006], which was not confirmed by the authors' own research.

CONCLUSIONS

No considerable diversification in the enzymatic activity of the soils, depending on their distance from the road edge, was found. The mean activity of catalase and dehydrogenases in the soils located 5 m from the road edge was, respectively, 4 and 7% greater than the activity of the soils located 200 m from the road edge. The mean arylsulfatase activity in the soils located 5 m from the road edge was 3% lower than in the soils located at a distance of 200 m.

A significantly positive correlation was found between the catalase and arylsulfatase activities, and the dehydrogenase activity in the soils. The arylsulfatase activity was positively correlated with the concentration of hydrogen ions in the soil and with the contents of organic carbon and total nitrogen in the soil.

Acknowledgements

The research results carried out within the subject No. 3101 were financed from the subsidy for science granted by the Polish Ministry of Science and Higher Education.

REFERENCES

1. Allef K., Nannipieri P. 1995. Methods in applied soil microbiology and biochemistry (enzyme activites). Academic Press, London.
2. Barabasz W., Vorisek K. 2002. Bioróżnorodność mikroorganizmów w środowiskach glebowych (Biodiversity of micro-organisms in soil environments). [In:] Aktywność drobnoustrojów w różnych środowiskach (Activity of micro-organisms in different environments). W. Barabasz (Ed.), Akademia Rolnicza, Kraków, 23–34.
3. Bielińska E.J., Ligęza S., Kawecka-Radomska M. 2010. Wpływ długoletniej emisji azotowej na aktywność enzymatyczną gleb uprawnych (The in-

fluence of long-term nitrogen emission on the enzymatic activity of arable soils). Zesz. Probl. Post. Nauk Rol., 556, 795–802.

4. Brzezińska M. 2006. Aktywność biologiczna oraz procesy jej towarzyszące w glebach organicznych nawadnianych oczyszczonymi ściekami miejskimi (badania polowe i modelowe) (Impact of treated wastewater on biological activity and accompanying processes in organic soils (field and model experiments)). Acta Agrophys., 131, Rozprawy i Monografie, 1–164.
5. Brzezińska M., Włodarczyk T. 2006. Methods of soil catalase and dehydrogenase activity measurement. [In:] Selected methodological aspects of soil enzyme activity tests. S. Russel (Ed.), A.I. Wyczółkowski (Ed.), A. Bieganowski (Ed.), Institute of Agrophysics Polish Academy of Sciences, Lublin, 59–74.
6. Dao L., Morrison L., Zhang Ch. 2010. Spatial variation of urban geochemistry in a roadside ground in Galway, Ireland. Sci. Total Environ., 408(5), 1076–1084.
7. Filipek-Mazur B., Tabak M., Gorczyca O. 2013. Trace element content in meadow sward and soil along Road No. 957 passing through Zawoja village. Ecol. Chem. Eng. A, 20, 631–642.
8. Gostkowska K., Furczak J., Domżał H., Bielińska E.J. 1998. Suitability of some biochemical and microbiological tests for the degradation degree podzolic soil on the back ground of it differentiated usage. Pol. J. Soil Sci., 30(2), 69–78.
9. Koper J., Piotrowska A., Siwik-Ziomek A. 2008. Aktywność dehydrogenaz i inwertazy w glebie rdzawej leśnej w okolicy Zakładów Azotowych „Anvil" we Włocławku (Dehydrogenase and invertase activities in a rusty soil in the neighbourhood of the Włocławek nitrogen plant "Anwil"). Proc. ECOpole, 2(1), 197–202.
10. Kuziemska B. 2012. Aktywność dehydrogenaz w glebie zanieczyszczonej niklem (The activity of dehydrogenases in the soil contaminated with nickel). Ochrona Środowiska i Zasobów Naturalnych, 52, 103–112.
11. Niemeyer J.C., Lolata G.B., de Carvalho G.M., Dogueira M.A. 2012. Microbial indicators of soil Heath as tools for ecological risk assessment of a metal contaminated site in Brazil. Appl. Soil Ecol., 59, 96–105.
12. Siwik-Ziomek A. 2005. Zawartość siarki i jej frakcji oraz aktywność arylosulfatazy w glebie płowej po zmianie nawożenia (Sulphur content and its fractions arylsulphatase activity in lesive soil after changes fertilization). Ekol. Tech., 13(6), 233–239.
13. Siwik-Ziomek A., Koper J. 2013. Wpływ nawożenia obornikiem i doboru roślin na zawartość siarczanów(V) i aktywność arylosulfatazy w glebie płowej (Effect of manure fertilisation and plant selection on the content of sulphates(VI) and the activity of arylsulphatase in luvisol). Proc. ECOpole, 7(1), 241–246.
14. Szymczak J., Kłódka., Smolik B., Pawlica M. 2011. Wpływ soli kadmu na aktywność enzymów stresu oksydacyjnego w glebie i kukurydzy (*Zea mays* var. *Saccharata*) (Effect of cadmium salt on the activity of oxidative stress enzymes in soil and maize (*Zea mays* var. *Saccharata*)). Ochrona Środowiska i Zasobów Naturalnych, 48, 210–215.
15. Wang P., Zhao W. 2008. Assessment of ambient volatile organic compounds (VOCs) near major roads in urban Nanjing, China. Atmos. Res., 89(3), 289–297.
16. Wyszkowska J., Kucharski M., Kucharski J., Borowik A. 2009. Activity of dehydrogenases, catalase and urease in copper polluted soil. J. Elementol., 14(3), 605–617.

PHYSIOLOGICAL REACTION OF COMMON OSIER (*SALIX VIMINIALIS* L.) VAR. JORR TO THE PRESENCE OF LEAD IN THE SUBSOIL

Katarzyna Malinowska[1], Ryszard Malinowski[2], Andrzej Stera[1]

[1] Department of Plant Physiology and Biochemistry, Faculty of Environmental Management and Agriculture, West Pomeranian University of Technology in Szczecin, 17 Słowackiego Str., 71-434 Szczecin, Poland, e-mail: katarzyna.malinowska@zut.edu.pl

[2] Department of Soil Science, Faculty of Environmental Management and Agriculture, West Pomeranian University of Technology in Szczecin, 17 Słowackiego Str., 71-434 Szczecin, Poland, e-mail: ryszard.malinowski@zut.edu.pl

ABSTRACT

The effect of lead ions of the concentrations within the range of 15–1000 mg·dm^{-3} on the physiological reaction of common osier var. Jorr was examined. The content of assimilation pigments, the rate the CO_2 assimilation, transpiration, the indices of relative water content and the deficit of water saturation and the content of lead in the nutrient solution. The studied physiological parameters in common osier var. Jorr were differentiated by the rate of lead ions in the nutrient solution. The Jorr variety of common osier was characterised by good values of the determined physiological parameters under stressful conditions at a large accumulation of lead. This suggests that it shows quite a high tolerance to the stress caused by contamination of the subsoil with lead.

Keywords: lead, photosynthetic pigments, CO_2 assimilation, transpiration, common osier.

INTRODUCTION

Pollution of the natural environment, resulted from the industrial activity of man, needs intensive procedures in order to restore the original state of soil and air. One of the ways of reclamation of soil chemically contaminated is the use of phytoremediation abilities of some plants [Baran 2000]. The effectiveness and the duration time of land reclamation with the use of phytoremediation depend not only on the concentration of impurities in the soil but also on a proper selection of plant species for a given type of pollution. Despite the fact that a series of plant species show increased resistance to given toxic compounds, not all, however, can be used in land reclamation [Baran 2000, Zemleduch and Tomaszewska 2007]. The studies within the range of chemistry, physiology and biochemistry allowed for more accurate acquaintance with mechanisms and processes responsible for the taking and the accumulation of toxins by plants. They also made it possible to use physiological and biochemical parameters for the assessment of the usability of plants for reclamation of degraded terrains. Owing to this it is possible to use deliberately natural abilities of plants and plan optimal methods of remediation of pollution from degraded land [Zemleduch and Tomaszewska 2007].

Shrub willows (*Salix* sp.) are used for many purposes including land reclamation and biological restoration of chemically degraded areas. The broad application of phytoremediation and phytoextraction to contaminated soils result mainly from their abilities to remove hard to degrade biologically impurities as well as from their high tolerance to pollution and their pioneering character [Eltop et al. 1991, Šottniková. et al. 2003, Deng et al. 2006, Hermle et al. 2006, Wrzosek et al. 2008].

The aim of the studies was to determine the physiological reaction of common osier (*Salix viminalis*) var. Jorr under the condition of the subsoil contaminated with lead and to determine the usefulness of this form for the development of the polluted soils.

MATERIAL AND RESEARCH METHODS

Common osier (*Salix viminalis* L.) var. Jorr was used as the experimental material. 60 cuttings of willow obtained from the last year's shoots were used for the studies. Laboratory studies were carried out as a hydroponic culture with different doses of lead. Willow cuttings were divided into four groups of the same number and placed in containers filled with Hoagland full nutrient solution. Lead nitrate (III) was added to the nutrient solutions 14 days after the cuttings were placed in the hydroponics, when the plants had taken roots and had shot. Three levels of contamination of the nutrient solutions with lead ions were applied: 15, 100 and 1000 $mg \cdot dm^{-3}$. The plants placed in Hoagland full nutrient solution were the control object. The determination of physiological parameters was carried out four times: on the 7th, 14th, 21st and 28th day after the day the rates of lead ions were applied. The measurements of the parameters of gas exchange (A – CO_2 assimilation, E – transpiration, c_i – concentration of CO_2, g_s – stomatal conductance for water vapor, g_c – stomatal conductance for CO_2) were made (the measurement was replicated three times) on leaves using a gas analyser TPS-2, working in an open system with a chamber of PLC-4 type. The measurement of the concentration of CO_2 (c_i) was carried out on the basis of its absorption in infrared radiation. The rate of photosynthesis was calculated as a quotient of the concentration of CO_2 absorbed by a leaf, in a defined time in the measurement chamber, and the value of the surface of leaves. The rate of transpiration was determined on the basis of the volumetric humidity sensor readings. In the analyser cuvette the following conditions were established: a permanent inflow of carbon dioxide, humidity equal to the humidity of environment and lighting equal to 2053 PAR ($\mu mol \cdot m^{-2} \cdot s^{-1}$), supplied by means of a light unit attached to the cuvette. On the basis of the obtained results of the assimilation and transpiration rates, photosynthetic effectiveness of the use of water (ω_F) was calculated. The determination of assimilation pigments (of chlorophyll "a" and "b" and carotenoids) were carried out on the same leaves on which the parameters of gas exchange were determined. The content of chlorophyll was assessed using the method of Arnon et al. [1956] modified by Lichtenthaler [1987], whereas the content of carotenoids was determined by means of the Hager and Meyer-Bethenrath method [1966]. The indices of the relative water content (RWC) and the deficit of water saturation (WSD) were determined according to Bandurska [1991]. The content of lead in leaves was determined after previous wet mineralization in a mixture (2:1) of nitric (V) and chloric (VII) acids with a spectrometer of atomic absorption Unicam Solaar 929. The statistical analysis of the studies was carried out using Duncan's test at the level of significance $\alpha_{0.05}$. Pearson linear correlation coefficients between the analysed variables characteristic of gas exchange and between the concentration of lead in leaves and the rate of assimilation of CO_2 and transpiration and the content of assimilation pigments in leaves were calculated.

RESULTS AND DISCUSSION

The obtained results of the measurements of gas exchange, the analysis of the content of assimilation pigments and the indices determining water balance of a plant showed a differentiated effect of lead on the determined physiological parameters.

It was observed that the increasing concentration of lead ions in the nutrient solution resulted in a decrease in the content of photosynthetic pigments in the leaves of common osier var. Jorr. The largest rate of lead ions caused a decrease in the concentration of chlorophyll "a" by about a little over 56%, chlorophyll "b" by 60%, and carotenoids by about 50%. Whereas the rates of lead ions in the amount of 15 and 100 $mg \cdot dm^{-3}$ caused a decrease in the assessed pigments from 20% (carotenoids at the rates of lead ions 15 $mg \cdot dm^{-3}$) to 37.5% (chlorophyll "a" at the rate of Pb^{2+} 100 $mg \cdot dm^{-3}$) – Table 1. Malinowska et al. [2010a] observed a significant decrease in the content of chlorophyll "a", "b" and carotenoids in the leaves of common osier, clone Bjor and Tora, cultured in the nutrient solution with different rates of lead ions. A significant decrease in the concentration of assimilation pigments in common osier, Jorr, was observed under the influence of cadmium ions [13]. A decreased amount of chlorophyll "a" and carotenoids by 62% and chlorophyll "b" by 53% were also obtained by Malinowska et al. [2010b] in common osier, Bjor, after the ap-

Table 1. The toxic impact of lead on the content of photosynthetic pigments [mg·g^{-1} FW], CO_2 assimilation [μmol·m^{-2}·s^{-1}], transpiration [mmol·m^{-2}·s^{-1}] and water use photosynthetic efficiency (ω_f) of common osier clone Jorr

Dose of Pb^{2+} [mg · dm^{-3}]	Chlorophyll a (% of control)	Chlorophyll b (% of control)	Carotenoids (% of control)	CO_2 assimilation (% of control)	Transpiration (% of control)	Water use photosynthetic efficiency (ω_f)
Clone Jorr						
0	2.95±0.12 (100)	1.11±0.11 (100)	1.15±0.15 (100)	7.77±0.26 (100)	0.81±0.09 (100)	9.59
15	1.94±0.15 (65.9)	0.82±0.09 (74.0)	0.92±0.11 (79.9)	3.93±0.18 (50.6)	0.61±0.04 (75.3)	6.44
100	1.84±0.14 (62.5)	0.72±0.09 (65.0)	0.83±0.14 (72.1)	3.14±0.16 (40.4)	0.48±0.10 (59.3)	6.54
1000	1.29±0.10 (43.7)	0.44±0.08 (39.5)	0.58±0.10 (50.3)	1.86±0.11 (23.9)	0.24±0.07 (29.6)	7.75
$LSD_{0,05}$ dose:	0.203	0.096	0.112	0.693	0.234	

plication of cadmium ions in the amount of 280 mg·dm^{-3}. Similar reactions to heavy metals were observed by Malinowska and Smolik [2011], Becerril et al. [1988].

A significant effect of lead on the rate of assimilation and transpiration in clone Jorr of common osier was observed. In the case of both processes a significant decrease in the rate was shown along with the increase in the rate of the rate of lead ions in the nutrient solution. The intensity of the process of photosynthesis and transpiration in clone Jorr at the presence of the highest rate of lead ions constituted 24% and 30%, respectively, in relation to the control plant (Table 1).

The decrease in the intensity of photosynthesis in common osier, var. Bjor and Tora, cultivated in the solution with the addition of lead ions was shown by Malinowska et al. [2010a], Malinowska et al. [2010b]. In our previous studies that cadmium caused a significant decrease in these processes in basket willow, clone Jorr. Photosynthetic effectiveness of water use is often a decisive indicator of plant productivity under the stressful conditions [Górny and Garczyński 2002]. The calculated index was differentiated depending on the applied rates of lead ions. A decrease in this parameter by about 33% in relation to the control plants was observed when the rates of 15 and 100 mg·dm^{-3} were applied. Whereas the application of the highest rate of lead ions caused an increase in the effectiveness of use of water by 14% in relation to the other rates applied. An increase in the value of this parameter results, first of all, from a low rate of transpiration of the studied variety of willow (Table 1). Toxic effect of ions of heavy metals concerns different types of blockades of key metabolic reactions by destabilization of the functional groups and structural changes of the enzymes and transportation proteins [Schützendübel et al. 2001, Verma and Dubey 2003] and it influences the lipid composition of membranes [Hall 2002]. This results, first of all, in disturbances to the processes of gas exchange, transpiration and stomatal conductance. There is a decrease in the effectiveness of the light-dependent reactions and light-independent („dark") biochemical reactions [Zenk 1996, Panda et al. 2003].

On the basis of the values of the correlation coefficient a negative significant relation between the content of total chlorophyll and the concentration of lead in the leaves of common osier, clone Jorr. The correlation coefficient (r) of this relation was r = - 0.5706. The significant linear correlation negative between the CO_2 assimilation and the content of Pb^{2+} in the leaves took place in common osier Jorr. The value of the correlation coefficient (r) for this relationship was r = - 0.6766 (Figure 1).

The indicators of the changes in water balance in a plant are RWC and WSD indices. Increasing rates of lead ions caused a decrease in the content of water in the leaves of the studied variety of common osier. The largest decrease in the index of relative water content by 23.6% was observed when the largest rate of lead ions in relation to the control plants was applied (Figure 2). The observed changes in the intensity of the studied physiological parameters could result, under unfavourable conditions, in both stress and repair mechanisms [Starck 2002].

On the basis of the parameters of the gas exchange the analysis of the linear correlation was carried out between the assimilation of CO_2 (A) and transpiration (E) and the concentration of carbon dioxide in intercellular spaces (c_i), stomatal

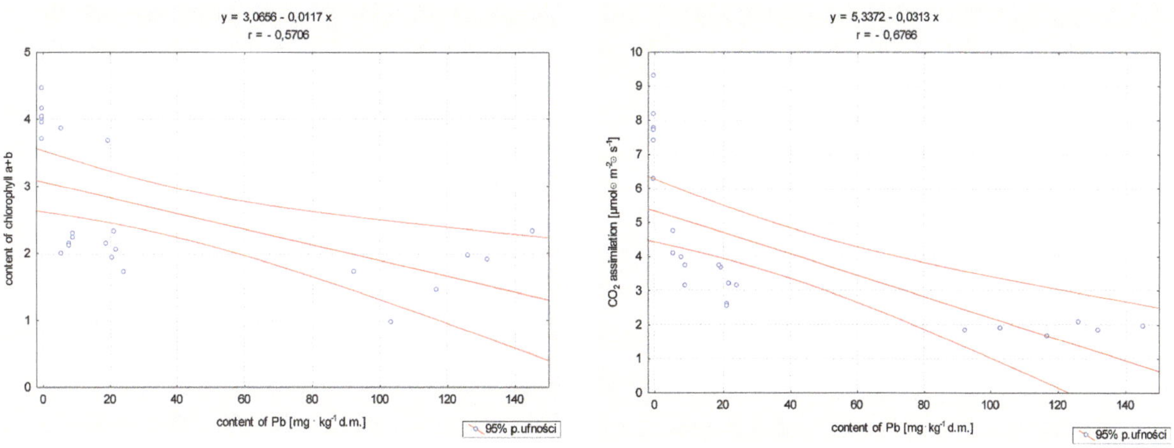

Figure 1. Correlation between the content of photosynthetic pigments [mg·g^{-1} FW], CO_2 assimilation [μmol·m^{-2}·s^{-1}] and the content of lead in the leaves of common osier

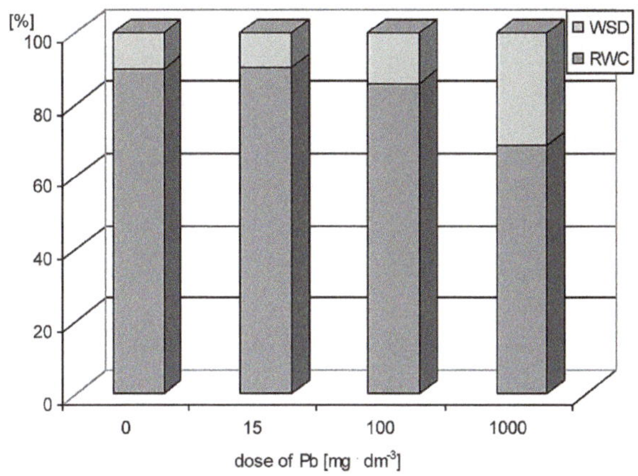

Figure 2. Water indices (%) of common osier – of clone Jorr in relation to a dose of lead

conductance for CO_2 (g_c) stomatal conductance for water vapor (g_s).

In the carried out studies the CO_2 assimilation and transpiration of common osier var. Jorr was limited to a large extent. Significant linear positive relationships between the parameters of gas exchange occurred in the studied variety of common osier both under control conditions and under conditions of contamination with lead ions. The relationship between these parameters increased along with the increase in the rate of Pb^{2+} (Table 2). The analysis of correlation coefficients for the relationship between transpiration and stomatal conductance for the water vapor showed the largest significant coefficient of correlation in Jorr when the highest rate of lead ions was applied (Table 2). Straight line relationship between transpiration and conductance of stomatal apparatuses in common osier under stressful conditions were shown by Wróbel et al. [2006]. The closing of stomatal apparatuses is a response of plants to many stress factors. It runs according to a complicated mechanism related to the phosphorylation of some proteins and the increase in the amount of abscisic acid and it leads to accumulation of H_2O_2 and activation of the calcium channel in membranes of the stomatal cells. This results in a decrease in the rate of transpiration [Mott and Parkhurst 1991, Hinckly et al. 1994, Laloi et al. 2004].

Table 2. Correlation between the parameters of gaseous exchange processes of common osier clone Jorr

Dose of Pb^{2+} [mg · dm^{-3}]	Variables		Correlation coefficients (r)
	y	x	
0	CO_2 assimilation (A)	Stomatal conductance for CO_2 (g_c)	r = 0.434*
	Transpiration (E)	Stomatal conductance for water vapor (g_s)	r = 0.496*
15	CO_2 assimilation (A)	Stomatal conductance for CO_2 (g_c)	r = 0.422*
	Transpiration (E)	Stomatal conductance for water vapor (g_s)	r = 0.564*
100	CO_2 assimilation (A)	Stomatal conductance for CO_2 (g_c)	r = 0.685*
	Transpiration (E)	Stomatal conductance for water vapor (g_s)	r = 0.703*
1000	CO_2 assimilation (A)	Stomatal conductance for CO_2 (g_c)	r = 0.733*
	Transpiration (E)	Stomatal conductance for water vapor (g_s)	r = 0.794*

The obtained results of the studied physiological parameters can be useful for the evaluation of the resistance of the studied clone of willow to the stress caused by lead, and their usefulness for reclamation of the antropogenically degraded land. The Jorr variety of common osier was characterised by good values of the determined physiological parameters under stressful conditions at a large accumulation of lead. This suggests that it shows quite a high tolerance to the stress caused by contamination of the subsoil with lead.

CONCLUSIONS

1. The rate of the assimilation CO_2, transpiration, and the content of assimilation pigments in common osier was differentiated by the rate of lead ions in the nutrient solution.
2. The CO_2 assimilation and transpiration of common osier var. Jorr were significantly limited stomatally.
3. A significant correlations were dependence was shown between the assimilation of CO_2, the content of total chlorophyll and the concentration of lead in the leaves of common osier, clone Jorr.
4. The addition of lead ions to the nutrient solution caused a decrease in the index of relative water content (RWC) and a deficit of water saturation in the leaves of the studied variety of common osier.
5. The obtained results suggest that the determined physiological parameters can be used for a thorough evaluation of the influence of common osier var. Jorr on the stress caused by the presence of lead ions in the subsoil.

REFERENCES

1. Arnon D.J., Allen M.B., Whatley F. 1956. Photosynthesis by isolated chloroplast. IV General concept and comparison of three photochemical reactions. Biochim. Biophys. Acta. 20, 449–461.
2. Bandurska H. 1991. The effect of proline on nitra te reductase activity in water – stressed baryle leaves. Acta Physiol. Plant., 1, 3–11.
3. Baran S. 2000. Ocena stanu degradacji i rekultywacji gleb. Lublin.
4. Becerril J.M., Munoz-Rueda A., Aparicio-Tejo P., Gonzales-Murua C. 1988. The effects of cadmium and lead on photosynthetic elektron transport In clover and lucerne. Plant Physiol. Biochem., 26, 357–363.
5. Deng H., Ye Z.H., Wong M.H. 2006. Lead and zinc accumulation and tolerance in populations of six wetland plants", Enviromental Pollution., 141, 69–80.
6. Eltop L., Bron G., Joachim O., Brinkmann K. 1991. Lead tolerance of *Betula* and *Salix* in themining area of mechernich. Plant and Soil., 131, 275–285.
7. Górny A.G., Garczyński S. 2002. Genotypie and nutrition-dependent variation in water use efficiency and photosythetic activity of leaves in winter wheat (*Triticum aesativum* L.). J. Appl. Genet., 43(2), 145–160.
8. Hager A., Mayer-Berthenrath T. 1966. Die Isolierung und quanttative Bestimung der Carotenoide und Chlorophyll von Blatern, Algen und isolierten Chloroplasten mit Hilfe Dunnschicht-chromatographischer Methoden. Planta., 69, 198–217.
9. Hall J.L. 2002. Cellular mechanism for heavy metal detoxification and tolerance. J. of Exp. Bot., 53, 1–11.
10. Hermle S., Günthardt-Goerg M., Schulin R. 2006. Effects of metal-contaminated soil on the performance of Young trees growing in model ecosystems under field conditions. Enviromental Pollution, 144, 703–714.
11. Hinckly T.M., Braatne J.H. 1994. Stomata. [In:] R. Wilkinson, K. Marcel (Eds.), Plant – environment interaction. New York, 323–355.
12. Laloi C., Apel K., Danon A. 2004. Reactive oxygen signaling: the latest news. Curr. Opin. Plant Biol., 7, 323–328.
13. Lichtenthaler H.K. 1987. Chlorophylls and carotenoids: Pigments of photosynthetic biomembranes. Methods Enzymol., 148, 350–380.
14. Malinowska K., Wróbel J., Malinowski R., Stera A. 2010a The toxic impact of lead on the physiological parameters of basket willow (*Salix viminalis* L.). Agricultural Sciences, 2(4), 45–49.
15. Malinowska K., Mikiciuk M., Berdzik A. 2010b. Zmiany wybranych parametrów fizjologicznych wierzby wiciowej (*Salix viminalis* L.) wywołane zróżnicowanym stężeniem kadmu w podłożu. Ochr. Środ. i Zasob. Natur., 42, 24–32.
16. Malinowska K., Mikiciuk M., Wróbel J., Czyżyk E. 2011. Influence of cadmium on physiological parameters of clone Jorr of basket willow (*Salix viminalis* L.) from aquatic cultures. Ecol. Chem. and Eng. A., 18(5-6), 763–770.
17. Malinowska K., Smolik B. 2011. Physiological and biochemical activity of spring wheat (*Triticum vulgare*) under the conditions of stress caused by cadmium. Envir. Prot. Eng., 37(1), 73–81.
18. Mott K.A., Parkhurst D.F. 1991. Stomatal responsem to humidity In Ir and helom. Plant Cell Environ., 14, 509–515.
19. Panda S., Chaudhury I., Khan M. 2003. Heavy metals induce lipid peroxidation and affect antioxi-

dants in wheat leaves. Biol. Plant., 46(2), 289–294.

20. Schützendübel A., Schwanz P., Teichmann T., Gross K., Langenfeld-Heyser R., Godbol D.L., Polle A. 2001. Cadmium-induced changes in antioxidative systems, hydrogen peroxide content, and differentiation in Scots Pine roots. Plant Physiol., 127, 887–898.
21. Šottniková A., Lunáčková L., Masarovičová E., Lux A., Streško V. 2003. Changes in the rooting and growth of willows and populars induced by cadmium. Biologia Plantarum., 46(1), 129–131.
22. Starck Z. 2002. Mechanizmy integracji procesów fotosyntezy i dystrybucji biomasy w niekorzystnych warunkach środowiska. Zesz. Prob. Post. Nauk Rol., 481, 113–123.
23. Wróbel J., Mikiciuk M., Stolarska A. 2006. Wpływ warunków zasolenia gleby na aktywność wymiany gazowej u trzech klonów wierzby wiciowej (*Salix viminalis* L.). Zesz. Prob. Post. Nauk Rol., 509, 269–281.
24. Wrzosek J., Gawroński S., Gworek B. 2008. Zastosowanie roślin energetycznych w technologii fitoremediacji. Ochr. Środ. i Zasob. Natur., 37, 139–151.
25. Zemleduch A., Tomaszewska B. 2007. Mechanizmy, procesy i oddziaływanie w fitoremediacji. Kosmos, Prob. Nauk Biol., 56(3-4), 393–407.
26. Zenk M. 1966. Heavy metal detoxification in plants – a review. Gene., 179, 21–30.
27. Verma S., Dubey R.S. 2003. Lead toxicity induces lipid peroxidation and alters the activities of antioxidant enzymes in growing rice plants. Plant Sci., 164(4), 645–655.

21

EFFECT OF FLY ASHES AND SEWAGE SLUDGE ON Fe, Mn, Al, Si AND Co UPTAKE BY GRASS MIXTURE

Jacek Antonkiewicz[1]

[1] Department of Agricultural and Environmental Chemistry, University of Agriculture in Krakow, 21 Mickiewicza Ave., 31-120 Krakow, Poland, e-mail: rrantonk@cyf-kr.edu.pl

ABSTRACT

Application of sewage sludge for environmental management of fly ashes landfill site affects chemical composition of plants. The aim of the present investigations was learning the effect of growing doses of municipal sewage sludge on the yield and uptake of Fe, Mn, Al, Si and Co by grass mixture used for environmental management of fly ashes landfill. The experimental design comprised of 5 objects differing by a dose of municipal sewage sludge supplied per 1 hectare: I. control, II. 25 t d.m., III. 50 t d.m., IV. 75 t d.m. and V. 100 t d.m. Application of sewage sludge resulted in the increase in yield. The content of analyzed elements in the grass mixture depended significantly on sewage sludge dose. Increasing doses of sewage sludge caused marked increase in Mn and Co contents, while they decreased Fe, Al and Si contents in the grass mixture. It was found that growing doses of sewage sludge caused an improvement of Fe to Mn ratio value in the grass mixture. Assessing the element content in the grass mixture in the view of forage value, it was found that Fe and Mn content did not meet the optimal value. Si content in plants was below the optimal value.

Keywords: fly ashes, municipal sewage sludge, grass mixture, yield, Fe, Mn, Al, Si, Co, uptake, quantitative ratios.

INTRODUCTION

In view of their heavy metal concentrations, a considerable amount of generated municipal sewage sludge qualifies for environmental management [Rozporządzenie... 2010]. In case of environmental management of fly ashes in landfill sites, the problem of creating environmental barrier relies on securing appropriate conditions for the development of vegetation introduced to the landfill [Antonkiewicz 2009, 2011]. The optimal technological solution seems to be covering the top layer of fly ashes landfill with municipal sewage sludge or its mixing with the top ash layer [Samaras et al. 2008, Klimont 2011]. Sewage sludge used in ashes landfill joins volatile parts of the ash preventing its dusting and supplies nutrients, including microelements, to plants used for the reclamation. Fly ashes subjected to biological reclamation also provide a valuable source of nutrients enabling growth and development of plants, particularly at their early development stages [Gilewska, Przybyła 2001, Bielińska et al. 2009, Rosik-Dulewska et al. 2011].

Presented research aimed to identify the impact of increasing doses of municipal sewage sludge on the yield and Fe, Mn, Al, Si and Co uptake by a grass mixture used for environmental management of fly ashes landfill site.

The present paper on the yielding and uptake of Fe, Mn, Al, Si and Co is a continuation of research on the effect of municipal sewage sludge on the yield, content and uptake of heavy metals by grass mixture [Antonkiewicz 2009].

MATERIALS AND METHODS

The research on the effect of increasing doses of municipal sewage sludge on the yield and microelement uptake by a grass mixture was conducted in 2005–2007 as a field experiment, on fly ashes landfill site owned by Dwory S.A. Chemical Enterprise in Oświęcim. A one factor field experiment was conducted in a randomized block design. The plot area was 10 m^2. The landfill possesses also a closed ash quarter filled with fly ashes, i.e. an ash-slag mixture from wet deposition of fly ashes with catalogue number 10.01 80 [Rozporządzenie... 2001]. The deposited ash, i.e. ash and slag mixture was supplied by hydrotransport and deposited in the ash quarter. The experimental design comprised 5 objects differing by a dose of supplied municipal sewage sludge per 1 hectare: I. control, II. 25 t d.m., III. 50 t d.m., IV. 75 t d.m. and V. 100 t d.m. The maximum dose of municipal sewage sludge – 100 t d.m.$\cdot$$ha^{-1}$ could be applied to the ground reclamation for agronomic and non-agronomic purposes, including for example reclamation of fly ashes landfill sites, in compliance with the Regulation of the Minister of Environment on municipal sewage sludge dated 1 August 2002, in force at that time [Rozporządzenie... 2002].

Municipal sewage sludge was applied to the surface two weeks prior to the grass mixture sowing. Previous research has shown that legumes (*Fabaceae*) were covering fly ashes less than ash-sludge mixtures [Antonkiewicz, Radkowski 2006], so in the field experiment set up on fly ashes landfill a mixture composed exclusively from grass plants was used. The grass mixture sown there was composed of *Festuca rubra* L., 'Brudzyńska' cv. (20%), *Festuca arundinacea* (Schreb.), 'Skarpa' cv. (25%), *Festuca pratensis* (L.), 'Skiba' cv. (25%), *Poa pratensis* (L), 'Skiz' cv. (5%) and *Lolium perenne* (L.), 'Solen' cv. (25 %).

The grass mixture was sown by hand in the first decade of August 2005 in the amount of 53 kg$\cdot$$ha^{-1}$. The norm of grass sowing was increased by 50% because at the beginning grasses germinate poorly in alkaline soil pH [Rogalski, Kardyńska 1999]. At the beginning, the sown grass mixture was sprinkled with water to the depth of 2 cm as the substratum was drying up. 3 weeks after sowing, plant emergencies were watched. No mineral NPK fertilization was applied in the field experiment. In the first year of vegetation the plants were cut once, in the first decade of October (1 regrowth), whereas in subsequent years it was cut once, in the first decade of September. A single harvest of the mixture was due to the conditions on the experiment site, among others intensive dusting of the landfill, which reduced the plant yielding, especially in the first and second year of the experiment.

After harvest the grass mixture was dried at 70 °C in a drier with forced air flow and the yield of dry mass was determined. Samples of the plant material were dry-mineralized in a muffle furnace at 450 °C and the obtained ash was dissolved in 20% nitric acid. The paper presents average yields of the grass mixture from respective years 2005–2007 and mean weighted contents of the elements for the investigated period. Approximate to total element contents were assessed in the ash and sewage sludge after organic matter incineration and digestion in the mixture of $HClO_4$ and HNO_3 (3:2, *v/v*) acids [Ostrowska et al. 1991]. Concentrations of Fe, Mn, Al, Si and Co in the ash and sludge, as well as in the plant material mineralisates were assessed by means of atomic emission spectrometer with inductively coupled argon plasma (ICP-AES).

RESULTS AND DISCUSSION

Basic physicochemical properties of wastes were presented in Table 1. Analysis of granulation revealed that considering the granulometric composition, the deposited fly ash matched ordinary dust [Systematyka... 1989]. The content of organic carbon and total nitrogen assessed in fly ash reveal the presence of unburned coal residue [Gilewska, Przybyła 2001, Su, Wong 2003].

The contents of Fe, Mn, Al, Si and Co in fly ash were higher by 108%, 40%, 117%, 165% and 298%, respectively, in comparison with these elements' concentrations in the municipal sewage sludge (Table 1).

Table 1. Properties of waste used in field experiment

Parameter		Unit	Ash	Sludge
pH	in H_2O	pH	8.83	6.72
	in KCl 1 mol$\cdot$$dm^{-3}$		8.74	6.25
Texture		-	Silt	-
			Content	
Dry mass		[%]	77.15	20.56
Organic carbon		[g $\cdot$ kg^{-1} d.m.]	40.36	268.46
Total nitrogen			1.55	38.91
Available form of	P	[mg $\cdot$ kg^{-1} d.m.]	145.98	648.96
	K		121.62	412.48
	Mg		453.97	865.87
Total content of	Iron		64436.0	31000.0
	Manganese		1236.35	569.47
	Aluminium		22655.0	8562.0
	Silicon		3563.0	895.0
	Cobalt		13.36	9.56

Municipal sewage sludge applied on fly ash landfill site revealed a slightly acid pH, whereas total organic carbon and nitrogen content was 6.6 and 25-fold higher in comparison with fly ash (Table 1). Municipal sewage sludge was a rich source of available macroelements, because of very high contents of available P, K and Mg forms. Assessment of municipal sewage sludge revealed that in view of heavy metal concentrations it met the requirements for application in agriculture and land reclamation for agronomic and non-agronomic purposes [Rozporządzenie... 2002, Antonkiewicz 2009, Rozporządzenie... 2010].

Yield of grass mixture

The effect of increasing doses of municipal sewage sludge on the yield of grass mixture cultivated on fly ash landfill site was presented in the previous paper [Antonkiewicz 2009]. The yield of grass mixture obtained in the field experiment was presented for individual years and as an average value for the period of investigations, i.e. 2005–2007 (Table 2).

Average yield of grass mixture dry mass obtained in the years 2005–2007 on fly ashes landfill fluctuated from 0.69 to 4.03 t d.m.·ha^{-1} depending on the object (Table 2). The lowest yield of the mixture was obtained on the control where no sewage sludge was used. Obtained yield of the grass mixture cultivated solely on fly ash may be explained by disadvantageous physicochemical properties of the substratum and nutrient deficiency in comparison with municipal sewage sludge treatments (Table 1).

Application of municipal sewage sludge on fly ashes landfill resulted in a significant increase in grass mixture yield. The smallest dose of sewage sludge, i.e. 25 t d.m.·ha^{-1} caused an increase in the grass mixture yield by 94% in comparison with the values obtained in the control. Doubling the sewage sludge dose (50 t d.m.·ha^{-1}) led to an increase in yield by 230% as compared with the control. Successive doses of sewage sludge also positively affected the amount of yield. The largest yield of grass mixture was obtained on the object where 100 t d.m.·ha^{-1} of sewage sludge was applied. The yield of grass mixture from this object was over 480% higher in comparison with the values obtained in the control. The research conducted on fly ashes landfill site demonstrated the best fertilizer value of the object where 100 t d.m.·ha^{-1} of sewage sludge was applied.

Table 2. Yield of grass mixture [t d.m.·ha^{-1}]

Object	Sludge dose [t d.m.·ha^{-1}]	Year of vegetation 2005	2006	2007	Average
I	Control	0.09	0.88	1.10	0.69
II	25	0.27	1.62	2.10	1.33
III	50	0.44	2.64	3.76	2.28
IV	75	0.78	3.36	4.71	2.95
V	100	1.13	4.58	6.39	4.03
RSD [%]		76.62	55.51	57.99	58.44
$LSD_{(\alpha=0.05)}$		0.10	0.27	0.41	0.220

Explanation: RSD – relative standard deviation.

The yield of the grass mixture obtained in respective years of experiment 2005–2007 was greatly diversified (Table 2). The lowest yields were produced in the first year of investigations, which should be explained by a long period of grass mixture emergencies and incomplete period of its vegetation. Much larger grass mixture yield was obtained in the second year and the largest in the third year of the research. Depending on the object, the yields obtained in the third year of research were between 5.7 and 12.2 times higher in comparison with those in the first year of vegetation. The highest yield increment, in comparison with the first year of vegetation, was registered in the control object and the lowest with the highest dose of sewage sludge. The highest yield increment was noted for grass mixture cultivated exclusively on fly ash, after the plant spreading.

Element content in the grass mixture

Fly ash and municipal sewage sludge were a potential source of Fe, Mn, Al, Si and Co for the grass mixture cultivated on the landfill (Table 1). The element content in grass mixture, depending on sewage sludge dose, ranged as follows: 159.65–1072.91 mg Fe, 21.93–47.68 mg Mn, 84.21–123.69 mg Al, 55.11–80.51 mg Si and 0.09–0.29 mg Co·kg^{-1} d.m. (Figure 1). Research has shown that the content of these elements in the grass mixture was diversified. The highest diversification was observed for Fe (RSD = 59%) and Mn (RSD = 46%), while the lowest for silicon and iron (RSD = 14%). The experiment demonstrated that municipal sewage sludge applied on the fly ashes landfill site led to a significant increase in Mn and Co contents in the grass mixture (Figure 1). The highest increment of Mn and Co content in the grass mixture was registered at the dose of 100 t d.m.·ha^{-1}. Increase in Co content at the application of the highest dose of sewage sludge exceeded 243%, whereas for Mn the increase was over 117% in relation to the control.

An opposite relationship was noted for Fe, Al and Si. The highest Fe, Al and Si content was noted in the grass mixture cultivated solely on fly ash, on the control not fertilized with sewage sludge. Increasing doses of sewage sludge caused a marked decrease in Fe, Al and Si content in the grass mixture. The content of the above mentioned elements in the grass mixture after the application of the highest sludge dose decreased 6.72, 1.47 and 1.46-fold, respectively, in comparison with the control.

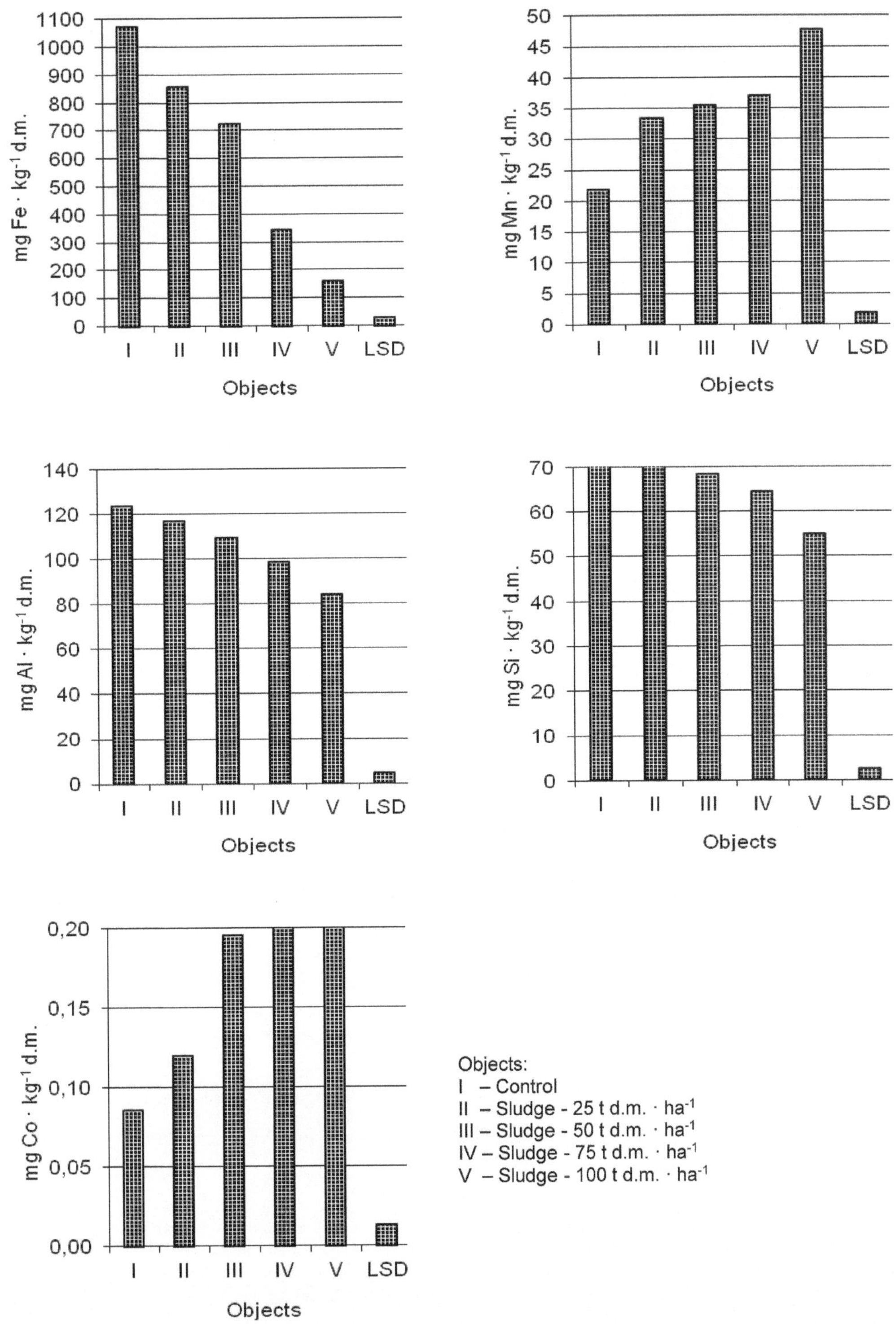

Figure 1. Content of elements in grass mixture

Element uptake

The paper presents a summary uptake of elements for the three-year research period (2005–2007). The element quantity absorbed from the respective objects (from the landfill) depended on the crop yield and individual element concentration in the yield (Figure 2). Considering the analyzed elements, Fe was absorbed in the greatest amounts, irrespective of the object, then Al, Si and Mn, whereas Co uptake was the lowest. The highest iron uptake by grass mixture was registered on the object where 50 t d.m.·ha^{-1} of sewage sludge was applied, whereas the subsequent increasing doses of sewage sludge caused a diminished uptake of this microelement. A dose of 100 t d.m.·ha^{-1} of municipal sewage sludge applied on the landfill significantly and to the greatest extent

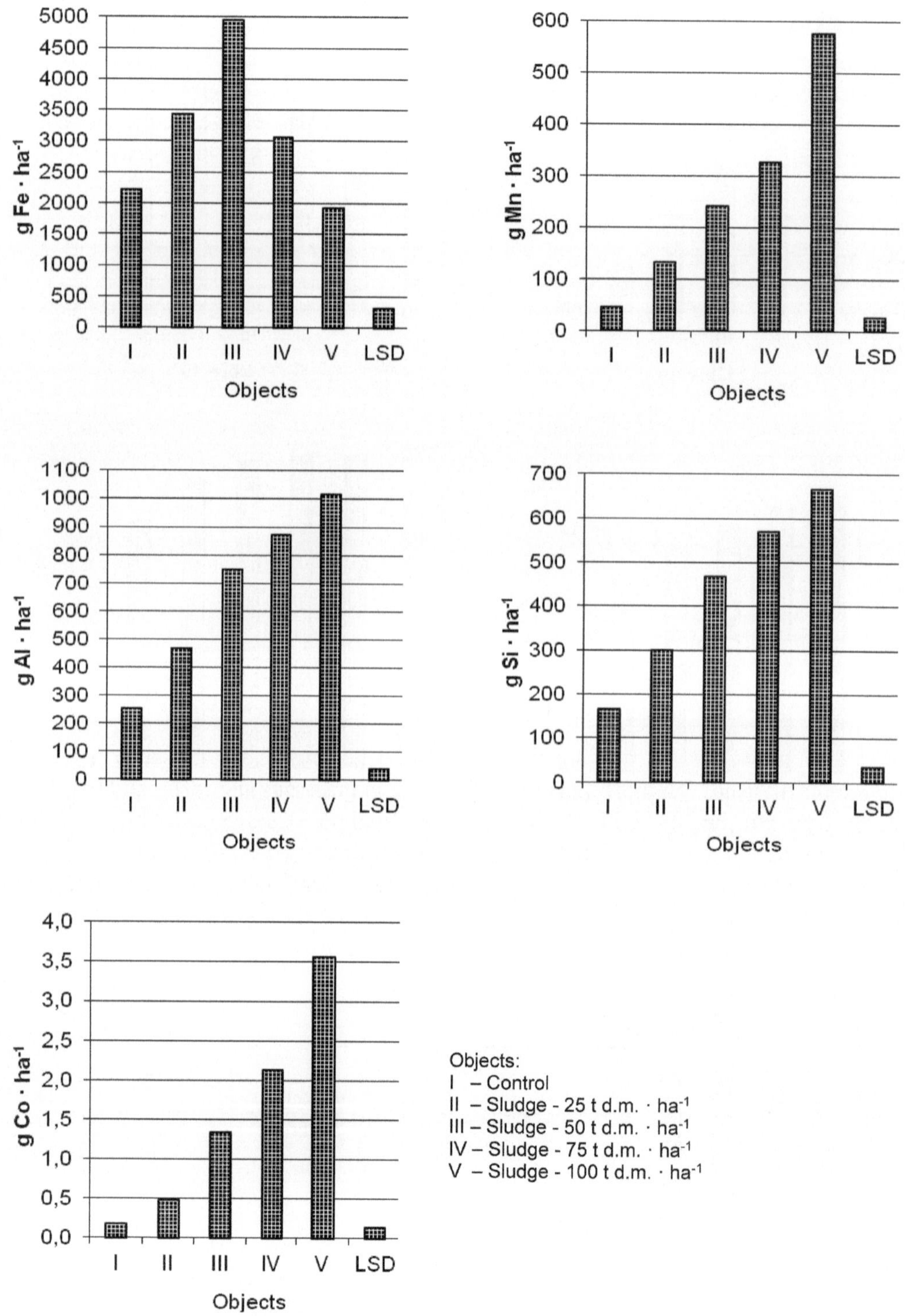

Figure 2. Uptake of elements by grass mixture

increased Mn, Al, Si and Co uptake by grass mixture. Increased absorption of Mn, Al, Si and Co following this dose application was 11.74, 2.98, 3.01 and 19.04-fold, respectively, in comparison with the control, i.e. fly ash not receiving sewage sludge. The research has shown that the lower dose of sewage sludge, the lower uptake of these elements by grass mixture. Reduced element uptake by the grass mixture was connected mainly with diminishing yield (Table 2). The lowest removal of these elements was noted in the control. The lowest uptake of microelements is explained by the fact that the grass mixture cultivated on the ash in the control produced lower yield and in case of Mn and Co low contents of these metals were noted in plant yield.

Weight relationships between elements in the grass mixture

For the cognitive reasons of the paper, the relationships between the analyzed elements in the grass mixture were presented. Increased doses of sewage sludge applied on fly ashes landfill clearly modified relationships between the studied elements (Table 3).

A considerable variation range of mutual relationships between the elements in the grass mixture was observed in the experiment, which indicates that grasses cultivated on wastes reveal a notable tolerance to quantitative relationships between these elements. The highest variation was assessed for Fe:Mn and Fe:Co ratios, whereas the lowest one for Al:Si ratio.

Application of increasing doses of municipal sewage sludge on fly ashes landfill caused a decline in values of Fe:Mn, Fe:Al, Fe:Si, Fe:Co and Mn:Co weight ratios in the grass mixture in comparison with the control. An opposite relationship, i.e. increasing values of weight ratios in comparison with noted in the control was registered for Mn:Al and Mn:Si ratios. In case of Al:Si ratio it was found that sewage sludge addition to fly ashes did not have any marked influence on this ratio value in the grass mixture (Table 3).

A proper growth of forage crops requires maintaining relatively stable and determined Fe:Mn ratio. It was assumed that the optimum Fe:Mn ratio is 1.5–2.5:1. According to Falkowski et al. [2000] below this ratio value 1.5, symptoms of manganese toxicity and iron deficiency are apparent, whereas when the Fe:Mn ratio exceeds 2.5, a manganese deficiency and iron excess appear in a plant.

Analyzing the value of Fe:Mn ratio in the grass mixture, it was determined that increasing doses of municipal sewage sludge caused an improvement of the ratio value between these elements. After the application of the highest dose of sewage sludge (100 t d.m.·ha^{-1}) the value of this ratio was most approximate to the optimal, however it remained below the optimal value. The value of Fe:Mn ratio in the grass mixture cultivated on fly ashes landfill evidenced Fe excess, but a deficiency of Mn. Assessing element content in the grass mixture from the point of forage value, it was found that it did not meet the optimal value for Fe and Mn.

Fly ashes landfills require a necessary reclamation through initiation of pedogenic processes and forming a vegetation cover using a method of a proper plant species selection [Antonkiewicz, Radkowski 2006, Klimont 2007]. In environmental management of fly ash landfills grasses are regarded as pioneer plants [Góral, Rola 2001]. Author's own research has shown that introduction of municipal sewage sludge causes a significant increase in grass mixture yield, and therefore environmental management of fly ash landfill site.

According to Gilewska and Przybyła [2001], application of municipal sewage sludge on ash rock improves its physical and chemical properties through increasing the content of among

Table 3. Relation between elements in grass mixture

Object	Sludge dose	Ratio							
	[t d.m. · ha^{-1}]	Fe : Mn	Fe : Al	Fe : Si	Fe : Co	Mn : Al	Mn : Si	Mn : Co	Al : Si
I	Control	48.9	8.7	13.3	12528.7	0,18	0,27	256.1	1,54
II	25	25.6	7.3	11.4	7148.0	0,29	0,44	279.0	1,56
III	50	20.4	6.6	10.6	3706.1	0,33	0,52	181.9	1,60
IV	75	9.3	3.5	5.4	1441.6	0,38	0,58	154.5	1,53
V	100	3.3	1.9	2.9	542.2	0,57	0,87	161.9	1,53
RSD [%]		82.1	50.1	50.3	96.3	41.4	40.5	27.6	1.93

Explanation: RSD – relative standard deviation.

others, microelements. Studies on efficiency of sewage sludge in reclamation of ash landfill sites were also conducted by Baran et al. [2008], who demonstrated an advantageous effect of municipal sewage sludge on the yield and chemical composition of plants.

The present paper focuses on such elements as iron, manganese, aluminum, silica and cobalt, which occur in plants in diversified quantities, but are crucial for their growth and development [Falkowski et al. 2000]. Therefore, for cognitive reasons the element content in the grass mixture was assessed in view of their fodder value. The following optimal values were assumed, below which their deficiency in forage plants occurs: 40–80 mg Fe, 40–70 mg Mn and 0.08–0.16 mg Co·kg^{-1} d.m. Animals fed with fodder abundant in silicon may deposit this element as urinary stones, therefore attempts were made to establish a limit of its safe content in animal feed, regarding 9.0 g Si·kg^{-1} d.m. as permissible content [Falkowski et al. 2000, Wiśniowska-Kielian, Lipiński 2007].

If one assumes the upper limit of Fe content in forage crops as 80 mg·kg^{-1} d.m., the grass mixture cultivated solely on fly ash contained over 13-fold more of this element in comparison with the optimal value. Increasing doses of municipal sewage sludge significantly diminished this microelement content in the grass mixture. The grass mixture cultivated on the highest dose of sewage sludge (100 t d.m.·ha^{-1}) contained over twice more of Fe than the upper limit of the optimal content of this microelement. Mn content in the grass mixture cultivated only on the ash ranged below the optimal, whereas increasing doses of sewage sludge resulted in a growing content of this element, and grass biomass obtained on the highest dose of sewage sludge (100 t d.m.·ha^{-1}) revealed the optimal content of this element. Co content in the grass mixture was above, while Si below the optimal value.

Assessment of chemical composition of the obtained grass mixture meant for reclamation of fly ash landfill site was not positive. The mixture should not be destined for animal feed due to the risk of heavy metals present in it in quantities exceeding the norm, which excludes it's forage use. The obtained yield of the grass mixture should be destined for biomass, used as energy source or for compost production. Moreover, industrial utilization of the plant biomass guarantees excluding heavy metals from the food chain.

Scientific literature [Kalembkiewicz, Sočo 2004, Wojcieszczuk et al. 2009] shows that fly ashes constitute not only a source of heavy metals, but are a valuable source of microelements, although the element solubility and bioavailability is generally low regarding their total content [Su, Wong 2003]. The factor limiting their availability is alkaline pH of ashes. For increasing microelement solubility, and therefore their availability, application of municipal sewage sludge is recommended, which is also a valuable source of microelements for plants [Kalembasa, Wysokiński 2008, Papadimitriou et al. 2008].

It results from the Author's own research that relationships between the elements in the grass mixture were clearly modified by municipal sewage sludge. A change of weight relationships between the elements in the grass mixture was probably caused by the effect of sludge on concentrations of individual elements in the soil solution, and therefore limiting or increasing their availability for grasses [Kalembkiewicz, Sočo 2004], due to acidification and increasing contents of these microelements in the substratum [Antonkiewicz 2009].

CONCLUSIONS

1. Growing doses of sewage sludge significantly increased yields of the grass mixture cultivated on fly ash landfill site.
2. Increasing does of sewage sludge resulted in a marked growth of Mn and Co content in the grass mixture cultivated on fly ash landfill site, whereas they declined its concentrations of Fe, Al and Si.
3. It was determined that sewage sludge supplement on fly ashes landfill led to a greater Fe, Mn, Al, Si and Co uptake by the grass mixture.
4. It was assessed that Fe and Co contents in the grass mixture were above the optimal value, whereas Mn and Si contents were below the optimal value.
5. Applied municipal sewage sludge obviously modified values of weight ratios between the analyzed elements in the grass mixture.

REFERENCES

1. Antonkiewicz J. 2009. An assessment of the environmental usability of furnace ashes and municipal sewage sludges. Zesz. Nauk UR w Krakowie, 454, Seria rozprawy 331, p. 131 [In Polish].

2. Antonkiewicz J. 2011. Assessment of bioavailability of heavy metals in wastes used for biological reclamation of hazardous waste landfill. Zesz. Nauk. UR w Krakowie, 484, Seria rozprawy 358, p. 119 [In Polish].
3. Antonkiewicz J., Radkowski A. 2006. Usability of selected grass and legume species for biological reclamation of ash dumps. Annales UMCS, sec. E, 61, 413-421 [In Polish].
4. Baran S., Wójcikowska-Kapusta A., Żukowska G., Milczarek T. 2008. Influence of the addition of kompost from sewage sludge and sludge-ashes on sorption properties of the soil-less ground formation. Zesz. Probl. Post. Nauk Roln., 533, 49-58. [In Polish].
5. Bielińska E.J., Baran S., Stankowski S. 2009. Assesment concerning usability of fluidal ashes from hard coal for agricultural purposes. Inżynieria Rolnicza, 6(115), 7-15 [In Polish].
6. Falkowski M., Kukułka I., Kozłowski S. 2000. Właściwości chemiczne roślin łąkowych. Wyd. AR Poznań, p. 132 [In Polish].
7. Gilewska M., Przybyła Cz. 2001. Utilization of sewage sludge in recultivation of ash disposal sites. Zesz. Probl. Post. Nauk Roln., 477, 217-222 [In Polish].
8. Góral S., Rola S. 2001. Trawy na popiołach elektrociepłowni nawożonych osadami ściekowymi. Inż. Ekol., 3, 146-150 [In Polish].
9. Kalembasa S., Wysokiński A. 2008. The influence of the addition of different quantities of calcium oxide and ash from brown coal to the waste activated sludges on the content of boron and molybdenum in the tested plants. Zesz. Probl. Post. Nauk Roln., 533, 189-196 [In Polish].
10. Kalembkiewicz J., Sočo E. 2004. Sequential extraction of Cr, Fe, Co, and Ni from industrial ash. Ecol. Chem. Eng., 11(4-5), 347-352.
11. Klimont K. 2007. Ocena przydatności wybranych gatunków roślin użytkowych do rekultywacji terenów zdewastowanych przez przemysł i gospodarkę komunalną. Problemy Inżynierii Rolniczej, 2, 27-36 [In Polish].
12. Klimont K. 2011. Reclamation efficiency of the sewage sludge on soil-less substrate of post-flotation lime and incineration ashes. Problemy Inżynierii Rolniczej, 2, 165-176 [In Polish].
13. Ostrowska A., Gawliński S., Szczubiałka Z. 1991. Methods of analysis and assessment of soil and plant properties. A Catalogue. Wyd. IOŚ, Warsaw, p. 334 [In Polish].
14. Papadimitriou C.A., Haritou I., Samaras P., Zouboulis A.I. 2008. Evaluation of leaching and ecotoxicological properties of sewage sludge-fly ash mixtures. Environ. Res., 106, 340-348.
15. Rogalski M., Kardyńska S. 1999. Initial growth and development of *Festulolium* and *lolium perenne* on ashes from the Dolna Odra power plant. Fol. Univ. Agric. Stetin. 197, Agricultura 75, 263-266 [In Polish].
16. Rosik-Dulewska Cz., Karwaczyńska U., Ciesielczuk T. 2011. Possibilities of utilization of organic and mineral waste according to rules of environmental protection. Rocznik Ochrona Środowiska, 13, 361-376 [In Polish].
17. Rozporządzenie Ministra Środowiska z dnia 1 sierpnia 2002 r. w sprawie komunalnych osadów ściekowych. Dz. U. RP 2002, Nr 134, poz. 1140. (Nie obowiązuje) [In Polish].
18. Rozporządzenie Ministra Środowiska z dnia 13 lipca 2010 r. w sprawie komunalnych osadów ściekowych. Dz. U. RP, Nr 137, poz. 924 [In Polish].
19. Rozporządzenie Ministra Środowiska z dnia 27 września 2001 r. w sprawie katalogu odpadów. Dz. U. RP 2001, Nr 112, poz. 1206 [In Polish].
20. Samaras P., Papadimitriou C.A., Haritou I., Zouboulis A.I. 2008. Investigation of sewage Sludge stabilization potential by the addition of fly ash and lime. J. Hazard. Mater., 154, 1052-1059.
21. Su D.C., Wong J.W.C. 2003. Chemical speciation and phytoavailability of Zn, Cu, Ni and Cd in soil amended with fly ash-stabilized sewage sludge. Environ. Int., 29, 895-900.
22. Systematyka gleb Polski. PTG. 1989. Roczn. Glebozn., 40(3/4), 1-150 [In Polish].
23. Wiśniowska-Kielian B., Lipiński W. (Red.). 2007. Ocena składu chemicznego roślin. Wyd. Polskie Towarzystwo Inżynierii Ekologicznej, Krajowa Stacja Chemiczno-Rolnicza. Kraków-Warszawa-Wrocław, 2007, p. 57 [In Polish].
24. Wojcieszczuk T., Meller E., Sammel A., Stankowski S. 2009. The content and dissolubility of some chemical components in coal ashes of different origin. Zesz. Probl. Post. Nauk Roln., 535, 483-493 [In Polish].

22

ASSESSMENT OF PAHS AND SELECTED PESTICIDES IN SHALLOW GROUNDWATER IN THE HIGHEST PROTECTED AREAS IN THE OPOLE REGION, POLAND

Mariusz Głowacki[1], Tomasz Ciesielczuk[1]

[1] Department of Land Protection, Opole University, Oleska 22, 45-052 Opole, Poland, e-mail: tciesielczuk@uni.opole.pl

Abstract

The ground water quality was determined after the analyses of water samples from 18 wells. The wells were in the Groundwater Area with the Highest Protection (Triassic water, Opole region, Poland), rural build up. The water table level was low: 0.5 – 18.0 m below the ground surface level (except for one artesian well). The following parameters were determined: pH, EC, colour, ammonium, nitrite, nitrate, dissolved orthophosphate, total phosphorus, dissolved oxygen, BOD, COD-Mn, COD-Cr, humic substances, chloride, sulphate, total hardness, alkalinity, dry residue PAHs (16 compounds), pesticides (6 compounds), however, only selected data were presented in this paper. In all the analysed water samples chloro-organic pesticides were observed. The analysed water contained heptachlor in the highest concentrations of 15.97 mg/dm^3. Good quality water must not include concentrations higher than 0.5 mg/dm^3 of heptachlor. However, the concentration was circa 32 times higher than this value. The second pesticide determining poor water quality is dieldrin. This compound in the investigated groundwater was 1.94 mg/dm^3 – 4 times higher than the limit for acceptable quality ground water. The concentration of pesticides also changed over the course of the research; the concentration in the analysed groundwater in the same well changed quite dramatically over a period of 1 year. Although PAHs and pesticides are potentially toxic for biological organisms they do exist in the environment as a product of the natural biological transformation of organic matter. The noted concentrations and compositions of PAH compounds were different to natural PAHs. It confirms the fact that agricultural activity influences groundwater quality.

Keywords: reservoir, groundwater, PAH, pesticides.

INTRODUCTION

Not only is waste disposal considered a critical factor in environmental degradation. Agricultural activities can adversely affect the environmental quality as a result of the emissions of nutrients and organic compounds [Fianko et al. 2009, Jayasekera et al. 2011]. It is also known as a non-point-source pollution [Ritter et al. 2002]. Persistent organic pollutants in a form of PAHs are present in all natural environmental components [Ake et al. 2003, Owabor et al. 2010, Pranagal and Oleszczuk 2007]. They may be of anthropogenic (mainly combustion processes) or natural (volcano eruptions, organic matter decomposition) origin [Lazzari et al. 1999, Thiele and Brummer 2002, Wilcke et al. 1999]. The highest amount can be traced in soil (about 80%) and bottom sediments [Oleszczuk et al. 2007]. PAHs are bad solvents in water but they can translocate in this environment because other organic compounds, existing as effluents, can take a role of a solvent [Włodarczyk-Makuła et al. 2003]. Pesticides used in agricultural areas can also be present in ground water and soil [Geyikci and Büyükgüngör 2011, Swarcewicz and Gregorczyk 2011]. This is a result of high stability, low microbial degradation and the sorption of these compounds in soil. Migration of PAHs and pesticides with humic substances in vertical water streams of the soil profile is a fundamental process. This could be a primary source of groundwater contamination by organic compounds [Pisarek and Głowacki 2005]. The compounds contained in groundwater, sometimes used as drinking water, are dangerous to human health because of the possibility of a mutagenic and carcinogenic reaction [Owabor et al. 2010, Ritter et al. 2002].

The aim of this paper is to determine the concentration of PAHs and pesticides in groundwater as a result of co-transport with humic substances contained in soil solutions.

MATERIALS AND METHODS

The research area was localized to the biggest ground water reservoir in the Opole region (south-west part of Poland) – The Main Groundwater Reservoir No. 333 (MRGW No 333) [Stasko and Wcislo 2006]. This region is situated in the western part of the Silesian Upland with post-volcanic rocks. An erupting volcano went through Triassic limestone sediments and formed a basaltic cone on towering limestone. Nowadays, this area is protected and known as "St. Anna Mountain" Landscape-Protection Park. The towering of limestone layers and their penetration by non-porous, little eroding basaltic rocks created good conditions for the karst phenomena [Flaczyk 1987]. The main water level of the reservoir is in the Triassic limestone [Kryza and Stasko 2000]. Total water capacity MRGW No. 333 was calculated at about 250 000 m^3/d, the reservoir being about 40 km long [Szczygieł et al. 2006]. The bottom of the mountain and the adjacent areas are used for agricultural purposes - the farmers grow various plants and breed stock animals.

The research from wells situated on the Main Groundwater Reservoir is presented in Figure 1.

Polycyclic aromatic hydrocarbons (PAHs) were examined in groundwater samples taken from wells localized on the main groundwater reservoir MRGW No 333 (Main Reservoir of Groundwater No 333) lower Silesia region (Poland). Wells 7 and 13 were removed from calculations due to incomplete results.

Preparation of Groundwater

Water samples were taken from wells and collected in amber glass bottles. Preservation was achieved by an addition of methanol. All samples were stored in a dark place without access to sunlight before analysis. The PAH extraction from the samples was performed with dichlomethane (GC grade) by 4-hour shaking. Dichlomethane – water sample ratio was 1:10 (v/v). Pesticide extraction from the samples was performed with hexane (GC grade) by 4-hour shaking. Hexane

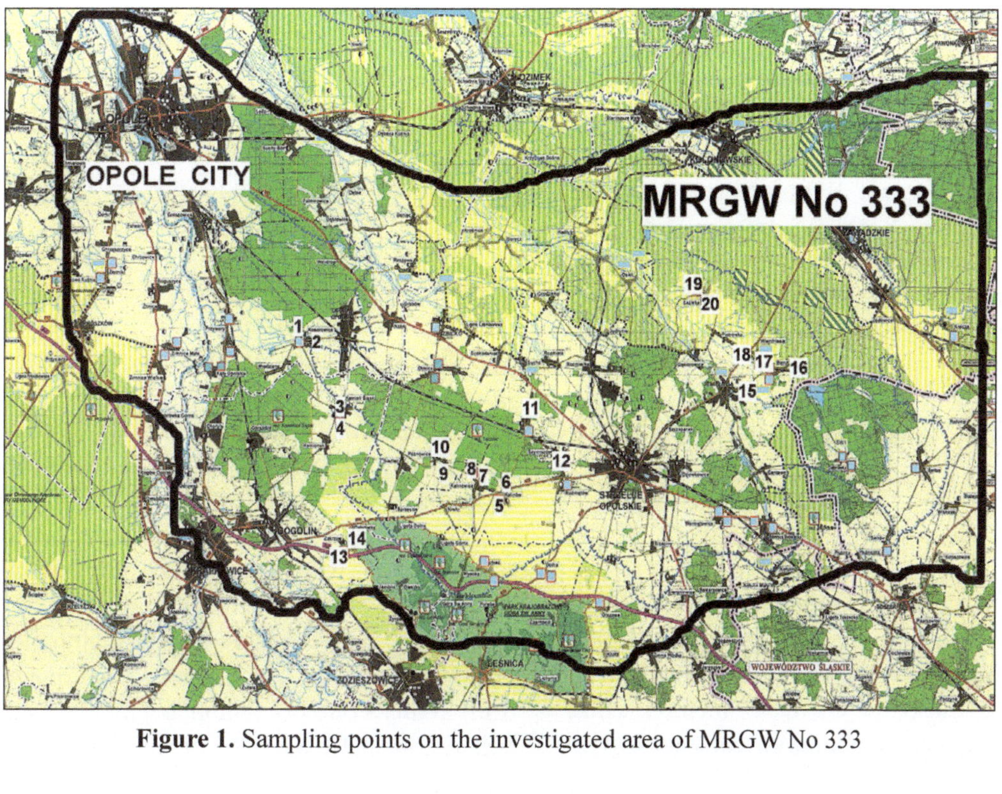

Figure 1. Sampling points on the investigated area of MRGW No 333

– water sample ratio was 1:10 (v/v). After that, the shaken organic layers were separated in separatory funnels, dried by anhydrous Na_2SO_4 and evaporated to about 0.5ml volume. All samples were determined in two replicates.

Samples analysis

In each sample 16 single compounds were determined which are recommended to be monitored by the U.S. Environmental Protection Agency (US-EPA). They include: naphthalene (NAPH), acenaphthylene (ACY), acenaphthene (ACE), fluorene (FLU), phenantrene (PHE), anthracene (ANT), fluoranthene (FLA), pyrene (PYR), benzo[a]anthracene (B[a]ANT), chrysene (CHR), benzo[b,k]fluoranthene (B[b,k]FLA), benzo[a]pyrene (B[a]PYR), dibenzo[a,h]anthracene (D[a,h]ANT), indeno[1,2,3-c,d]pyrene (Ind[1,2,3-c,d]P), benzo[g,h,i]perylene (B[g,h,i] PER). PAHs were determined in unpurified extracts by the GC-FID method using a ZB-5 (5% diphenyl–95% dimethylsiloxane) capillary column with constant carrier (He) flow through the column in the quantity of 1 ml/min. The temperature of the chromatograph (Varian CP 3800) injector was 300 °C while the FID detector was 320 °C. The temperature program of the stove began at 100 °C which lasted for 2 minutes, and then with a stable increase of 5 °C/min the temperature reached 300 °C which lasted for 10 minutes. Certified PAHs standards (2000 mg/cm^3 of each compound; LGC Promochem Corporation, Teddington, UK) were used in order to determine the calibration curve [Rosik-Dulewska et al. 2008a]. Recovery levels for this procedure were low for naphthalene (62–74%), and higher (83–94%) for the rest of the individual PAHs. The detection limit was 0.005 mg/dm^3 for each compound of PAHs group. The uncertainty of the results was calculated as a standard deviation value. Moreover, particular compound concentration ratios: ANT/(ANT+PHE), BaA/(BaA+CHR), FLT/(FLT+PYR) were calculated for estimation: petrogenic and pyrogenic compounds – from burning solid fuels.

Pesticides determination

In each sample 6 single compounds (Heptachlor, Aldrine, Dieldrine, DDE, DDD, DDT) were determined. Pesticides were determined in unpurified extracts by GC-ECD method using DB-1701 capillary column with constant carrier (He) flow through the column in quantity of 1ml/min. The temperature of the chromatograph injector was 250 °C while the ECD detector was 310 °C. The temperature program of the stove began at 150 °C which lasted for 1 minute, and then with a stable increase of 5 °C/min the temperature reached 305 °C that lasted for 10 minutes. Certified pesticides standards (500 mg/cm^3 of each compound; LGC Promochem Corporation, Teddington, UK) were used in order to determine calibration curve.

Basis analysis

The conductivity and pH was determined with the use of electronic instruments: pH-meter and conductivity-meter immediately after sampling. Temperature and deep horizon groundwater was determined in the well with a thermometer and a specialized meter with a water table detector on the end.

RESULTS AND DISCUSSION

In autumn, the water table was situated at 0.99 m under ground level (ugl) (well 9) to 9.0 m ugl (well 15). The mean depth of the water table during autumn was 3.48 m ugl. In winter the water table was measured from 0.7 m ugl (well 18) to 8.78 m ugl (well 15). The mean depth of the water table during autumn was 3.35 m ugl. Well 8 is particularly noteworthy. This is an artesian spring with an outflow of 0.7 m above ground. It is a rare phenomenon in the Lower Silesia Region. The temperature of the investigated water was high during autumn 10.3–19.0 °C and low during winter 3.5–10.0 °C. Values of pH and EC were typical for unpolluted ground waters (Table 1).

The investigated waters initially created a shallow water layer. It was mostly located from 1.0 to 3.5 meters under ground level. Additionally, the character of the soil and rocks – full of caverns – made it possible for effective transmission of rainwater (also with pollutants) directly and quickly to the water layer [Aghazadeh and Mogaddam 2011]. Only three wells (no. 14 to 16) were deeper – the water table being under 6 meters ugl. The pH values were typical for well-buffered waters and these values were close to existing data from the Opole region [Rosik-Dulewska et al. 2008b]. In this area, it manifests itself in rocks rich in calcium and magnesium carbonates, so pH data from other areas are lower

Table 1. Main water samples' parameters

Nr of well	Deep to horizon water [m ugl]		pH		EC [μS*cm^{-1}]		Temperature [°C]	
	Min–max	average	Min–max	median	Min–max	average	Min–max	average
1	1.10–1.12	1.05	7.02–7.10	7.05	760–872	806	5.5–17.0	9.9
2	0.81–1.15	0.99	7.42–7.68	7.59	103–415	287	4.5–17.0	10.2
3	0.83–1.59	1.23	7.07–7.68	7.11	524–726	606	3.1–16.0	5.3
4	2.08–2.48	2.30	6.90–7.21	6.95	609–951	833	6.4–16.0	10.7
5	2.10–2.76	2.35	7.17–7.95	7.36	462–836	643	5.7–16.0	10.0
6	1.80–2.28	2.06	7.00–7.34	7.22	695–1145	867	7.5–16.0	12.0
8	dnp	dnp	7.04–7.40	7.25	649–774	702	9.4–11.0	10.2
9	0.76–1.12	0.92	7.00–7.33	7.16	580–764	656	6.6–19.0	12.0
10	2.03–2.12	2.07	7.04–7.30	7.23	695–857	767	8.2–14.8	11.0
11	3.85–4–64	4.04	7.85–8.33	8.16	372–490	420	7.0–13.2	9.9
12	1.09–1.70	1.38	7.31–8.11	7.88	363–420	382	5.5–13.2	9.0
14	7.12–7.85	7.49	7.08–7.68	7.41	709–895	786	8.9–11.3	10.2
15	8.78–9.46	9.07	6.89–7.49	7.27	816–895	872	9.1–10.3	9.6
16	6.60–6.85	6.73	6.68–7.71	7.21	460–535	506	7.6–11.3	9.5
17	0.98–1.63	1.32	6.68–7.86	7.47	502–620	561	2.6–13.0	8.1
18	0.51–1.15	0.85	6.26–7.62	7.35	424–619	547	2.1–13.1	7.4
19	2.74–3.04	2.89	6.24–7.02	6.67	439–524	502	7.5–13.4	10.3
20	2.25–2.30	2.27	6.88–7.30	7.09	292–528	410	6.5–12.5	9.7

Table 2. Content of PAHs in water samples taken in autumn (A) and winter (W) [mg/dm^3]

Well	1		2		3		4		5		6	
Compound	A	W	A	W	A	W	A	W	A	W	A	W
NAPH	0.294	0.693	0.671	1.106	0.764	2.045	0.011	1.042	0.705	1.957	0.341	1.187
ACY	0.023	0.013	0.019	0.228	0.025	0.128	<0.005	0.017	0.028	0.045	0.024	0.019
ACE	0.050	0.093	0.030	<0.005	0.043	2.212	0.016	<0.005	0.041	0.158	0.035	0.056
FLU	0.019	0.090	0.032	0.487	0.015	1.013	0.009	0.047	0.075	0.377	0.122	0.546
FEN	0.106	0.060	0.032	0.322	0.116	1.238	0.038	0.139	0.095	0.474	0.140	0.632
ANT	0.178	0.052	0.103	0.473	0.150	0.100	0.058	0.132	0.169	0.026	0.160	0.037
FLA	0.330	0.117	0.223	0.128	0.353	0.718	0.129	0.044	0.559	0.084	0.392	0.032
PYR	0.129	0.135	0.427	0.402	0.022	0.065	0.010	<0.005	0.022	0.022	<0.005	0.027
B(a)ANT	0.124	0.044	0.081	0.019	0.111	0.104	0.105	0.120	0.197	0.034	0.117	0.059
CHR	0.071	<0.005	0.191	0.145	0.064	<0.005	0.033	<0.005	0.097	<0.005	0.069	<0.005
B(b)FLU	0.602	0.385	0.111	<0.005	0.599	<0.005	0.185	<0.005	0.712	0.035	0.544	0.058
B(k)FLU	0.753	0.028	1.278	<0.005	0.659	0.055	0.253	0.281	0.667	0.104	0.188	0.166
B(a)PYR	<0.005	0.050	0.809	0.032	<0.005	<0.005	<0.005	<0.005	0.037	0.040	0.018	<0.005
Ind(1,2,3-cd)P	0.026	0.180	<0.005	0.121	0.113	0.254	0.016	<0.005	0.034	0.096	0.022	0.114
D(a,h)ANT	<0.005	<0.005	0.022	0.310	0.160	<0.005	0.049	<0.005	0.073	<0.005	0.220	<0.005
B(g,h,i)PER	<0.005	<0.005	0.032	<0.005	<0.005	<0.005	<0.005	<0.005	<0.005	<0.005	<0.005	<0.005
Mean value	0.169	0.121	0.254	0.236	0.199	0.496	0.057	0.114	0.219	0.216	0.149	0.183

Table 3. Content of PAHs in water samples taken in autumn (A) and winter (W) [mg/dm^3]

Well / Compound	8		9		10		11		12		14	
	A	W	A	W	A	W	A	W	A	W	A	W
NAPH	0.639	1.151	1.103	0.812	0.926	0.832	0.391	0.627	0.634	0.863	0.113	0.651
ACY	<0.005	0.022	0.031	0.043	0.021	0.050	0.012	0.168	0.013	<0.005	0.004	<0.005
ACE	0.008	0.123	0.059	0.068	0.033	<0.005	0.010	0.005	0.018	<0.005	0.012	<0.005
FLU	0.047	0.796	0.030	0.025	0.081	0.016	0.059	0.659	0.039	0.182	0.003	0.337
FEN	0.104	0.825	0.124	0.038	0.084	0.035	0.047	0.124	0.035	1.722	0.015	0.194
ANT	0.115	0.030	0.242	0.055	0.205	0.070	0.080	0.183	0.281	0.340	<0.005	0.024
FLA	0.393	0.021	0.393	0.092	0.313	0.107	0.183	0.122	0.097	0.005	0.010	0.027
PYR	0.220	0.036	<0.005	0.007	<0.005	0.041	0.025	0.018	0.016	0.106	0.007	0.006
B(a)ANT	0.524	0.097	0.231	0.139	0.128	0.178	0.023	0.017	0.047	0.039	<0.005	0.013
CHR	0.447	<0.005	0.010	0.036	0.040	0.020	0.005	<0.005	0.289	0.044	<0.005	0.015
B(b)FLU	0.026	0.038	0.199	0.063	0.595	0.069	0.128	0.230	0.748	0.026	0.423	0.042
B(k)FLU	0.876	0.177	0.318	0.153	0.152	0.188	0.193	0.551	0.633	0.119	0.586	0.068
B(a)PYR	1.067	0.069	0.085	0.056	0.063	0.055	0.065	<0.005	0.151	0.029	<0.005	<0.005
Ind(1,2,3-cd)P	0.779	0.184	0.028	0.148	0.014	0.207	0.034	<0.005	0.168	0.080	<0.005	<0.005
D(a,h)ANT	0.126	0.133	0.192	0.086	0.090	0.116	0.035	<0.005	1.190	<0.005	<0.005	0.074
B(g,h,i)PER	0.460	<0.005	<0.005	<0.005	<0.005	<0.005	<0.005	<0.005	<0.005	0.231	<0.005	<0.005
Mean value	0.364	0.231	0.190	0.114	0.172	0.124	0.081	0.169	0.272	0.237	0.073	0.091

Table 4. Content of PAHs in water samples taken in autumn (A) and winter (W) [mg/dm^3]

Well / Compound	15		16		17		18		19		20	
	A	W	A	W	A	W	A	W	A	W	A	W
NAPH	0.064	0.648	0.274	0.586	0.527	0.685	0.603	1.479	1.034	1.657	0.551	1.157
ACY	<0.005	0.019	0.006	0.004	0.020	0.004	0.014	0.019	0.026	0.018	0.039	0.007
ACE	0.048	0.112	0.021	0.022	0.033	0.021	0.035	0.157	0.020	0.126	0.034	0.207
FLU	0.039	2.026	0.002	0.392	0.010	0.324	0.014	0.611	0.016	0.031	0.034	0.056
FEN	0.060	0.238	0.004	0.564	0.063	0.413	<0.005	0.489	0.051	0.209	0.042	0.107
ANT	<0.005	0.039	0.020	0.122	0.093	0.043	<0.005	0.719	0.093	0.049	0.068	0.041
FLA	0.007	0.027	0.057	0.023	0.223	0.007	0.040	0.023	0.200	0.017	0.227	0.013
PYR	0.084	0.007	0.008	0.008	<0.005	0.008	0.094	0.015	<0.005	0.043	<0.005	0.032
B(a)ANT	<0.005	0.036	0.013	0.047	0.052	0.052	<0.005	0.114	0.090	0.080	<0.005	0.023
CHR	0.115	<0.005	0.024	0.046	0.025	0.099	0.022	0.054	0.072	<0.005	<0.005	<0.005
B(b)FLU	0.235	0.069	0.042	0.147	0.280	0.117	0.012	<0.005	0.165	0.106	0.104	0.116
B(k)FLU	0.840	0.091	0.135	0.208	0.315	0.266	0.149	0.298	0.366	0.163	0.212	0.143
B(a)PYR	<0.005	<0.005	<0.005	0.048	<0.005	0.051	0.126	0.128	<0.005	<0.005	<0.005	<0.005
Ind(1,2,3-cd)P	0.264	<0.005	0.037	0.058	<0.005	0.201	<0.005	<0.005	0.081	0.151	<0.005	<0.005
D(a,h)ANT	0.177	<0.005	0.026	0.543	<0.005	1.009	<0.005	<0.005	<0.005	<0.005	0.145	<0.005
B(g,h,i)PER	<0.005	<0.005	<0.005	<0.005	<0.005	<0.005	<0.005	0.071	<0.005	<0.005	<0.005	<0.005
Mean value	0.121	0.207	0.042	0.176	0.103	0.206	0.069	0.261	0.138	0.166	0.091	0.119

[Fianko et al. 2009]. The electrically conductivity (EC) is lower than in existing data from the polluted areas and shows that these waters are not polluted with electrolytes [Aghazadeh and Mogaddam 2011, Ciesielczuk et al. 2006a]. The temperature of shallow water is more dependent on atmospheric conditions. Only three samples – wells 14 to 16 were less dependent, particularly during the winter time when the temperature was between 7.6–9.1 °C.

The investigated area is not under industry pressure so organic contaminants mostly come from agricultural activity eg. fertilizing with organic fertilizers and emissions from individual heating systems. Noted concentrations of particular compounds are showed in Tables 2–4.

The sum of 16 PAHs analyzed compounds was high – minimal values of 0.65 mg/dm^3 were noted in the sample from well 16 in the autumn and 1.452 mg/dm^3 in the winter (well 14). Maximum sum values were noted in sample 8 (autumn) 5.832 mg/dm^3 and 7.931 mg/dm^3 in well 3 (winter). The Ministry of Environment regulations [Regulation 2008] classify this water as "V" – the poorest quality class. However, this regulation does not state what compounds should be examined in the samples. Naphthalene is easily soluble in water – so its concentration was relatively high – and made over 50% of the total content. In general, samples taken during the winter contained lower amounts of PAHs than in the autumn samples. Data from the literature showed that soil eluate may contain 0.3–2.0 mg/dm^3 of particular compounds from the PAHs group. However, there were 3–4-ring compounds. The rest of the compounds – 4 and 5 ring – were noted in trace amounts [Goodarzi and Mukhopadhyay 2000]. Especially high concentrations were observed in groundwater polluted by petroleum products [Ciesielczuk et al. 2006b]. Also, elution of PAHs from soil was relatively high [Czop and Wandrasz 2007]. Calculated PAH ratios: ANT/(ANT+PHE), BaA/(BaA+CHR), FLT/(FLT+PYR) showed both sources to be equal: petrogenic and pyrogenic for these compounds. There was no correlation between the sum of 16 and between the sum of 6 PAHs (according to Ministry regulations) and the depth of the water table. High concentrations of investigated PAH compounds was probably an effect of the influence of the well area and water administration. Apart from PAHs, chlorinated pesticides were also observed (Table 5).

Pesticide analyses covered six compounds. In the case of heptachlor, aldrin and dieldrin, high

Table 5. Content of pesticides in water samples taken in autumn (A) and winter (W) [mg/dm^3]

Well / Compound	Heptachlor		Aldrin		Dieldrin		DDE		DDD		DDT	
	A	W	A	W	A	W	A	W	A	W	A	W
1	0.453	1.056	0.123	–	0.176	–	0.054	–	0.342	–	0.363	–
2	0.701	0.411	0.600	–	0.806	–	–	–	–	–	–	–
3	0.833	0.263	0.198	–	1.160	–	–	–	–	–	–	–
4	1.009	0.767	0.214	–	1.215	–	–	–	0.432	–	0.746	–
5	15.969	0.292	6.406	–	1.751	–	1.038	–	1.586	–	1.054	–
6	0.870	1.346	0.415	–	1.994	–	–	–	–	–	–	–
8	0.315	0.155	0.186	–	0.954	–	0.033	–	–	–	–	–
9	1.152	0.878	0.452	–	0.656	–	0.021	–	0.025	–	0.038	0.017
10	0.995	0.560	0.159	–	0.332	–	–	–	–	–	–	–
11	0.735	0.338	0.385	–	0.418	0.021	–	–	–	–	–	–
12	0.683	0.452	0.752	–	0.084	–	0.009	–	0.080	–	–	–
14	0.334	0.363	0.954	–	0.060	–	0.058	0.008	0.007	–	0.075	–
15	0.128	0.225	0.605	–	0.175	–	0.020	–	–	0.009	0.043	–
16	0.373	0.385	0.857	–	0.659	–	–	–	–	–	–	–
17	0.485	0.452	0.907	–	0.445	–	–	–	–	–	–	–
18	1.083	1.350	1.150	0.052	0.338	–	0.064	–	0.546	–	–	0.023
19	0.787	0.681	0.858	–	0.549	–	–	–	–	–	–	–
20	0.564	0.405	0.762	–	0.587	–	–	–	–	–	–	–

concentrations in investigated samples were noted. Other authors found much lower concentrations of these compounds [Geyikci and Büyükgüngör 2011]. Heptachlor was the most noted and the autumn mean concentration reached 1.53 mg/dm^3, for aldrin 0.888mg/dm^3 and for dieldrin 0.687 mg/dm^3. During winter the concentrations of aldrin and dieldrin were low and rarely exceeded the detection limit. Also, the presence of pesticides which had not been in use for around 20 years such as DDT was noted. The pollutants DDT, and metabolites, were observed in lower concentrations and only in some samples. The sample most polluted by these compounds was observed in well 5 – analoguous to high concentrations of heptachlor, aldrin and dieldrin. During winter, the determined amount of DDT and its metabolites (DDE, DDD) were lower and mostly under the detection limit. This confirmed high stability and environmental pollution from this product. There were no observed correlations between pesticides' concentration and the well depth.

CONCLUSION

The water from the investigated wells was shallow groundwater – 1st level water table, therefore, the pollution risk from external sources is highly possible. The samples contained a wide range of concentrations of PAHs and pesticides. In winter higher amounts of PAHs were observed, however, they were mostly low toxicity 2 and 3 ring compounds, in contrast to autumn with lower summary values but higher amounts of 5 and 6-ring PAHs. The sum of PAHs was high and investigated water samples were in the IV and V – poor quality class (according to Ministry regulations). The pesticides content was also wide ranging and high – showing great pressure from human agricultural activities. The highest noted concentration of heptachlor exceeded the maximum accepted level by 32 times. The most possible source of PAHs and pesticides in investigated water samples was soil eluate which leaks to ground water as an effect of rainfall.

Acknowledgments

This investigation was financed by Polish Ministry of Science and Higher Education Grant No 2592/P01/2007/32 from the science supporting funds in years 2007–2010.

REFERENCES

1. Aghazadeh N., Mogaddam A.A. 2011. Investigation of hydrochemical characteristics of groundwater in the Harzandat aquifer, Northwest of Iran. Environmental Monitoring and Assessment 176, 183-195.
2. Ake C.L., Wiles M.C., Huebner H.J., McDonald T.J., Cosgriff D., Richardson M.B., Donnelly K.C., Phillips T.D. 2003. Porous organoclay composite for the sorption of polycyclic aromatic hydrocarbons and pentachlorophenol from groundwater. Chemosphere 51, 835-844.
3. Ciesielczuk T., Rosik-Dulewska Cz., Nabzdyjak T. 2006a. Assessment of aliphatic and polycyclic aromatic hydrocarbons in ground water on the fuel storage area of airport (in polish), Ecological Chemistry and Engineering Vol. 13, No S4, 531-538.
4. Ciesielczuk T., Kusza G., Poluszyńska J. 2006b. Assessment of PAHs and total content of organic matter in Landfill Leachate and Groundwater, Ecological Chemistry and Engineering 11(13), 1225-1230.
5. Czop M., Wandrasz J.W. 2007. Analysis of mobility Polycyclic Aromatic Hydrocarbons in soils (in polish). Environmental Protection and Natural Resources, IEP, Warsaw 31, 144-148.
6. Fianko J.R, Osae S., Adomako D., Achel D.G. 2009. Relationship between land use and groundwater quality in six districts in the eastern region of Ghana. Environmental Monitoring and Assessment 153, 139-146.
7. Flaczyk Z. 1987. The environmental conditions of agricultural activity (in Polish), IUNG, Pulawy, pp. 82.
8. Geyikci F., Büyükgüngör H. 2011. Monitoring of organochlorine pesticides in the surface waters from Mid-Black Sea Region, Turkey Environmental Monitoring and Assessment 173, 127-137.
9. Goodarzi F., Mukhopadhyay (Muki) P.K. 2000. Metals and polyaromatic hydrocarbons in the drinking water of the Sydney Basin, Nova Scotia, Canada: a preliminary assessment of their source, International Journal of Coal Geology 43, 357-372.
10. Jayasekera D.L., Kaluarachchi J.J., Villholth K.G. 2011. Groundwater stress and vulnerability in rural coastal aquifers under competing demands: a case study from Sri Lanka. Environmental Monitoring and Assessment 176, 13-30.
11. Kryza J., Stasko S. 2000. Groundwater Flow Rate and Contaminant Migration in Fissure-Karstic Aquifer of Opole Triassic System Due to Man Activity. Environmental Geology 39, 3/4, 384-389.
12. Lazzari L., Sperni L., Salizzato M., Pavoni B. 1999. Gas chromatographic determination of organic micropollutants in samples of sewage sludge and compost: behavior of PCB and PAH during composting, Chemosphere 38, 1925-1935.

13. Oleszczuk P., Baran S., Baranowska E., Pranagal J. 2007. Content of Polycyclic Aromatic Hydrocarbons in long-term flooded soil. Ecological Chemistry and Engineering 14(S1), 109-116.

14. Owabor Ch.N., Ogbeide S.E., Susu A.A. 2010. Estimation of transport and degradation parameters for naphthalene and anthracene: influence of mass transfer on kinetics. Environmental Monitoring and Assessment 169, 607-617.

15. Pisarek I., Glowacki M. (2005) Some properties of aquatic humic substances from underground water of Opole region. Polish Journal of Environmental Studies 14, V: 29-32.

16. Pranagal J., Oleszczuk P. 2007. Content and migration of Polycyclic Aromatic Hydrocarbons in different soils used. Ecological Chemistry and Engineering 14(S2), 233-244.

17. Regulation of Polish Ministry of Environment from 23 of July 2008 No 143 pos. 896 (in Polish)

18. Ritter L., Solomon K., Sibley P. 2002. Sources, pathways, and relative risks of contaminants in surface water and groundwater: a perspective prepared for the walkerton inquiry. Journal of Toxicology and Environmental Health, Part A, 65, 1-142.

19. Rosik-Dulewska Cz., Ciesielczuk T., Ramus K. (2008a) Changes in the Polycyclic Aromatic Hydrocarbons (PAHs) content in an urban waste composting process. (In) Management of Pollutant Emission from Landfills and Sludge; Pawłowski, Dudzińska & Pawłowski (eds) Taylor&Francis Group London, 85-89.

20. Rosik-Dulewska Cz., Karwaczyńska U., Ciesielczuk T. 2008b. The impact of municipal landfill on the concentration of heavy metals in genetic soil horizons. (In) Management of Pollutant Emission from Landfills and Sludge; Pawłowski, Dudzińska & Pawłowski (eds) Taylor&Francis Group London, 117-124.

21. Stasko S., Wcislo M. 2006. Finite-difference method limitation in case of groundwater resource evaluation by modelingin karstic-fracture system, Geologos 10, 241-251.

22. Swarcewicz M.K., Gregorczyk A. 2011. The effects of pesticide mixtures on degradation of pendimethalin in soils. Environmental Monitoring and Assessment.

23. Szczygiel I., Kryza J., Fic A. 2006. Aquitard conductivity calibration of the Triassic aquifer numerical model of the Strzelce Opolskie area, Geologos 10, 272-283.

24. Thiele S., Brummer G.W. 2002. Bioformation of polycyclic aromatic hydrocarbons in soil under oxygen deficient conditions. Soil Biology & Biochemistry 34, 733-735.

25. Wilcke W., Muller S., Kanchanakool N., Niamskul Ch., Zech W. 1999. Polycyclic aromatic hydrocarbons in hydromorphic soils of the tropical metropolis Bangkok. Geoderma 91, 297-309.

26. Włodarczyk-Makuła M., Wiśniowska E., Janosz-Rajczyk M. 2003. Carcinogenic PAHs content in over-sludge water. Ecological Chemistry and Engineering 10(5), 467-476.

23

EFFECT OF DIFFERENT SUBSTANCES ON SOME PROPERTIES OF SOIL CONTAMINATED WITH HEATING OIL

Mirosław Wyszkowski[1], Veranika Sivitskaya[2]

[1] Department of Environmental Chemistry, University of Warmia and Mazury in Olsztyn, 4 Łódzki Sq., 10-727 Olsztyn, Poland, e-mail: miroslaw.wyszkowski@uwm.edu.pl

[2] Department of Environmental Chemistry, University of Warmia and Mazury in Olsztyn, 4 Łódzki Sq., 10-727 Olsztyn, Poland, e-mail: veranika.sivitskaya@uwm.edu.pl

ABSTRACT

The aim of the study was to determine the effect of the application of different substances (nitrogen, compost, bentonite, zeolite and calcium oxide) to soil on its selected properties after prior contamination with heating oil: 0, 5, 10, 15 and 20 g kg^{-1} d.m. of soil. Heating oil contamination and the application of different substances had a significant effect on the tested soil properties. In the series without additives, heating oil caused an increase in soil pH and a decrease in hydrolytic acidity, the total exchange bases and cation exchange capacity. Bentonite and calcium oxide had the strongest effect of all the substances on soil properties. They induced a rise in soil pH, an increase in the total exchange bases and cation exchange capacity and a decrease in hydrolytic acidity. The effect of other substances, particularly nitrogen and compost, on the tested soil properties was significantly lower. On the soil contaminated with heating oil, the effect of individual substances on the degree of base saturation was relatively small because it did not exceed 8% in all of the test series.

Keywords: heating oil contamination, nitrogen, compost, bentonite, zeolite, calcium oxide, soil properties.

INTRODUCTION

Intensive development of industry, construction and the increase in diverse needs of population in developed countries involves the need to obtain, produce, transport and store rising quantities of fuels [Nadim et al. 2000, Wyszkowski et al. 2004]. Unfortunately, this is not harmless to the environment because of the entailed possibility of its pollution, which is the result of different kinds of events, mainly failures [Nadim et al. 2000]. One of these fuels is heating oil, most often used for heating buildings. If the soil environment is contaminated, the effects can be substantial and sometimes even hard to reverse. Oil derivatives have a very strong effect not only on physicochemical soil properties [Baran et al. 2002, Kucharski, Jastrzębska 2005, Ziółkowska, Wyszkowski 2010], but they also cause a loss of microbiological and biochemical balance in the soil [Wyszkowska, Wyszkowski 2010], which often becomes unfit for plant cultivation. Therefore, it becomes necessary to make an attempt to restore soil balance, e.g. by the application of different kinds of substances, bioremediation and phytoremediation or the use of more costly *ex situ* methods.

The aim of the study was to determine the effect of the application of different substances (nitrogen, compost, bentonite, zeolite and calcium oxide) to soil on its selected properties after prior contamination with heating oil.

MATERIALS AND METHODS

The study was performed in the greenhouse of the University of Warmia and Mazury in Olsztyn (north-eastern Poland) on soil with the granulometric composition of loamy sand. The soil properties were as follows: pH at 1 mole

KCl · dm^{-3} – 4.52; hydrolytic acidity (HA) – 25.4 mmol(+) · kg^{-1}; the total exchange bases (TEB) – 29.1 mmol(+) · kg^{-1}; cation exchange capacity (CEC) – 54.5 mmol(+) · kg^{-1}; the degree of cation saturation (BS) – 53.4%; C_{org} content – 11.28 g · kg^{-1}; the content of available forms of: phosphorus – 71.88 mg P · kg^{-1}; potassium – 118.60 mg K · kg^{-1} and magnesium – 104.18 mg Mg · kg^{-1}. The soil was contaminated with heating oil in the amounts of: 0; 5; 10; 15 and 20 g · kg^{-1} DM. The study was conducted in five series: without the substances and with the application of nitrogen (200 mg N · kg^{-1} soil in the form of 46% urea), compost (3%), bentonite and zeolite (2% relative to soil mass) and 50% calcium oxide in a dose corresponding to one full hydrolytic acidity. Macro- and micronutrients were also applied to all the pots in the same amounts (in mg · kg^{-1} soil): N – 100 ($CO(NH_2)_2$), P – 30 (KH_2PO_4); K – 100 (KH_2PO_4 + KCl); Mg – 50 ($MgSO_4 \cdot 7H_2O$); Mn – 5 ($MnCl_2 \cdot 4H_2O$); Mo – 5 [$(NH_4)_6Mo_7O_{24} \cdot 4H_2O$]; B – 0.33 ($H_3BO_3$). During the establishment of the experiment, heating oil, compost, bentonite and calcium oxide, as well as macro- and micronutrients in a form of aqueous solutions, were mixed with 9 kg soil and transferred to polyethylene pots. Spring barley (*Hordeum vulgare* L.) cv. Rebel was sown in the prepared pots as well as white mustard (*Sinapis alba* L.) cv. Bardena after its harvest in the flowering phase. The moisture content was maintained at 60% of the capillary water capacity during the study. Soil samples for the tests were collected at the time of white mustard harvest in the flowering phase.

The following were determined in the soil, after prior drying and screening through a 1 mm mesh sieve: pH – by the potentiometric method in an aqueous KCl solution with a concentration of 1 mole · dm^{-3}, hydrolytic acidity (HA) and the total exchange bases (TEB) – by the Kappen method [Lityński et al. 1976]. Cation exchange capacity (CEC) and the degree of soil base saturation (BS) were computed based on hydrolytic acidity (HA) and the total exchange bases (TEB) by the following formulas: CEC = TEB + HA; BS = TEB · CEC^{-1} · 100. Moreover, the following were determined in the soil before the experiment was established: organic carbon (C_{org}) content – by the Tiurin method [Lityński et al. 1976], available phosphorus and potassium content – by the Egner-Riehm method [Lityński et al. 1976], available magnesium content – by the Schachtschabel method [Lityński et al. 1976]. A two-factor analysis of variance (ANOVA) from the Statistica package [StatSoft, Inc. 2010] and Pearson's simple correlation coefficients (to determine the relationship between soil heating oil contamination and the tested soil properties) were used for statistical analysis of the obtained test results.

RESULTS AND DISCUSSION

Heating oil contamination and the application of different substances had a significant effect on the tested soil properties (Tables 1 and 2). In the series without additives, pH was positively correlated with increasing heating oil doses. The pH rose under increasing heating oil doses in the soil with a dose of 20 g · kg^{-1} soil. In the series with the respective additives, the relationship between heating oil doses and pH was not always so clearly directed as in the series without additives. In the series without additives, hydrolytic acidity in the studied soil decreased under increasing heating oil doses. The highest heating oil dose (20 g · kg^{-1} soil) caused a decrease by 22% compared to the object without the tested oil derivative. The effect of increasing heating oil doses on the total exchange bases and cation exchange capacity was negative. No relationship was detected between increasing heating oil doses and the degree of base saturation for the soil under white mustard.

Contamination of soils with oil derivatives has a significant effect on soil properties. Experiments by Kucharski and Jastrzębska [2005] indicate that heating oil caused a decrease in pH, the sum of base cations, cation exchange capacity and the degree of sorption complex base saturation, which is partly consistent with the results of own research. In a study by Baran et al. [2002], point sources of contamination with oil derivatives induced an increase in pH, the sum of base cations and cation exchange capacity.

All the alleviating substances applied to the soil caused changes in the tested properties in the objects contaminated with heating oil (Figure 1).

Comparing the effect of the substances in the soil contaminated with heating oil in the amount of 5–20 g · kg^{-1} soil, CaO and, to a slightly lower degree, bentonite, showed the most stimulating effect on soil pH compared to the series without additives. Bentonite and calcium oxide, as the only substances, contributed to a decrease in hydrolytic acidity in the soil by 39% and 31%, compared to the series without additives (Figure 1). Zeolite and

Table 1. pH and hydrolytic activity in soil after plants harvest

Dose of heating oil in g · kg^{-1} of soil	Kind of substance neutralizing effect of heating oil						
	Without additions	Nitrogen	Compost	Bentonite	Zeolite	CaO	Average
pH							
0	5.34	4.76	5.11	6.84	5.35	6.33	–
5	5.20	5.60	5.25	6.58	5.34	6.52	–
10	5.32	5.27	5.53	6.71	5.32	7.01	–
15	5.40	5.23	5.34	6.65	5.46	6.95	–
20	5.86	5.27	5.46	6.55	5.36	6.83	–
r	0.768**	0.342	0.747**	-0.683*	0.444	0.779**	–
LSD for: heating oil dose – 0.13**; kind of neutralizing substance – 0.14**; interaction – 0.31**							
Hydrolytic activity [mmol · kg^{-1} of soil]							
0	31.0	24.9	29.6	17.5	30.1	24.3	26.2
5	28.4	23.9	26.1	16.4	30.2	19.2	24.0
10	26.6	28.4	32.6	16.7	31.2	18.4	25.6
15	28.0	29.9	30.9	15.6	28.4	19.1	25.3
20	24.3	28.5	29.8	17.2	29.3	17.5	24.4
Average	27.7	27.1	29.8	16.7	29.8	19.7	25.1
r	-0.883**	0.814**	0.337	-0.305	-0.514	-0.820**	-0.419
LSD for: heating oil dose – 0.8**; kind of neutralizing substance – 0.8**; interaction – 1.9**							

Explanation: r – correlation coefficient; significant for: ** – P≤0.01, * – P≤0.05.

Table 2. Total exchange bases (TEB), cation exchange capacity (CEC) and base saturation (BS) in soil after plants harvest

Dose of heating oil in g · kg^{-1} of soil	Kind of substance neutralizing effect of heating oil						
	Without additions	Nitrogen	Compost	Bentonite	Zeolite	CaO	Average
Total exchange bases (TEB) [mmol · kg^{-1} of soil]							
0	105.6	96.2	77.5	117.2	119.6	100.0	102.7
5	94.4	87.4	93.6	130.0	115.1	101.5	103.7
10	85.8	88.9	90.3	134.6	106.8	112.9	103.2
15	86.9	85.3	81.7	132.5	102.6	115.4	100.7
20	96.7	75.2	92.7	132.7	105.0	116.8	103.2
Average	93.9	86.6	87.2	129.4	109.8	109.3	102.7
r	-0.496	-0.920**	0.410	0.756**	-0.914**	0.943**	-0.262
LSD for: heating oil dose - n.s.; kind of neutralizing substance - 6.7**; interaction - 14.9**							
Cation exchange capacity (CEC) [mmol · kg^{-1} of soil]							
0	128.5	121.4	100.9	129.4	142.2	116.5	123.1
5	117.2	106.8	116.7	144.4	140.4	116.5	123.7
10	107.5	109.9	111.1	151.5	134.2	128.5	123.8
15	106.9	107.5	101.1	147.5	133.0	128.4	120.7
20	117.3	99.5	113.3	144.1	132.5	131.8	123.1
Average	115.5	109.0	108.6	143.4	136.4	124.3	122.9
r	-0.586	-0.859**	0.205	0.616*	-0.941**	0.924**	-0.391
LSD for: heating oil dose - n.s.; kind of neutralizing substance - 7.1**; interaction - 15.8*							
Base saturation (BS) [%]							
0	82.1	79.2	76.7	90.6	84.1	85.8	83.1
5	80.5	81.8	80.2	90.0	82.0	87.1	83.6
10	79.8	80.9	81.3	88.9	79.6	87.9	83.0
15	81.3	79.0	80.7	89.8	77.2	89.9	83.0
20	82.4	75.6	81.8	92.1	79.3	88.6	83.3
Average	81.2	79.3	80.1	90.3	80.4	87.9	83.2
r	0.215	-0.674*	0.840**	0.381	-0.856**	0.857**	-0.127
LSD for: heating oil dose - n.s.; kind of neutralizing substance - 1.1**; interaction - 2.4**							

Explanation: r – correlation coefficient; significant for: ** – P≤0.01, * – P≤0.05.

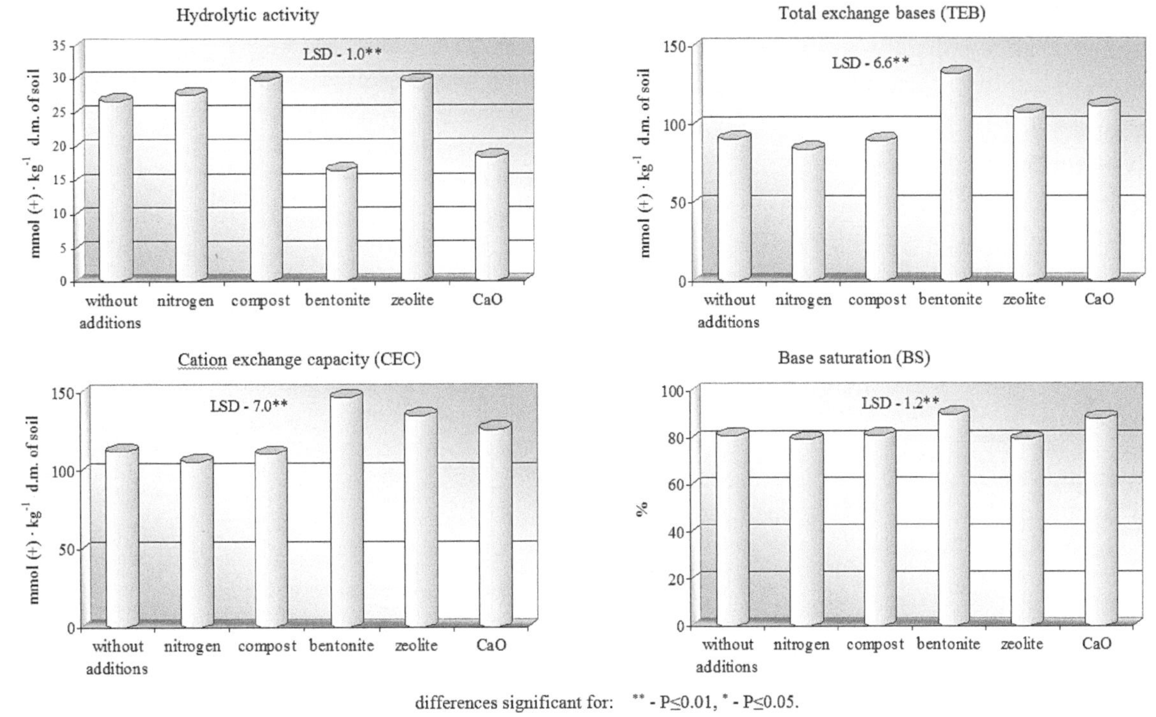

Figure 1. Soil properties after plants harvest (average from series contaminated with 5–20 g heating oil per kg of soil)

compost caused an increase in hydrolytic acidity in the soil, but only by 11%. The addition of nitrogen had no substantial effect on the tested property.

In the soil contaminated with heating oil, bentonite and calcium oxide showed the most stimulating effect and contributed to an increase in the total exchange bases by 46% and 23% and in cation exchange capacity by 26% and 13% compared to the series without additives (Figure 1). An increase in the total exchange bases by 18% and cation exchange capacity by 20% compared to the series without additives was recorded after the application of zeolite.

In the soils contaminated with heating oil, the effect of individual substances on the degree of base saturation was relatively low because it did not exceed 5% in all the test series (Figure 1). An exception was the application of bentonite and calcium oxide, which increased the value of the tested index by 8% and 7%, respectively, compared to the series without the alleviating substances.

The application of neutralizing additives to soil usually has a positive effect on the properties of soil contaminated with small quantities of oil derivatives. Organic substances act particularly favorably, as indicated by the studies of Riffaldi et al. [2006], Quintern et al. [2006], Wyszkowski and Ziółkowska [2009]. Organic substances, especially compost, have a positive effect on the sorption properties of soil and accelerate the decomposition of oil derivatives [Ziółkowska, Wyszkowski 2010]. The positive effect of bentonite and CaO on soil pH was confirmed by Wyszkowski and Ziółkowska [2011] who found that in soil contaminated with gasoline and diesel fuel these additives contributed to an increase in pH, the total exchange bases and exchange capacity. Bentonite, calcium oxide and compost decreased hydrolytic acidity in contaminated soil and contributed to a rise in the degree of base saturation. This creates more optimal conditions for a proper element cycle in the soil environment and plant growth [Ziółkowska, Wyszkowski 2010]. When soil is highly contaminated with oil derivatives, plant cultivation is not possible [Ogboghodo et al. 2004].

CONCLUSIONS

1. Heating oil contamination and the application of different substances had a significant effect on the tested soil properties.
2. In the series without additives, heating oil caused an increase in soil pH and a decrease in hydrolytic acidity, the total exchange bases and cation exchange capacity.
3. Bentonite and calcium oxide had the strongest effect of all the substances on soil properties. They induced a rise in soil pH,

an increase in the total exchange bases and cation exchange capacity and a decrease in hydrolytic acidity.

4. The effect of the other substances, particularly nitrogen and compost, on the tested soil properties was significantly lower. On the soil contaminated with heating oil, the effect of individual substances on the degree of base saturation was relatively small because it did not exceed 8% in all of the test series.

Acknowledgements

The study was completed as a part of research funded by the Polish National Science Centre (NCN) No. N N305 305340.

REFERENCES

1. Baran S., Bielińska E.J., Wójcikowska-Kapusta A. 2002. Kształtowanie się aktywności enzymatycznej w glebach zanieczyszczonych produktami ropopochodnymi. Acta Agroph., 70, 9–19.
2. Kucharski J., Jastrzębska E. 2005. Effects of heating oil on the count of microorganisms and physico-chemical properties of soil. Polish J. Environ. Stud., 14(2), 189–198.
3. Lityński T., Jurkowska H., Gorlach E. 1976. Analiza chemiczno-rolnicza. PWN, Warszawa, 129–132.
4. Nadim F., Hoag G.E., Liu Sh., Carley R.J., Zack P. 2000. Detection and remediation of soil and aquifer systems contaminated with petroleum products: an overview. J. Petroleum Sci. Eng., 26, 169–178.
5. Ogboghodo I.A., Erebor E.B., Osemwota I.O., Isitekhale H.H. 2004. The effects of application of poultry manure to crude oil polluted soils on maize (Zea mays) growth and soil properties. Environ. Monit. Assessm., 96(1-3), 153–161.
6. Quintern M., Lein M., Joergensen R.G. 2006. Changes in soil-biological quality indices after long-term addition of shredded shrubs and biogenic waste compost. J. Plant Nutrit. Soil Sci., 169(4), 488–493.
7. Riffaldi R., Levi-Minzi R., Cardelli R., Palumbo S., Saviozzi A. 2006. Soil biological activities in monitoring the bioremediation of diesel oil-contaminated soil. Water Air Soil Pollut., 170(1-4), 3–15.
8. Statsoft Statistica 10.0 (data analysis software system). www.statsoft.com; 2012.
9. Wyszkowska J., Wyszkowski M. 2010. Activity of dehydrogenases, urease and phosphatases in soil polluted with petrol. J. Toxicol. Environ. Heal. A, 73(17), 1202–1210.
10. Wyszkowski M., Wyszkowska J., Ziółkowska A. 2004. Effect of soil contamination with diesel oil on yellow lupine field and macroelements content. Plant Soil Environ., 50, 218–226.
11. Wyszkowski M., Ziółkowska A. 2011. Effect of compost, bentonite and CaO on some properties of soil contaminated with petrol and diesel oil. Ecol. Chem. Eng., A, 18 (9-10), 1373–1381.
12. Wyszkowski, M., Ziółkowska, A. 2009. Role of compost, bentonite and calcium oxide in restricting the effect of soil contamination with petrol and diesel oil on plants. Chemosphere 74, 860–865.
13. Ziółkowska A., Wyszkowski M. 2010. Toxicity of petroleum substances to microorganisms and plants. Ecol. Chem. Eng. S, 17(1), 73–82.

24

POSSIBLE USE OF SPIDER WEBS FOR THE INDICATION OF ORGANIC ROAD POLLUTANTS

Justyna Rybak[1]

[1] Department of Environmental Protection, Wroclaw University of Technology, Wybrzeże Wyspiańskiego 27, 50-370 Wrocław, Poland, e-mail: justyna.rybak@pwr.wroc.pl

ABSTRACT

The results of the research assessing the possibility of using spider webs for the indication of organic pollutants (PAHs) in the long period of time in the air polluted with road traffic emissions are presented. It was proved that webs could be used for the indication of pollutants of this type. Because of the great availability and dense structure of webs belonging to representatives of Agelenidae family, they can serve as a very effective tool. What is more, the secluded location of webs is favorable and influences the accuracy and credibility of results, since deposited pollutants are not washed off with rain, neither are they decomposed by microorganisms or sunlight. The influence of co-existing contaminants is also minimized.

Keywords: PAHs, road pollutants, spider webs, indication.

INTRODUCTION

Already for a long time, the biomonitoring methods support the instrumental monitoring techniques of mosses, vascular plants and lichens are usually used to evaluate the pollution of atmospheric aerosol [Ciesielczuk et al. 2012]. Spider webs absorb air pollution thus they can also serve as a perfect tool in the environment assessment. Advantages of the use of webs are as follows: a low cost of samples collections, great availability of the research material, sheltered location of webs preventing them from destruction by weather conditions (rain, wind, sun), non-invasiveness of studies, an easy sample collection [Rybak 2012, Rybak et al. 2012, Rybak and Olejniczak 2013]. The ability of webs to accumulate toxins enables a long-term assessment of air pollution level in the selected site, which is an additional advantage, in contrast to standard measurements which provide information about the temporary state of the environment only. Thus, spider webs can serve as a perfect indicator of air pollution level [Rybak 2012, Rybak et al. 2012, Rybak and Olejniczak 2013, Xiao et al. 2006].

The aim of studies was to test the hypothesis if spider webs are suitable for the assessment of air contamination with polycyclic aromatic hydrocarbons in a long period of time. PAH concentrations are dependent on the nature of emissions to the atmosphere, as well as on rates of wet and dry deposition, sunlight, wind and chemical transformations. Thus, the accumulation rates depends on the site location. Research of this type focusing on application of spider silk for the indication of pollutants were done several times only [Hose at al. 2002, Rybak 2012, Rybak et al. 2012, Rybak and Olejniczak 2013, Xiao et al. 2006]. Listed authors (with the exception of Rybak and Olejniczak 2013) were not focused on PAHs. The work of Rybak and Olejniczak [2013] proved that spider webs of Agelenidae family are good for indication of road traffic emissions concerning PAHs (of high molecular weight mainly) previously defined as markers of traffic contamination (as benzo[*a*]pyrene, pyrene, fluoranthene and phenanthrene and others such as benzo[*a*] anthracene and indeno[*1,2,3-c,d*]pyrene that may also serve this role). The cited work focused on the possibility of indication with the use of spider webs (only from Agelenidae family exposed for

60 day at sampling sites) and, what is more, on finding possible PAHs sources. In the presented work the accumulation of PAHs is studied in 5 different spider species in long time period (the webs were exposed at sampling sites during 10, 60 and 90 days).

MATERIAL AND METHODS

The research were conducted in 2010 (spring, summer, autumn), 8 sites were chosen for studies within the city of Wrocław. As many as 5 species from Agelenidae family: *Malthonica silvestis* (L. KOCH, 1872), *Malthonica ferruginea* (PANZER, 1804), Amaurobiidae family: *Amaurobius ferox* (WALCKENAER, 1830) and finally Theridiidae: *Theridion mystaceum* L. KOCH, 1870 and *Theridion melanurum* HAHN, 1831 were selected. They were not present at all sites. Species that do not eat their own old web were chosen. Hose's studies (2002) proved that some pollutants (mainly heavy metals) are deposited on the web, but they are not built in into the matrix of Cribellatae webs. Probably, these observation could be also true for other spider species. What is more, spider silk is a protein product dependent on food spiders food. Hence, the contaminated food can affect the contamination of predators body and, in consequence, it might be present in their webs, as spiders prey can derive from a larger area than their victims. But spiders from Agelenidae, Amaurobiidae and Theridiidae families feed mainly on non-flying insects and other invertebrates (some beetles, springtails, ants and earthworms) which do not travel such long distances as flying insects do [Park and Moon 2012].

Samples were collected in the city of Wrocław (Lower Silesia, SW Poland). Two types of webs were used for the analysis: dry silk of Cribellatae representatives such as Amaurobiidae and sticky or dry of Ecribellatae such as Theridiidae and Agelenidae. For chemical analyses only newly constructed webs were used after 10 and 60 and 90 days of exposure. Webs of similar size and weight were used [Hose at al. 2002]. They were collected from walls of hydrotechnical building protected from rain in a great distance from the main routes (reference site 1), from walls and fences of housing estate away from heavy traffic (reference site 2), from the wall of an underground car park about 15 meters (site 3) and about 70 meters away (site 4) from a highly traffic road, from railway viaducts walls situated in a quiet area but in the vicinity of heavy traffic roads (site 5 and site 6) and from fences and hedges along a busy street contributing highly to the worst pollution in the city of Wrocław by road emissions [WIOŚ 2012], samples were collected about 10 meters (site 7) and about 70 meters (site 8) away from the roadside [Rybak 2012, Rybak et al. 2012, Rybak and Olejniczak 2013]. Figure 1 presents the map of sampling sites.

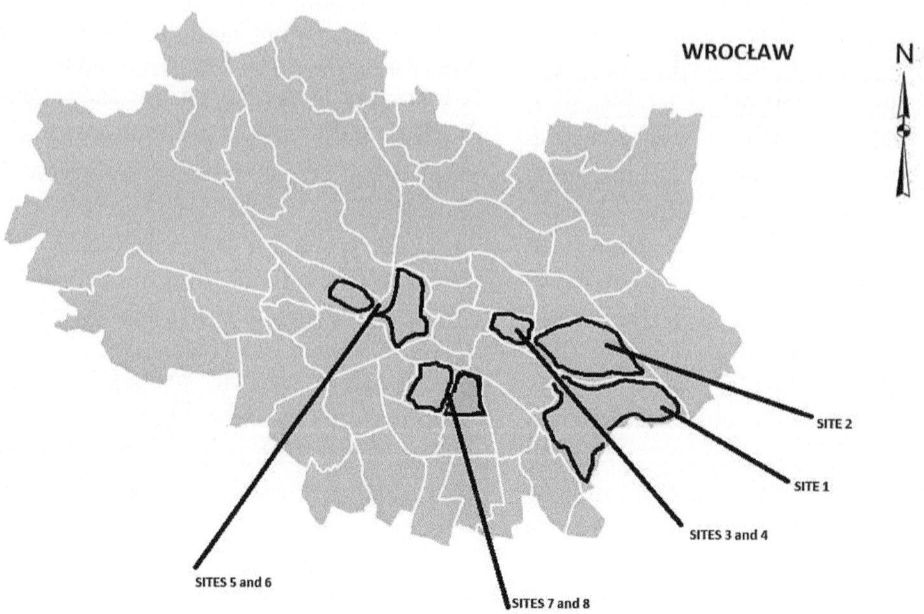

Figure 1. The map of sampling sites (map modified on the basis: [http://commons.wikimedia.org/wiki/File:Wroc%C5%82aw_Huby.png, [Siliesiac] (Category:Districts of Wrocław), licencja: [CC-BY-SA 3.0 Deed] (link do: http://creativecommons.org/licenses/by-sa/3.0/)

The silk was collected into clean glass phials with glass, sterile baguettes and then frozen until their use for chemical analyses. PAHs content was determined with chromatograph GC-MS of the absolute calibration for PAHs standards and average sensitivity 0.1 ng/50 μl (examined solution, i.e. 0,002 μg/ml). The samples were extracted with three portions (1 ml) mixtures of methylene chloride with methanol (9:1 v/v). Extracts after spinning (10 000 rpm) were evaporated to a dry residue in the stream of nitrogen (temp. 30 °C) and then dissolved in 50 μl of methylene chloride and analyzed. A qualitative and quantitative analysis of organic pollutants included in samples was done by Chemical Department in the University of Wrocław [Rybak and Olejniczak 2013]. Sixteen polycyclic aromatic hydrocarbons were determined in the samples according to US EPA recommendation. Table 1 shows names and corresponding recovery rates (RR) and standard deviations (SD_{RR}) [Rybak & Olejniczak 2013].

Non-parametric (Wilcoxon signed-rank Test, Mann-Whitney U-Test) statistical tests were used because the data were only partly normally distributed. Statistical tests were performed to determine differences in PAHs content between months of observation and between distance from emission source and among spider webs types. The probability level (p) of all tests was 0.05.

Table 1. Names, recovery rates RR and standard deviation (SD_{RR}) for corresponding PAHs in the web samples (n= 6) [Rybak & Olejniczak 2013]

PAH	RR [%]	SD_{RR}
Naphthalene	75.3	12.4
Acenaphthylene	89.5	5.3
Acenaphthene	98.7	8.9
Fluorene	99.4	12.1
Phenanthrene	87.3	13.4
Anthracene	88.7	6.2
Fluoranthene	98.1	16.1
Pyrene	91	8.2
Benzo[*a*]anthracene	89.2	14.2
Chrysene	93.1	12.5
Benzo[*b*]fluoranthene	95.3	7.8
Benzo[*k*]fluoranthene	96.4	6.5
Benzo[*a*]pyrene	90.1	9.9
Indeno[*1,2,3-c,d*]pyrene	99.6	15.7
Dibenzo[*a,h*]anthracene	95.4	11.8
Benzo[*g,h,i*]perylene	82.7	12.5

RESULTS AND DISCUSSION

PAHs content at sites in 2010 for the genus *Malthonica* (Agelenidae family), *Amaurobius ferox* (Amaurobiidae family), for the genus *Theridion* (Theridiidae family) are shown in Tables 2, 3, 4 and 5.

PAH concentrations according to Wilcoxon signed-rank Test for June, July and October 2010 differed significantly (for *Malthonica* genus: June-July $Z = -5{,}009$; June-October $Z = -5.25$; July-October $Z = -5.84$; $p = 0{,}001$; for *Amaurobius ferox*: June-July $Z = -4.45$; June-October $Z = -3.79$; July-October $Z = -3.79$; $p = 0{,}001$; for *Theridion* genus: June-July $Z = -4.23$; June-October $Z = -4.45$, July - October $Z = -4.45$ $p = 0{,}001$). PAHs concentrations according to Mann-Whitney U-Test were not significantly different between site 3 and 4 ($U = 165.0$, $p = 0.56$), but they were significantly different between site 7 and 8 ($U = 13$, $p = 0.01$) proving that the average concentrations of PAHs decrease with the distance from roadside.

As it was shown in Tables 3 and 4 the differences between the same sites (site 5 and 6) and different species (dry silk of *Amaurobius ferox*, and sticky silk of genus *Theridion*) were not significantly different ($p > 0.05$ Mann-Whitney U-Test). The highest concentrations of PAH were observed at sites with the heaviest traffic. The total PAH concentration, Σ PAH in web samples exposed for 10 days in reference site 2 (ΣPAH =787 ng/g) was comparable with the mean ΣPAH of 827 ng/g determined in the moss *Pseudoscleropodium purum* exposed for 2 months in Santa Cruz city on the Tenerife islands [Ares et al. 2009], but the total concentrations of PAHs in the reference site 1 exposed for 10 days (ΣPAH = 85 ng/g) are comparable with those in control samples of mosses *Pleurozium schreberi* exposed for 7 days (ΣPAH= 99 ng/g) and collected in the areas of Kotorz village in Poland, characterized with little pollution [Ciesielczuk et al. 2013]. The highest values of PAHs were obtained for highly traffic oriented site 7, webs exposed for 90 days contained 11786 ng/g ΣPAH. Such a result is not comparable with others obtained with the use of mosses and lichens or vascular plants, as these biomonitors were not exposed for such a long time. In this context spider webs seem to be more efficient when evaluating long-term air condition, then other biomonitors dependent on water and sun. In general, roadway tunnel, viaducts and car

Table 2. Mean concentrations and standard deviation of PAHs at sites in 2010, the samples were collected in June (I data series for 6 sites) after 10 days of exposure for genus *Malthonica.* L. d. = below detection limit. Concentrations are given in ng/g dry weight (n = 5)

PAH [ng/g]	Site no. 1(i)	Site no. 2(i)	Site no. 3(i)	Site no. 4(i)	Site no. 7 (i)	Site no. 8(i)
Naphthalene	L. d.	5.1±2.2	24.92±1.1	1.4±0.12	12.97±0.3	3.18±0.04
Acenaphthylene	L. d.	L. d.	L. d	L. d	34.71±4.64	15.34±1.8
Acenaphthene	L. d.	L. d.	L. d.	19.35±3.1	11.03±2.3	L. d.
Fluorene	L. d.	10.83±7.8	L. d.	34.45±7.9	18.23±4.3	L. d.
Phenanthrene	L. d.	8.9±3.5	140.92±35.8	129.99±23.1	200.11±13.2	10.21±2.8
Antheracene	L. d.	13.6±1.6	65.16±12.3	20.61±2.5	411.75±19.2	211.22±12.1
Fluoranthene	16.18±3.1	341.8±61.5	144.77±9.0	170.13±54.9	534.42±13.4	307.6±11.7
Pyrene	34.33±2.2	28.12±7.4	29.85±12.0	27.5±11.5	604.36±11.23	345.32±12.8
Benz[a]anthracene	L. d.	11.55±6.1	174.62±12.2	98.88±15.7	715.12±21.6	245.36±33.7
Chrysene	11.05±1.1	16.21±1.8	119.4±6.9	89.26±9.9	664.9±23.1	165.12±13.1
Benzo[b]fluoranthene	L. d.	17.76±4.1	121.84±15.8	107.45±43.8	300±12.7	121.21±18.6
Benzo[k]fluoranthene	L. d.	12.12±2.7	L. d.	L. d.	L. d.	L. d
Benzo[a]pyrene	24.41±7.3	28.67±9.5	L. d.	41.34±13.91	915.43±12.3	343.12±16.3
Indeno[1,2,3-c,d]pyrene	L. d.	L. d	179.1±21.3	167.66±27.36	555.35±12.8	332.66±32.11
Dibenz[a,h]anthracene	L. d.	19.12±5.2	L. d.	L. d.	306.57±26.1	105.83±9.1
Benzo[g,h,i]perylene	L. d.	273.46±17.5	L. d	L. d.	567.16±23.3	326.5±56.1
Sum of PAHs	85.97±13.7	787.24±128.7	1000.58±126.4	908.02±213.79	5852.11±200.47	2532.67±220.25

Table 3. Mean concentrations and standard deviation of PAHs at sites in 2010, the samples were collected in July (II data series for 4 sites) after 60 days and in October (III data series for 4 sites) after 90 days of exposure for genus *Malthonica.* L. d. = below detection limit. Concentrations are given in ng/g dry weight (n = 5)

PAH[ng/g]	Site no. 1 (II)	Site no. 2(II)	Site no. 3(II)	Site no. 7(II)	Site no. 1 (III)	Site no. 2(III)	Site no. 3(III)	Site no. 7(III)
Naphthalene	L. d.	9.8±4.7	2.9±1	27.14±7.71	L. d.	46.41±12.67	12.31±5.78	45.12±9.18
Acenaphthylene	L. d.	L. d.	L. d	91.74±6.31	L. d.	L. d.	L. d	221.65±128.91
Acenaphthene	L. d.	L. d.	69.86±1.9	57.87±12.61	L. d.	L. d.	110.71±23.08	137.16±54.11
Fluorene	L. d.	15.84±4.1	156.78±11.8	67.34±11.05	L. d.	61.27±23.86	273.91±32.17	168.23±47.51
Phenanthrene	L. d.	38.5±10	414.17±18.3	544.54±34.21	L. d.	159.67±28.13	503.22±127.83	714.86±141.28
Antheracene	L. d.	23.19±6.9	58.12±7.91	790.25±40.2	L. d.	76.11±18.19	108.27±19.17	1095.54±185.52
Fluoranthene	58.8±14.8	567.53±27.1	674.76±7.65	890.8±43.31	90.12±23.91	1062.63±48.92	903.11±123.5	1190.12±143.16
Pyrene	67.14±18.7	39.7±4.14	98.51±11.7	1005.16±33.6	158.41±34.63	103.62±17.13	180.43±55.12	1075.33±234.62
Benz[a]anthracene	L. d.	100.2±23.1	234.1±10.14	990.45±16.8	L. d.	159.98±90.6	404.87±154.08	990.45±16.8
Chrysene	48.14±18.6	36.32±5.7	321.56±19.2	1007.37±45.18	109.21±27.32	166.98±11	508.76±122.71	1142.71±167.16
Benzo[b]fluoranthene	L. d.	47.63±4.1	567.89±34.2	691.75±23.48	L. d.	199.12±23.67	718.73±187.06	792.53±123.9
Benzo[k]fluoranthene	L. d.	34.43±4.6	L. d.	L. d.	L. d.	140.78±41.87	L. d.	L. d.
Benzo[a]pyrene	145.67±12.3	103.56±7.1	190±12.33	1434.66±58.12	242.22±67.18	257.84±55.1	345.13±167.18	1523.78±169.23
Indeno[1,2,3-c,d] pyrene	L. d.	L. d	589.15±67.03	771.45±23.32	L. d.	L. d	719.63±168.14	883.31±134.41
Dibenz[a,h]anthracene	L. d.	49.12±12.2	L. d.	712.3±69.8	L. d.	204.13±56.13	L. d.	813.72±173.15
Benzo[g,h,i]perylene	L. d.	516.42±34.5	L. d.	898.56±12.13	L. d.	1061.23±43.02	L. d.	991.62±138.34
Sum of PAHs	319.7±64.4	1572.44±143.5	3374.9±202.16	9981.38±437.83	599.96±153.04	3699.77±470.29	4789.08±1215.82	11786.13±1867.27

Table 4. Mean concentrations and standard deviation of PAHs at sites in 2010, the samples were collected in June (I data series for 2 sites) after 10 days, in July (II data series for 2 sites) after 60 days and in the October (III data series for 2 sites) after 90 days of exposure for *Amaurobius ferox.* L. d. = below detection limit. Concentrations are given in ng/g dry weight (n = 5). No significant differences were found between concentrations of PAH for *Amaurobius ferox* and genus *Theridion* at sites (Table 4) 5 and 6 (p <0.05 Mann-Whitney U-Test)

PAH [ng/g]	Site no. 5 (I)	Site no. 6 (I)	Site no. 5 (II)	Site no. 6(II)	Site no. 5 (III)	Site no. 6(III)
Naphthalene	10.05±1.2	7.17±2.3	27.89±6.8	31.54±8.84	56.42±15.12	78.31±19.71
Acenaphthylene	14.02±4.32	L. d.	56.8±13.98	7.82±1.07	89.63±24.07	19.35±10.83
Acenaphthene	L. d.	L. d.	L. d.	L. d.	L. d.	L. d.
Fluorene	13.46±2.86	7.1±4.81	35.43±6.74	69.23±12.43	148.32±16.22	193.24±126.67
Phenanthrene	107.01±23.77	24.53±4.16	534.71±78.12	90.34±16.12	931.63±145.36	207.76±53.16
Antheracene	23.55±4.23	63.77±4.07	101.97±12.5	321.87±56.9	238.74±29.71	512.76±147.02
Fluoranthene	128.04±17.98	72.64±23.6	719.18±48.57	261.6±47.16	852.72±89.61	418.72±158.72
Pyrene	217.29±22.79	171.7±45.92	827.07±83	432.41±65.37	1116.18±193.28	602.3±197.03
Benz[a]anthracene	348.6±54.73	541.51±53.15	548.31±67.38	678.43±19.86	739.42±175.83	964.27±120.95
Chrysene	335.05±12.27	170.75±47.02	909.89±23.14	684.13±16.62	1093.76±132.44	975.24±182.41
Benzo[b]fluoranthene	44.39±43.96	272.64±13.09	178.19±19.24	608.1±65.38	369.86±122.31	851.14±78.22
Benzo[k]fluoranthene	71.96±13.6	L. d.	191.52±26.07	2.52±0.8	493.3±137.83	13.21±4.15
Benzo[a]pyrene	349.07±12.09	198.11±38.08	581.41±45.97	456.12±56.76	702.32±162.86	632.37±179.17
Indeno[1,2,3-c,d]pyrene	97.2±12.66	43.4±12.06	108.12±42.11	105.9±22.8	295.34±161.78	236.82±76.63
Dibenz[a,h]anthracene	135.05±18.94	47.17±2.76	167.89±34.8	181.54±48.12	289.19±146.17	472.28±127.52
Benzo[g,h,i]perylene	514.02±67.9	10.75±0.78	794.12±49.03	56.33±25	881.33±156.17	106.18±33.27
Sum of PAHs	2408.76±313.3	1631.24±251.7	5782.5±557.45	3987.88±406.47	8249.16±1339.12	6574.85±1515.46

Table 5. Mean concentrations and standard deviation of PAHs at sites in 2010, the samples were collected in June (I data series for 2 sites) after 10 days, in July (II data series for 2 sites) after 60 days and in the October (III data series for 2 sites) after 90 days of exposure for genus *Theridion.* L. d. = below detection limit. Concentrations are given in ng/g dry weight (n = 4)

PAH [ng/g]	Site no. 5 (i)	Site no. 6(i)	Site no. 5 (ii)	Site no. 6(ii)	Site no. 5 (iii)	Site no. 6(iii)
Naphthalene	18.52±2.56	18.05±3.14	34.12±6.12	39.16±2.85	157.18±32.82	80.54±23.92
Acenaphthylene	17.04±4.07	3.05±0.98	68.18±10.65	8.19±2.33	195.37±32.18	31.27±12.56
Acenaphthene	L. d.	L. d.	L. d.	L. d.	L. d.	L. d.
Fluorene	16.12±1.9	3.61±0.7	44.27±5.93	78.31±9.3	181.19±67.12	179.63±39.84
Phenanthrene	106.3±26.04	34.66±12.76	598.6±66.01	100.66±6.51	1074.03±213.8	269.71±69.62
Antheracene	31.85±12.55	75.81±1.64	115.09±45.4	373.51±65.23	226.1±155.32	691.72±126.48
Fluoranthene	125.93±27.04	102.35±23.14	690.23±61.31	251.69±41.49	1031.47±123.43	362.78±156.98
Pyrene	188.89±66.43	116.25±23.33	749.43±77.12	386.52±71.32	1050.64±186.23	889.43±160.44
Benz[a]anthracene	355.56±55.17	453.43±45.78	491.64±43.51	699.56±67.12	809.75±154.62	1409.51±186.19
Chrysene	407.41±100.67	214.44±56.89	897.32±44.12	707.3±45.13	1207.43±164.03	1432.06±87.31
Benzo[b]fluoranthene	59.26±12.77	178.7±34.45	199.2±18.07	643.24±34.36	439.37±40.09	1442.7±139.28
Benzo[k]fluoranthene	59.63±13.1	L. d.	187.31±35.12	3.16±1.1	406.42±73.31	12.26±6.79
Benzo[a]pyrene	205.56±67.89	166.68±28.9	600.2±56.89	512.1±66.93	1090.53±118.07	1067.18±165.06
Indeno[1,2,3-c,d]pyrene	105.7±20.5	36.61±13.42	191.37±32.2	107.23±12.65	467.48±64.46	268.34±59.17
Dibenz[a,h]anthracene	86.52±7.62	38.31±12.55	178.22±22.9	191.38±43.21	463.36±67.76	467.78±89.12
Benzo[g,h,i]perylene	270.04±24.09	8.05±2.37	762.13±56.18	65.1±18.07	1160.47±103.18	145.7±36.62
Sum of PAHs	2054.31±442.4	1450±260.05	5807.31±581.53	4167.11±487.6	9960.79±1596.42	8750.61±1359.38

park (sites 3, 4, 5, 6, 7 and 8) are characterized by the specific air circulation and are sheltered from the rainfall, which effectively protect walls before gradual removing of accumulated pollutants. Unfortunately, the silk of all used species, with the exception of the Agelenidae family, was rarely available in urban environment what seems to be the additional obstacle in the potential use of these webs for air monitoring. It seems that all Agelenids are very suitable for organic pollutants indication, since the representatives can be identified relatively easily and they produce an extensive, easily recognizable webs. Most importantly, they are active all-year-round, even in low temperatures and their webs are resistant to degradation of any type [Roberts 1995, Rybak 2012, Rybak et al. 2012, Rybak and Olejniczak 2013].

The concentration of PAHs depends on traffic intensity, the distance from the emission sources, meteorological conditions, the size of the particulates and their rates of deposition [Jones et al. 1989, Kumar et al. 2011, Oliveira et al. 2011]. The mechanism of PAHs accumulation depends upon their chemical and physical properties and the environmental conditions. PAHs dispersion in the atmosphere is divided between the gas phase (PAHs of high volatility) and particulate matter (the least volatile compounds with 5, 6 rings), PAHs of medium volatility (with 3 or 4 rings) are distributed between the gas and particulate phases [Harmens et al. 2013]. Compounds in gas phase are transported to the areas much distant from pollution sources, whereas solid particulate absorbed PAHs are deposited in great amounts near the emission source [Harmens et al. 2013, Thomas 1986]. The degradation rate of PAHs is generally very slow and depends upon several factors [Kumar et al 2011]. But they could easily decompose as result of the activity of microorganisms and sunlight [Chen et al. 2010]. However, photolysis is much lower for PAHs presented in the atmospheric particulate matter (PM) than in water or organic solvents [Tobiszewski and Namieśnik 2012]. As significant differences were observed in PAHs content depending on the exposure time, we can say that the location of sites (places sheltered from rain and sunlight) was the main factor stimulating the constant accumulation process. The pollutants of this type can migrate for significant distances. On the other hand, a decrease of PAHs content was observed near emission sources caused by a strong wind [Cabrerizo et al. 2012]. In the presented studies factors influencing or disturbing PAHs accumulation were eliminated (sheltered places), therefore, a statistically significant decrease of PAH concentrations was observed along with increasing the distance from the heavy traffic roadside (sites 7 and 8). In earlier conducted studies, the authors proved that active monitoring with the use of mosses in roadside tunnels is the most effective when monitoring road pollutants [Harmens et al. 2013, Zechmeister et al. 2006].

CONCLUSION

The results presented proved that webs can serve as an effective long term tool for the indication and evaluation of air pollutants, such as polycyclic aromatic hydrocarbons. The spider webs-based studies of air pollution are very small by scale compared with studies based on other bioindicators such as mosses, lichens and vascular plants. Because of their excellent accumulative properties resulting from webs high density, the webs of Agelenidae representatives can serve as an easy indication tool, applied almost all-year-round, what seems to be an additional advantage comparing to bioindicative methods often limited to the vegetative season. The recorded high concentrations of PAHs (with the exception of reference sites 1 and 2) were highly traffic oriented. All sampling sites were protected from rain washing giving them the possibility to accumulate PAHs in a long period of time (decomposition processes are limited). As the webs are not dependent on sunlight, their use seems to be favorable than other commonly used biomonitors, protecting PAHs from their fast degradation. What is more, the relation between concentrations of PAHs in atmospheric aerosol and accumulated in webs has never been studied, thus further studies are required.

Acknowledgments

The work was financed by government funds No. N N 305 096639 "Wykorzystanie właściwości kumulacyjnych sieci pajęczych do indykacji zanieczyszczeń komunikacyjnych".

REFERENCES

1. Ares A., Aboal J.R, Fernández J.A., Real C., Carballeira A. 2009. Use of the terrestrial moss *Pseudoscleropodium purum* to detect sources of small scale contamination by PAHs. Atmospheric Environment, 43: 5501–5509.

2. Cabrerizo A., Dachs J., Barcelò D., Jones K.C. 2012. Influence of organic matter content and human activities on the occurrence of organic pollutants in Antarctic soils, lichens, grass and mosses. Environmental Science and Technology, 46, 1396-1405.
3. Chen I., Zhang Y., Liu B. 2010. In situ simultaneous determination the photolysis of multi component PAHs adsorbed on the leaf surfaces of living *Kandelia candle* seedlings. Talanta, 83, 324-331.
4. Ciesielczuk T., Olszowski T., Prokop M., Kłos A. 2012. Application of mosses to identification of emission sources of polycyclic aromatic hydrocarbons. Ecological Chemistry and Engineering, 19 (4), 585-595
5. Foelix R.F. 1996. Biology of spiders. Oxford University Press, New York, 1-329.
6. Harmens H., Foan L., Simon V., Mills G. 2013. Terrestrial mosses as biomonitors of atmospheric POPs pollution: A review. Environmental Pollution, 173, 245-254.
7. Hose G.C., James, J.M., Gray, M.R. 2002. Spider webs as environmental indicators. Environmental Pollution, 120,725-733.
8. Jones, K.C., Stratford, J.A., Waterhouse, K.S., Vogt, N.B. 1989. Organic contaminants in Welsh Soils: polynuclear aromatic hydrocarbons. Environmental Science & Technology, 23(5), 540–550.
9. Kumar V., Kothiyal N.C. 2011. Distribution behavior of polycyclic aromatic hydrocarbons in roadside soil at traffic intercepts within developing cities. International Journal of Environmental Science and Technology, 8(1), 63-72.
10. Nyffeller M., Moor H., Foelix R.F. 2001. Spiders feeding on earthworms. Journal of Arachnology, 29, 119-124.
11. Oliveira C., Martins N., Tavares J., Pio C., Cerqueira M., Matos M., Silva H., Oliveira C., Camoes F. 2011. Size distribution of polycyclic aromatic hydrocarbons in a roadway tunnel in Lisbon, Portugal. Chemosphere, 83(11), 1588-1596.
12. Roberts M.J. 1995. Spiders of Britain and Northern Europe. Collins Field Guide. Harper Collins. London, 1-342.
13. Rybak J. 2012. Zastosowanie sieci pajęczych do oceny zawartości wybranych metali ciężkich w powietrzu na przykładzie Wrocławia. Ochrona Środowiska, 34(4), 47-50.
14. Rybak J., Sówka I., Zwoździak A. 2012. The preliminary assessment of use of spider webs for the indication of air contaminants. Environment Protection Engineering, 38(3), 175-181.
15. Rybak J., Olejniczak T. 2014. Accumulation of polycyclic aromatic hydrocarbons (PAHs) on the spider webs in the vicinity of road traffic emissions. Environmental Science and Pollution Research, 21(3), 2313-24.
16. Tobiszewski M., Namieśnik J. 2012. PAH diagnostic ratios for the identification of pollution emission sources. Environmental Pollution, 162, 110-119.
17. Wojewódzki Inspektorat Ochrony Środowiska (2012). Raport o stanie środowiska w województwie dolnośląskim w 2011 roku .Wrocław, 1-245.
18. Thomas W. 1986. Representivity of mosses as biomonitor organisms for the accumulation of environmental chemicals in plants and soils. Ecotoxicology and Environmental Safety, 11, 339-346.
19. Xiao-Li S., Yu P., Hose G.C., Jian C., Feng-Xiang L. 2006. Spider webs as indicators of heavy metal pollution in air. Bulletin of Environmental Contamination and Toxicology, 76(2), 271-277.
20. Zechmeister H.G., Dullinger S., Hohenwallner D., Riss A., Hanus-Illnar A., Sharf S. 2006. Pilot study on road traffic emissions (PAHs, heavy metals) measured by using mosses in a tunnel experiment in Vienna, Austria. Environmental Science and Pollution Research, 13, 398-405.

25

EFFICIENCY OF LEACHING TESTS IN THE CONTEXT OF THE INFLUENCE OF THE FLY ASH ON THE ENVIRONMENT

Jan Kalembkiewicz[1], Elżbieta Sitarz-Palczak[2]

[1] Department of Inorganic and Analytical Chemistry, Faculty of Chemistry, Rzeszow University of Technology, 6 Powstancow Warszawy Ave., 35-959 Rzeszow, Poland, e-mail: kalembic@prz.edu.pl

[2] Department of Inorganic and Analytical Chemistry, Faculty of Chemistry, Rzeszow University of Technology, 6 Powstancow Warszawy Ave., 35-959 Rzeszow, Poland, e-mail: epalczak@prz.edu.pl

ABSTRACT

The leachability of heavy metals (Cu, Pb and Zn) from the coal fly ash samples was studied. The investigations executed using three leachability tests (USEPA, TCLP, ASTM). The effect of different parameters was determined (the diameter of ash grains, the kind of leaching solutions, pH of leaching solutions, the volume ratio of leaching solutions to the mass of ash samples, and the leaching time) on the leachability of the heavy metals from fly ash samples. Moreover, the influence of pH and changes in the redox potential of the leaching solutions as well as the presence of organic compounds that could potentially form complexes with metals and solubility changes of metals. The concentration of the metals studied (Cu, Pb and Zn) in all obtained solutions was determined by FAAS method. On the basis of the research investigations, optimal leachability conditions for Cu, Pb and Zn from fly ash in this study have been determined. We have also found that pH of the leaching solution, and the presence of organic compounds which have a potential capacity of complexion the metals are the important factors in determining the solubility of Cu, Pb and Zn.

Keywords: heavy metals, fly ash, leaching test.

INTRODUCTION

Depositing waste at landfills and taking place in these uncontrolled processes physical, chemical, and biological properties pose a serious threat to the environment. For this reason, the waste from storage must be evaluated together with organic designation degree of toxicity [Witkowska-Kit 2000]. The degree of impact on the environment of waste containing contaminants in a form of toxic substances or harmful, depending on is not only the type and concentrations of these substances, but mainly their mobility. The mobility of pollutants contained in the waste, expressed as susceptibility to leaching them in water (also called leaching test), is the most important ecological sights indicator release for depositing waste at landfills. In Poland there is a lack of uniform standards and regulations relating to the preparation of aqueous extract and evaluate the reliability of the results obtained in the leaching test. Currently, two procedures are used to draw up the waste water. The first procedure is the result of the regulation of the Council of Ministers, 1997. The second procedure was established by Polish Standards [PN-Z-15009/1997]. The results obtained after the application of the above procedures do not indicate a total of the contents of the component in the waste, which can be subject to physicochemical change in the environment, give only an approximate indication in terms of assessing potential risks posed to the environment at the time referred to the waste.

Leaching tests can be divided into two main categories: 1. static tests and group and 2. dynamic tests and core barrels. In the static tests and group a quantity of the sample solution is specified and joined by a fixed period of time. The product of this connection is the leach substance. The main goal of the tests is a group that, at the end of the entire period of the process will be given a stable state of molecules. In the dynamic tests and column, the substance resulting in leaching constantly flows through the sample. Summary and comparison of group recruitment tests and column charts are contained in the following works [Zandi et al. 2007, Fällman, Aurell 1996, Egemen, Yurteri 1996, Lewin 1996, Wahlström 1996, Quevauviller et al. 1996].

Most waste in the field of employment is due to the structure of coal-fired power generation. As a result of burning coal solid waste (furnace waste) is produced: fly ash, slag, mixture of ash and slag, microspheres, boiler ash, flue gas desulfurization gypsum fluidized beds using limestone (wet method), wastes from flue gas desulfurization methods (dry and semi-dry methods), etc. [Galas et al. 2005]. Fly ash according to [Bilitewski et al. 2003] its the residues from the combustion of ground coal in amber furnaces, which are lifted of emissions and are captured by dust device. Ashes are contained in the mixture of dust-gas created from coal (hard coal and lignite).

Ash properties primarily depend on the chemical composition of non-flammable substances contained in different grades of coal burned [Pachowski 2002]. Fly ash, formed as a result of burning coal, may contain trace amounts of substances which during liquidation or disposal may be washed out, what poses a potential threat to the environment. Environmental conditions have a huge impact on the mobility of ash as well as on the physical and chemical properties. In order to simulate the behavior of the leached fly ashes in different environmental conditions and to reduce the discrepancies between the measurements made on the ground, and these have been developed in the laboratory, standard leaching tests. The most common tests used by eluting extraction procedures, i.e. the "Extraction" Procedure (EP) [Esakku et al 2008, Sophia, Swaminathan 2005], the "Toxicity by Characteristic Leaching Procedure" (TCLP) [Gitari et al. 2009, Lu et al. 2009, Jing et al 2006, Janus et al. 1998] and standardized extraction procedure developed by the American Society Of Testing And Materials (ASTM) [Dermata et al. 2004, Mangialardi 2003]. The leaching tests are widely used mainly as indicators to estimate the potential release of trace metals from the ash, in order to assess their impact on the environment in the process of rendering or deposition of fly ash under laboratory conditions [Sarode et al., 2010]. In addition, they allow the classification of trace elements of ash on the basis of their behavior during the leaching under different conditions and to identify the mechanisms of leaching [Fytiano, Tsaniklidi 1998, Hardaway et al. 1999, Janus et al. 1998, Praharaj et al. 2002, Sitarz-Palczak, Kalembkiewicz 2012].

Existing standard leaching schemes are insufficient because they do not depict any scenarios of waste liquidation or disposal , including ash. To reduce the surface storage of the ashes, new marketing options for their re-use are still searched for, which requires comprehensive and preceded by studies of their composition, physicochemical properties and susceptibility to further processing by various technologies and behavior in contact with various external factors.

The tests carried out in the framework of this work were to assess the impact on the environment from coal ash as they deposit in landfills and the certification of fitness for non-industrial utilization. To this end, studies were carried out regarding changes to the solubility of selected heavy metals (Cu, Pb and Zn) in the fly coal ash in anaerobic conditions.

MATERIALS AND METHODS

Apparatus

The flame atomic absorption spectrometer Perkin-Elmer 3100 Model (Shelton Instruments, CT USA) was used for Cu, Mn and Zn determination in the solutions. Measurements were performed at wavelength 324.8 nm (Cu), 217 nm (Pb) and 213.9 nm (Zn) using a yellow fuel-rich air-acetylene flame (acetylene flow velocity 2.0 $dm^3 \cdot min^{-1}$; air flow velocity 8.0 $dm^3 \cdot min^{-1}$) and burner height of 3 mm. The analytical lines were selected using a slit width 0.7 mm and followed by [Perkin-Elmer 1982]. Hollow cathode lamp at 10 mA (Cu and Zn), 20 mA (Pb) was used. The centrifuge tube-test Model WE 1 (Precision Engineering, Poland) – was used for the centrifugation of the soil extracts at the at appropriate rotary speed of 3000 rpm. The universal laboratory shaker Vibramax 100 Model (Heidolph Instruments, Germany) and a hot plate HP 88720-26 Model (Barnstead/Thermolyne, USA) were used for the extraction. The pH of the extraction solutions was determined with a pH meter CPI-551 Model (Elmetron, Poland) supplied with a glass combination electrode (Eurosensor, Poland).

Reagents and solutions

All chemicals and reagents were of analytical grade or higher purity and were obtained from POCH, Gliwice, Poland. The solutions were prepared by dissolving appropriate compounds in double-distilled water from the Water Purification System (SolPure 7 POLL LAB, Poland). The reagents were prepared and stored in clean polyethylene bottles. Glassware and plastic ware (PE) used throughout the experimental work were previously soaked in 10% nitric acid bath overnight and washed thoroughly in double distilled water. Standard solutions were prepared from standard solutions for atomic absorption (Sigma-Aldrich Chemie GmbH, Switzerland) – the concentration of all metals in standard solutions were the same and equals 1.000 $\mu g \cdot cm^{-3}$ in 1% of HNO_3. Working standard solutions containing Cu, Pb and Zn were prepared by serial dilution of the appropriately of standard solution for atomic absorption.

Sample collection and preparation

Coal fly ash samples used for the experiments were collected from the electric filter from Rzeszów S.A power-plant (Rzeszów, Poland). The 0.5 kg sample was prepared from overall air-dried 10 kg sample by a "quarterning" method – according to Polish Branch Standard [BN-81/0623-01] procedure. After that, the air-dried ash was sieved, initially through a laboratory sieve of 1 mm diameter, and then milled in an agate mortar to fine powder ($\varphi \leq 100$ μm) – according to Polish Standard [PN-77/G-04528/00].

Fly ash

Physicochemical characteristics of ash samples included determination of pH using potentiometric method [PN-ISO-10390]; determination of organic carbon content - Tiurin method [PN-95-Z-15005] and composition of granulometric by using the sieve methods [PN-R-04032]. The total content of metals is obtained after mineralization in a mixture of concentrated HNO_3 acid, HF and HCl under conditions described in the work [Sitarz-Palczak, Kalembkiewicz 2008].

Standard leaching tests

Three standard leaching tests (USEPA, TCLP, ASTM) were selected for this study to determine stability of trace elements (Cu, Pb and Zn) in fly ash samples under three different conditions (Table 1). Leaching tests were carried out based on the principles laid in standard procedures. To obtain USEPA, TCLP and ASTM extracts 5 g of fly ash was agitated in polyethylene bottle containing 100 cm^3 of extracting medium for the specified time periods of 24 and 18 h, respectively. After the required period of agitation, all extracts were stirred with a glass rod and filtered through 0.45 μm filter paper. The filtrates were acidified with a concentrated HNO_3 addition and transferred to a polyethylene bottle and stored in a cool dry place for further analysis by atomic absorption spectrophotometry (FAAS). All extractions were carried out as triplicate runs.

Effect of parameters of ash and leaching process

The effect of selected parameters on the output samples of ash (particle diameter ash) and leaching process (a type of leaching solution, the pH of the solution, the ratio of volume of leaching solution to mas of the sample and the influence of the ash leaching time) on the leaching of heavy metals from the ash samples was studied. In order to determine the effect of particle diameter ash and the type of leaching solution has three standard tests of leaching: USEPA TCLP, and ASTM (Table 1), with respect to the fraction of the ashes received after the sieve analysis. The proceedings were the same as in the case of a standard leaching procedures described above. In order to determine the effect of pH leaching solution (2.5% CH_3COOH) on the leachability of metals from the ash particle diameter $\varphi = 1.5$ mm done series of 2.5% CH_3COOH that pH values: 2; 3; 4; 7 and 12. Adjust the pH was carried out using 0.1 $mol \cdot dm^{-3}$ HCl and 0.1 $mol \cdot dm^{-3}$ NaOH. Then use the so prepared solutions 2.5% CH_3COOH fixed pH values were leaching of metals (Cu, Pb and Zn) under the conditions given in Table 1 – in relation to a standard TCLP procedure. To determine the influence of relative elution volume [cm^3] to mass [g] on the leaching Cu, Pb and Zn with samples of ash made a series of leaching tests according to the standard TCLP procedure, in which the ratio of L:S [cm^3:g] was, respectively: 1: 1; 2: 1; 5: 1; 10: 1, 20: 1 and 50: 1. In order to determine the influence of time on the leachability Cu, Pb and Zn from ash fractions with a diameter of $\varphi = 1.5$ mm using standard leaching test, the following conditions were established as optimal i.e. leach-

Table 1. Conditions of leaching tests

Tests conditions	USEPA [USEPA 1987]	TCLP [TCLP 1311 1990]	ASTM [ASTM D 3987-85 2001]
Leaching solution	0,5 mol/dm^{-3} nitric acid	2,5% acetic acid	Distilled water
Liquid to solid ratio [cm^3/g]	20:1	20:1	20:1
Leaching time [h]	24	24	24
pH of leachig solutions	0.6 ± 0.1	4.0 ± 0,1	6.3 ± 0.1
Temperature [^{0}C]	19 – 25	19 – 25	19 – 25
Number extraction	1	1	1

ing solutions 2.5% CH_3COOH (TCLP); pH = 2.5; the ratio of the volume of leaching solution to the mass of the sample L:S = 50:1 [cm^3:g]. Assuming the following leaching times: 0,5; 2; 4; 6; 8, 12 and 24 h. subsequent procedure was the same as in the case of leaching procedures described above.

Effects of selected organic compounds with a potential capacity of complexion the metals on the solubility of Cu, Pb and Zn in anaerobic conditions

In order to determine the effect of complexion substances on the contents of selected metals in the soluble forms contained in the ashes two series of complete extractions by shaking samples of ash were performed, with the complexion addition substances: citric acid or tartaric acid with the following concentrations: 10, 20, 40, 100 mg·dm^{-3} corresponding 6.25; 12.50; 25.00; 62,50 mg·kg^{-1} ash. The solubility change of Cu, Pb and Zn in anaerobic conditions was studied in a series of incubation of samples of ash with the addition of organic substances constituting a potential source of easily accessible coal they were glucose and tartaric acid with a concentration of 200 mg·dm^{-3}. Incubations are performed at in confinement without access of oxygen for a period of 7 days, stirring the sample for 10 minutes from time to time to ensure the balance between the permanent and the solution. After the incubation was carried out through the extraction conditions of shaking in leaching tests (USEPA and TCLP), which is given in Table 1.

RESULTS AND DISCUSSION

Basic physicochemical properties of ash were presented in Table 2. On the basis of the obtained results, it was concluded that coal ash shows slightly alkaline pH. Working with [Łączny 2002] it is known that the energy (fly ashes) compounds of heavy metals present in the waste are in slightly alkaline environment, because the ions of these metals in this environment are generally less mobile complex compounds. If intended for the storage of fly ash, or other energy waste will have basic character, the main threat to the environment will be releasing such compounds as sodium, potassium and calcium sulphate. So in the case of investigational ash (wood and coal) there is no risk of wash of heavy metal ions with alkaline base nature of these wastes. Organic carbon content in the ash residue of unburned carbon [Su, Wong 2003]. Based on the results of the analysis, the granulometric composition it was found that the greatest part is the fraction of particulate matter, which is consistent with the results obtained by [Antonkiewicz 2014]. The total content of trace metals in the ash is varied. Since they all occur at concentrations above 50 mg · kg^{-1} can they have a large impact on the environment (Table 2). In a view of the fact that ash is a heterogeneous material, elements contained therein are not evenly distributed. The content of organic carbon assessed in fly ash reveal the presence of unburned coal residues. The surface layer is enriched in

Table 2. Physicochemical properties of fly ash (the results presented reflect values of data taken from 10 samples)

Parameter	Value of parameter
pH:	
H_2O	9.1
1 mol·dm^{-3} KCl	8.9
Content of organic carbon [g·kg^{-1}]	53.9
Texture [%]:	
Clay (0 – 2 μm)	0.9
Silt (2 – 50 μm)	75.7
Sand (> 50 μm)	23.4
Total content [mg·kg^{-1}] of metal	
Cu	63.7 (± 3.9)
Pb	101.2 (± 4.6)
Zn	200.0 (± 4.1)

copper and zinc, while lead is distributed evenly between the sensor and the surface of the grains [Izquierdo, Quero 2012].

Standard leaching tests

In natural conditions (under the influence of atmospheric conditions and the presence of microorganisms – in landfills there are numerous physical, chemical and biological processes, not subject to the control. These processes give rise to, first and foremost, the aging of the deposited material, its weathered and leaching. Aging is generally to reduce the solubility, and weathered to the creation of partially soluble products. The leachability applies to an easily soluble metal form, present in the waste, i.e. first of all, salt, in which the composition has sulfates of calcium, potassium, sodium and magnesium. The immediate effect of leaching is an increase in permeability of the deposits, increased salinity of groundwater and surface water. The leachability of Cu, Pb and Zn in fly ash obtained after the application of three standard leaching tests (USEPA TCLP and ASTM) was presented in Table 3. The highest degree of leaching for Pb of 56.8% in relation to the total content of Pb in the fly ash, were obtained using the USEPA test; respectively 22.3% was obtained by means of TCLP procedure. While the ASTM test using distilled water was obtained the lowest value of leaching equal to 2.5% in relation to the total contents of Pb in the fly ash. The degree of leaching for copper was 62.7% using the USEPA test and only 19.6% in the case of the TCLP test in relation to the total content of Cu in the fly ash. The leachability of copper in the case of ASTM test was below the limit of detection for this metal, as determined by FAAS method. Zinc showed the greatest percentage of leachability using the USEPA procedures and it was 35% in relation to the total content of the metal in the ashes. and 14.6% after applying the TCLP test. The leachability of Zn in ASTM test does not exceed the limit of detection for the determination of this metal by the FAAS method. In summary, containing a test in which the largest were the concentration of soluble forms in respect of all designated metals in coal ash is the USEPA. In the light of the above, it can be concluded that this test shows the highest efficiency in relation to the determined metals Cu, Pb and Zn. The USEPA procedure is more effective due to aggressive conditions during the leaching of metals. Taking into account the value of the degree of leaching of individual elements, it can be concluded that the degree of leaching of Cu, Pb and Zn is decreasing in the following series: Cu > Pb > Zn. In spite of all processes taking place in the environment it is suggested that the most appropriate and best simulating natural conditions is ASTM procedure, in which the eluting solution is distilled water. However, research carried out for the TCLP test may be helpful in the diagnosis of possible risks which might occur in the event of "acid rain".

Statistical analysis of the results obtained on the concentrations of metals in the soluble forms, and their total content in the ash was carried out using a linear regression. Figure 1 shows the obtained relationship between leachability of Cu, Pb and Zn in relation of the total content of these metals in studied the ashes after application of the TCLP test. On the basis of the obtained result, it was found that an important relationship between the soluble forms of metals and their total content in fly ash prove the obtained correlation values of the coefficients, which are : Zn – 0.68, Cu – 0.83, Pb – 0.82. The results of the studies on the content of the soluble forms Cu, Pb and Zn in fly ash al-

Table 3. Summary of the concentrations of Cu, Pb and Zn in the soluble forms and relative leaching of these metals by USEPA TCLP and ASTM procedures (concentration of metal in the fly ash and the confidence interval were given for n = 10 at p = 95%)

Procedure	Concentration of metals [mg·kg^{-1}]	Cu	Pb	Zn
USEPA	Leachability	40.0 (± 2.6)	57.5 (± 2.9)	70 (± 3.5)
	Relative leaching [%]	62.7	56.8	35.0
TCLP	Leachability	12.5 (± 0.7)	22.6 (± 1.1)	29.1 (± 1.6)
	Relative leaching [%]	19.6	22.3	14.6
ASTM	Leachability	< l. d.	2.5 (± 0.2)	< l. d.
	Relative leaching [%]	-	2.5	< l. d.
Leachability limit [Rozporządzenie...2006]		0.05	0.5	2.0

l. d. - limit of detection by FAAS method.

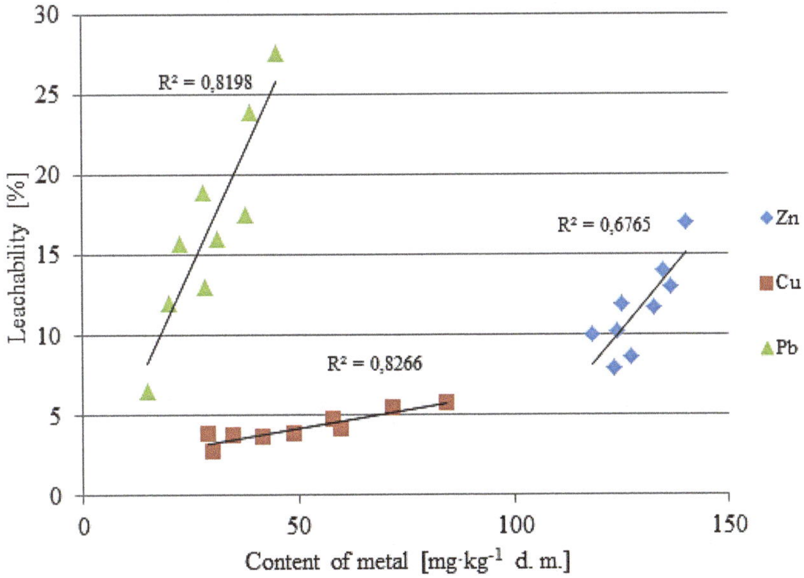

Figure 1. Relation between leachability of Zn, Cu and Pb and total concentration this metals in fly ash after the TCLP test used

low an approximate assessment of the risks posed to the water – soil environment at the time by a specified waste. For this purpose, the concentrations of trace investigated metals received in the soluble forms in the ash were compared with the limit values for wastewater entering into waters or into the ground, assembled in a Table 3. On the basis of the obtained results, it was found that the concentration of Zn obtained after leaching at the USEPA, TCLP and ASTM tests exceed acceptable concentrations for this metal for wastewater entering into waters or into the ground. The same situation occurs in the case of Cu and Pb after leaching with USEPA TCLP, tests from the fly ash. The obtained results, due to the heterogeneity of the test material, should be verified on the basis of the follow-up testing of leachability of metals from the ashes.

In assessing the dangers of fly ash containing pollution in a form of heavy metals, in addition to the degree of leaching, one should take into account the additional effect of the aging of waste. It can be both beneficial and detrimental. This is a very important ecological factor for the future in relation to the waste deposited in landfills. It may actually happen that the waste with low leaching today become more washed out after a few years, or waste with high leaching become harmless after a few years of storage. Linked to this need for constant monitoring of ash storage facilities, which, in the light of the obtained results pose a risk of contamination of the environment water near the landfill.

Effect of parameters of ash and leaching process

There is a number of factors that may affect the leach of each constituent elements in fly ash leaching tests. Some of these factors are related to the properties and composition of fly ash, others depend on conditions of leaching [Zandi et al., 2007]. Test conditions (LS ratio, leachant pH and leaching time) have a great impact on these results. The conditions for measuring potential leachability have not been clearly addressed in literature; there are a few contradictory studies on the principles of leaching and effects of control factors such as pH and LS ratio on fly ash leaching behavior [Fällman 1997].

Particle size of fly ash

Fly ash have poorly developed a pore structure. The degree of development of the structure of the surface of the grains and the size of the specific surface area to a large extent determine the susceptibility of the ashes on the leaching of soluble parts from them. Therefore, it can be seen, that both the grain diameter as well as their appropriate size will determine indirectly the leaching solutions on the tested metals. Figures 2 - 4 illustrate the effect of particle diameter on the leachability of soluble forms of Cu, Pb and Zn from the fly ash using the USEPA TCLP tests. On the basis of the results presented in Figures 2 - 4, it was concluded that particle diameter of fly ash has a significant effect on the leachability

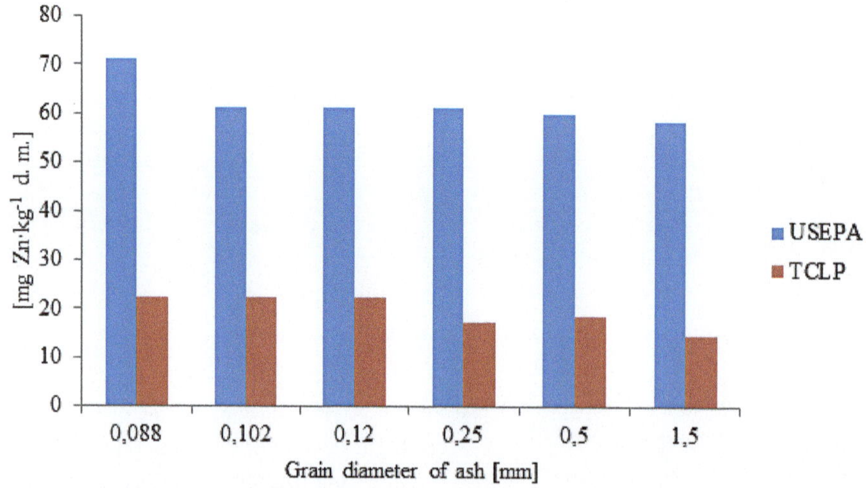

Figure 2. Content of Zn in the soluble forms in the fly ash with different grain diameter of ash obtained by the use of the leaching test (USEPA, TCLP) and determined by FAAS method

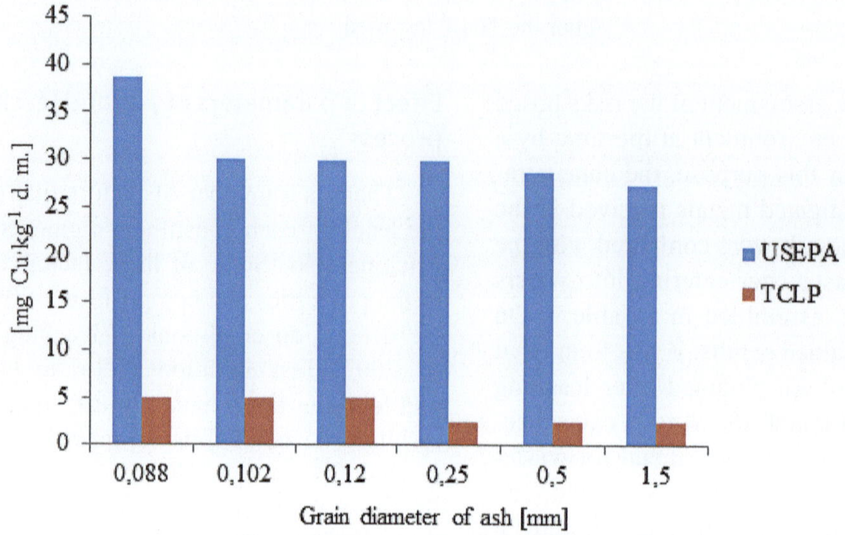

Figure 3. Content of Cu in the soluble forms in the fly ash with different grain diameter of ash obtained by the use of the leaching test (USEPA, TCLP) and determined by FAAS method

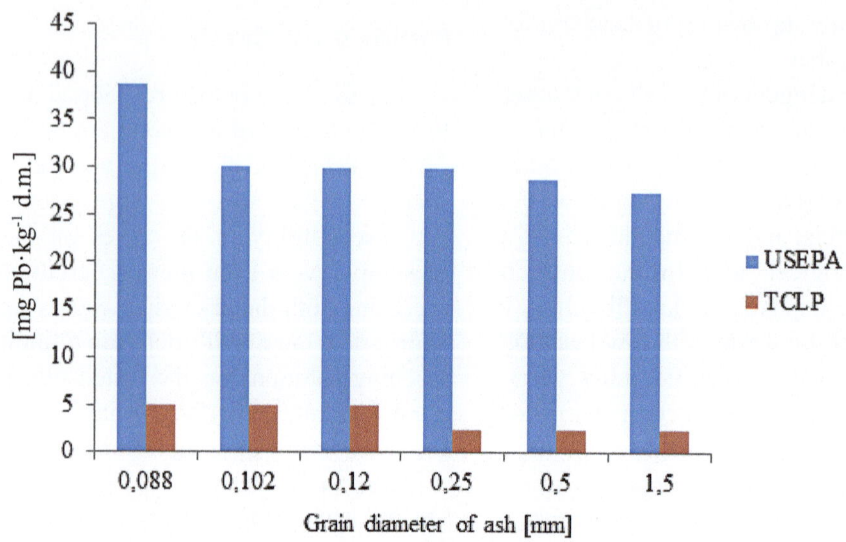

Figure 4. Content of Pb in the soluble forms in the fly ash with different grain diameter of ash obtained by the use of the leaching test (USEPA, TCLP) and determined by FAAS method

of soluble forms of Zn and Cu, as well as in the application of the tests USEPA and TCLP. With regard to both the highest of leaching metals were received by the ash of 0.088 mm diameter and solution of HNO_3 (USEPA) respectively, for was equal to 72 mg·kg^{-1}, while for Cu 38 mg·kg^{-1} in respect of the dry mass of the ash. The effect of particle of diameter of ash on leachability of Pb shows an inverse relationship, as in the case of other metals by applying USEPA and TCLP tests, namely the leachability of Pb increases with ash particle diameter. The leachability of lead using the USEPA test shows the highest concentration of soluble forms of metal particle diameter ash equal: 0.088 and 1.5 mm. The lowest Pb content in the soluble forms was received for grain diameter of ash equals 0.102 mm.

Leachability as a function of pH

The pH of the solution is very important as it determines the surface charge of the fly ash, and the degree of ionization and speciation of the elements in solution. The interactions between the charged ions in solution and the surface of fly ash particles contribute to release or adsorption of species. The results shown in Figure 5 indicate the trend of pH release of Cu, Pb and Zn. The concentration of soluble forms of all studied metals in fly ash decreases with increasing alkalinity of the leaching solution. pH in the range of 2 to 3 found the largest leachability for all heavy metals tested; respectively for Zn 25–15%, copper 31–19% and lead 36–26% in relation to the total contents of the metals in the fly ash. Low concentrations of these metals were observed in leachate solution (2.5% CH_3OOH) at high pH (> 9) but increased at pH < 9. The leaching trend for these metals indicates the dependence of the release of some metals on the dissolution of hydrous oxides of iron and manganese. Mn and Fe-oxides present in fly ash are formed during combustion of coal, and on hydrolysis form Mn and Fe hydroxides. Trace metals, such as Cu and Pb are associated with these hydroxides and on dissolution of these hydroxides, the trace metals are bound to be released into solution. Adsorption/desorption also plays a role in controlling the concentrations of heavy metals.

Fly ash material generates an alkaline pH within a few minutes of contact with water in most cases, due to higher quantities of soluble basic oxides than the soluble acid phases [Choi et al 2002, Reardon et al. 1995]. Several studies [Cho et al. 2005, Lee, Saunders 2003, Steenari et al. 1999] have shown that at higher pH, the surfaces of the fly ash particles are negatively charged. This would lead to increase in the adsorption of most heavy metals from the solution. The increase in adsorption of heavy metals ions in fly ash solutions at higher pH (> 10) may also be attributed to calcium content which provides alkalinity in the system raising the pH to strongly alkaline values, thereby facilitating the uptake of metal ions by the fly ash.

The pH of the environment has a great impact an chemical speciation of leachate elements from fly ash in the aqueous environment except for highly soluble elements which are mobile under both acidic and basic conditions. The difference

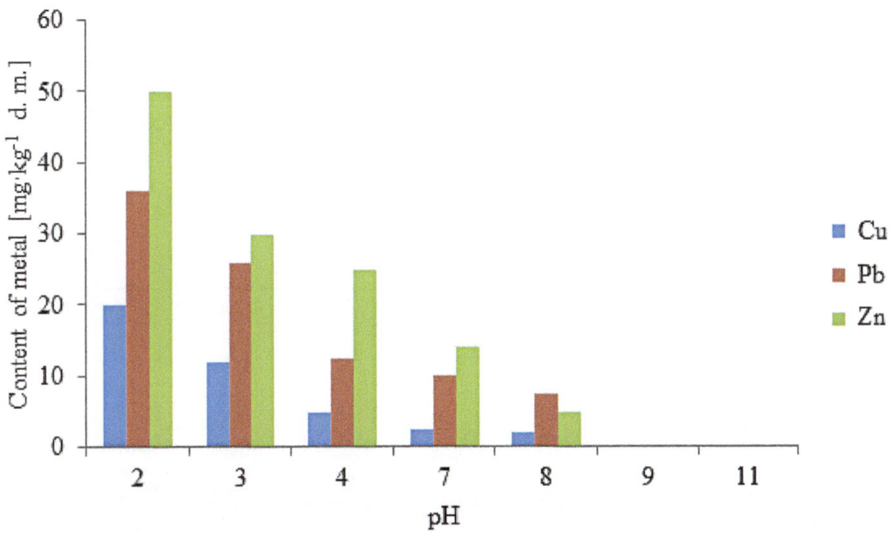

Figure 5. Change of content of Cu, Pb, Zn in the soluble forms in the fly ash under the influence of changes to the influence pH of leaching solution obtained after the TCLP procedure

between total quantity of element in the fly ash and the potentially leachable constituent is due to structure and mineralogy of the fly ash and associations of the element major insoluble phases present in the fly ash.

Solubility of metal compounds is an important property that allows metals to penetration into the structure of the porous adsorbent. The concentration of such speciation forms of metals as: MOH^+, $M(OH)_2$, $M(OH)_3^-$, $M(OH)_4^{2-}$ at pH 6 < are too small to affect the sorption process the dominant forms of metal M^{2+} (M = Cu, Zn, Pb) [Ho, et al., 2002]. This is confirmed by the data presented in Figure 6 the speciation graph of metal ions, whose curves are compatible with the data from literature [Zirino, Yamamoto, 1972, Cuppett et al. 2006, Cárdenas López et al. 2007, Deliyanni et al., 2007]. They have been generated on the basis of the program Medusa [http://www.kemi.kth.se/medusa/]. The speciation curves for the different ions of heavy metals indicate that different forms dominate in the solution within a certain range of pH; for example Zn^{2+} is a dominant form in solution at pH < 6. Above this pH in solution increases the concentration of $ZnOH^+$ and ZnO. With regard to the adsorption of Cu and Pb it should be kept in mind that these speciation forms of metals, or simple ions Cu^{2+} and Pb^{2+} have mainly to pH about 6, above which exponentially increases the concentration of CuO and $Pb(OH)_2$. In general, there are different mechanisms that may control the release of a particular constituent from a particular fly ash in an aqueous environment. Highly soluble elements available on the surface of the fly ash often fall into the category of available controlled leaching if the concentration released into the leachant is limited only by the amount of the element in the fly ash. These element are expected to become mobile soon after contact with aqueous environment and leach at high concentrations. Leachability of some other constituents may be solubility controlled where leachability of elements in the fly ash can be described by dissolution and precipitation processes.

Leachability as a function of redox potential

The leaching of some metals e.g., Fe, Cr, As can depend significantly on the redox conditions. The redox conditions can change due to the access of air to the sample during the storage or in performing the leaching test [Wahlström1996]. The results concerning the effect of the potential redox on the contents of metals in the soluble form in the fly ash are shown in Table 4. At high values of the potential (≈170 mV), the concentration of Cu, Pb and Zn in the soluble forms are highest and are respectively equal to: Cu - 37.5; Pb - 50 and Zn - 80 $mg \cdot kg^{-1}$. At the same time, as the high value of the potential suggests, these metals are present in the oxidized forms by taking the highest oxidation: Cu^{II}, Pb^{IV} and Zn^{II}. Along with the decline in the potential decrease the content of Cu, Pb and Zn in the soluble forms, which testifies to their presence in the reduced forms.

Leachability as a function of L:S ratio

The ratio of the volume of leaching solution to the mass of sample fly ash (L:S) is one of the more important factors affecting the leaching of trace elements from the ashes samples. The most commonly used ratio of the liquid phase to the solid phase (L:S) is located in the range (10–20): 1, but with a higher ratio, for example 50:1, the leaching substance is more easily mixed with fly ash, and trace components are increasingly leached away from the seeds of the ash. The results shown in Table 5 fully confirmed the correlation.

Table 4. Effect of the redox potential of leaching solutions on the content of Cu, Pb and Zn in the soluble form in the fly ash obtained after the TCLP procedure (concentration of metal in the fly ash and the confidence interval were given for n = 6 at p = 95%)

Value of SEM [mV] of leaching solutions	Value of SEM [mV] before leaching	Value of SEM [mV] after leaching	Concentration of soluble forms of metal [$mg \cdot kg^{-1}$]		
			Cu	Pb	Zn
189	178	170	37.5 (±5.5)	50.0 (±7.2)	70,0 (± 9.5)
162	159	140	20.5 (±2.8)	26.0 (±3.3)	55,0 (±5.8)
100	95	89	5.0 (±)	12.5 (±1.5)	25.0 (±6.5)
-25	-20	-10	2.5 (±1.1)	12.5 (±3.1)	7.5 (±1.8)
-96	-96	-88	< l. d.	5.0 (±1.9)	5.0 (±1.7)
-161	-167	-158	< l. d.	< l. d.	< l. d.

l. d. – limit of detection by FAAS method.

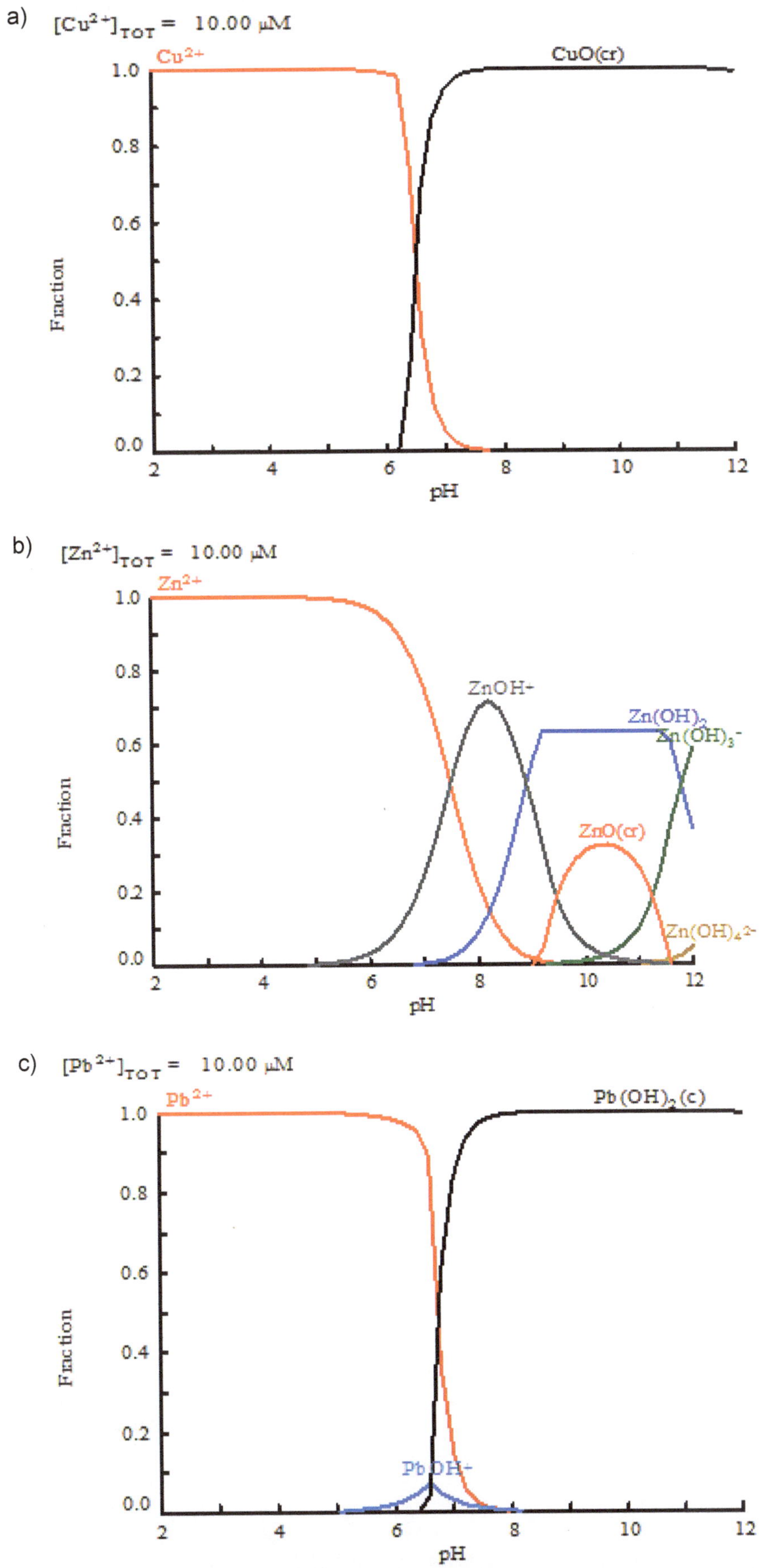

Figure 6. Speciation curves of a) Cu, b) Zn, c) Pb
Comments: (c) – solid substance , (cr) – crystalline form

Table 5. The content of Cu, Pb and Zn in the soluble forms in fly ash (φ =1,5 mm) obtained by a variety relations of volume of 2.5% acetic acid (pH = 2.5) to the mass of the sample – L:S (concentration of metal in the fly ash and the confidence interval were given for n = 6 at p = 95%)

Ratio L:S [cm³:g]	Concentration of soluble forms of metal [mg·kg⁻¹]		
	Zn	Cu	Pb
2:1	2.5 (± 0.2)	2.5 (± 0.2)	< l.d.
5:1	2.5 (± 0.1)	2.4 (± 0.1)	< l.d.
10:1	14.9 (± 0.7)	2.4 (± 0.5)	12.4 (± 0.9)
20:1	30.0 (± 1.6)	8.9 (± 0.6)	24.9 (± 1.3)
50:1	29.4 (± 1.6)	8.7 (± 0.7)	24.0 (± 1.6)

l. d. – limit of detection by FAAS method.

Leachability as a function of time

The execution time of the leaching test is an important parameter affecting the mobility and speciation of trace elements in fly ashes. The study of the effect of execution time of leaching of trace metals (Cu, Pb and Zn) from ash are essential from the perspective of the protection of the environment. The main objective of the research is to estimate the leaching time quantity of heavy metals in the soluble forms that are subject to migration, with the ash landfills in the wild in real life conditions. The results of these studies are necessary to assess the potential of the possible time of the release of the components of fly ash in the context of its liquidation in rendering specific conditions [Zandi et al., 2007]. The results obtained for this aspect of the study are shown in Figure 7. Time of leaching showed a significant impact on the concentration of Cu, Pb and Zn in soluble forms. Copper concentrations were relatively constant for the execution time of the leaching test from 0.5 to 12 h. For the leaching time equal to 24 h, the contents of this metal increases. Lead showed a tendency to increase the concentration over the execution time of leaching test from 15 to 25 mg·kg^{-1}. Zinc, like lead, showed an increase to the highest concentration with increasing execution time of the leaching test. It achieved the highest concentration after 24 hours of leaching test duration.

Effects of selected organic compounds with a potential capacity of complexion the metals on the solubility of Cu, Pb and Zn in anaerobic conditions

One of the effective ways of reducing the solubility of heavy metals is the use of probing by using specific groups of organic compounds, obtained by chemical synthesis or isolated from natural substances. For complexion compounds tartaric acid, citric acid, glucose were used. For complexion compounds of natural origin may be classified as humic substances. In order to determine the effect of selected organic compounds with a potential capacity of complexion of metals on the solubility of Cu, Pb and Zn in anaerobic conditions, the fly ash made with the addition of complexion reagents i.e. tartaric acid, citric acid and glucose. The results of the concentrations of studied metals in the soluble forms in fly ash containing tartaric acid in the concentration range 10–200 mg dm^{-3}, the obtained after applying the TCLP test are shown in Figure 8. The presence of tartaric acid affects the concentration of the form soluble forms of Pb. Namely, for Zn there was an increase in the concentration of metal in relation to the concentration of the metal in the soluble forms in the ash without tartaric acid, whereas for Pb it was reduced. The concentration of the soluble forms of Cu in the fly ash throughout the range of concentrations $C_4H_6O_6$ is the same.

Figure 9 illustrates the changes in concentrations of metals in the soluble forms contained in the ashes under the influence of different concentrations of citric acid in the range of 10-100 mg·dm^{-3} introduced into the fly ash. The addition of citric acid to the fly ash caused an increase in the concentration of soluble forms of Zn in the ash as compared with to the fly ash no the complexion substances addition. This is different in the case of the other elements; for Cu concentration in the soluble forms decreased by half the value of the concentration of copper in the soluble forms in the fly ash with no citric acid addition, whereas for Pb it was equal to zero. Regardless of the value of the concentrations $C_6H_8O_7$, the quantity of soluble forms of metal in the ash remains the same. Generally, the presence of citric acid is beneficial for leaching of Zn, as compared to other elements.

The concentrations of the investigated metals in solutions obtained after leaching containing fly ash with addition of 200 mg·dm^{-3} of glucose by TCLP procedure revealed no significant results. Comparing the results obtained with the concentrations of metals in the soluble form of fly ash with no glucose addition, an increased concentration of Zn can be seen . The contents of Cu and Pb in soluble forms has not changed significantly.

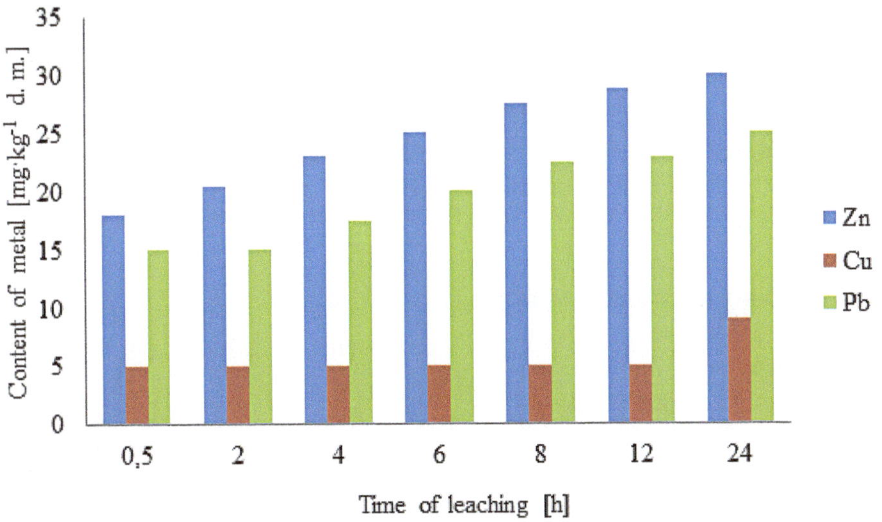

Figure 7. Relation between a time of leaching and the concentration of Zn, Cu and Pb in the soluble forms in the fly ash having a grain diameter φ = 1.5 mm, obtained by leaching with 2.5% CH_3COOH, pH 2.5

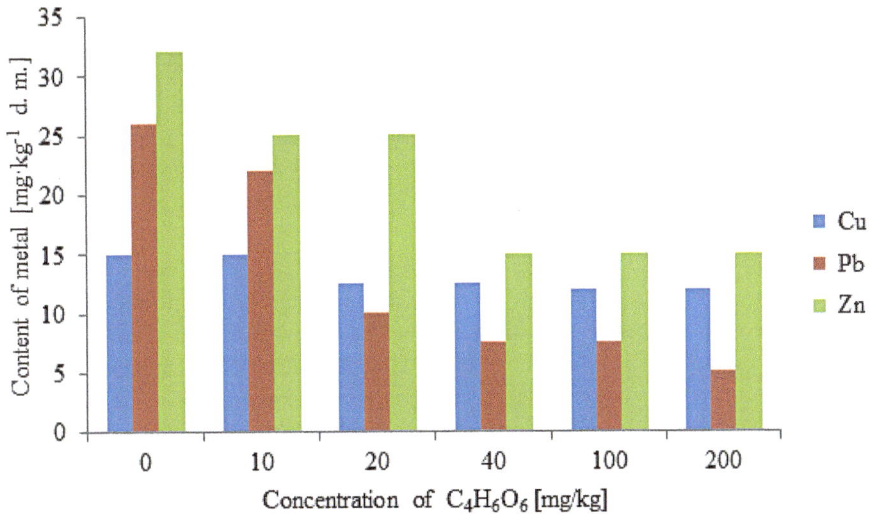

Figure 8. Changes of concentrations of soluble forms of Zn, Cu and Pb in the fly ash with the addition of tartaric acid, obtained by the use the TCLP test

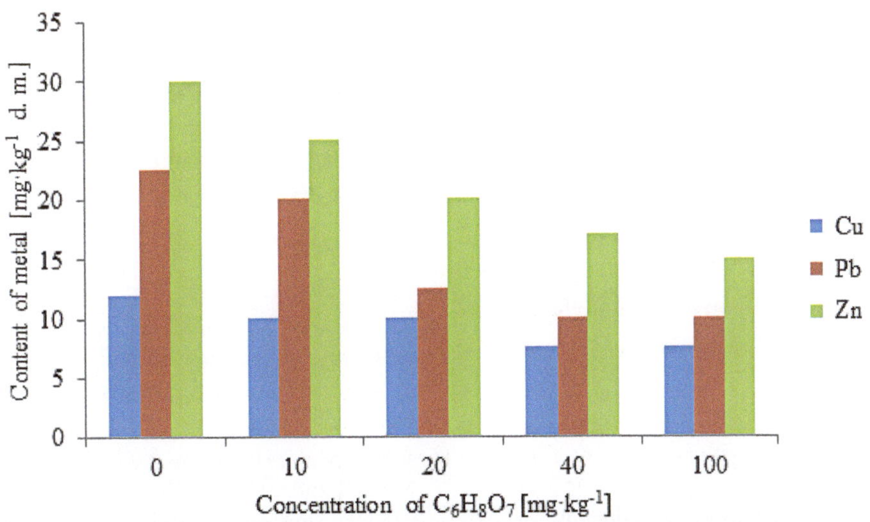

Figure 9. Changes of concentrations of soluble forms of Zn, Cu and Pb in the fly ash with the addition of citric acid, obtained by the use the TCLP test

CONCLUSIONS

1. The contents of metals in the soluble forms received after leaching the fly ash using the standard leaching testing exceeds the permissible concentrations of Cu, Pb and Zn in relation to the standards for waste entering into waters or into the ground.
2. Considering, the solubility and mobility of heavy metals the fly ash can pose a threat to water and soils in the area of storage and storage creates the risk of inorganic contamination of the environment.
3. With regard to the ageing of inorganic waste, such as fly ash, it should be pointed out that monitoring both the composition and mobility of heavy metals present in the fly ashes is important.
4. Depending on the chemical composition of the leaching solutions and the diameter of the grains examined fly ash, exhibits a different level of mobility of Cu, Pb and Zn in the soluble forms depending on the conditions of the leaching process.
5. On the basis of the conducted research it can be concluded that the optimal method for assessment of the impact of surface ash storage on the environment is the TCLP test made under the following conditions: particle diameter ash φ = 1.5 mm; pH 2.5% CH3COOH = 2.5; L:S ratio = (20) 50: 1; leaching time = 24 h.
6. In addition, it was found that the presence of organic compounds with a potential complexion of metals has a significant impact on the concentration of Cu, Pb and Zn in soluble forms. Generally, the addition of tartaric or citric acid in the fly ash causes a decrease in the concentrations of metals in the soluble forms, with regard to ash without the above mentioned acids added after being applied of the TCLP test in optimal conditions.
7. Presence of glucose in fly ash caused an increase in the concentration of the soluble forms of Zn received by TCLP procedure compared to the concentrations of the metal in the soluble forms in fly ash with no glucose addition. The contents of Cu and Pb in the soluble forms has not been a significant change.

REFERENCES

1. Witkowska-Kit B., Zasucha J. 2000. Metody badań szkodliwego oddziaływania odpadów na środowisko, Utylizacja Odpadów Przemysłowych i Komunalnych, 2, 5–7 [In Polish].
2. Rozporządzenie Rady Ministrów z dnia 30.12.1997 r. w sprawie opłat za składowanie odpadów. Dz. U. RP, Nr162, poz. 1135 [In Polish].
3. Polska Norma PN-Z-15009/1997. Odpady stałe – Przygotowanie wyciągu wodnego [In Polish].
4. Zandi M., Russel N.V. 2007. Design of a leaching test framework for coal fly ash accounting for environmental conditions. Environ. Monit. Assess., 131, 509–526.
5. Fällman A.M. Aurell B. 1996. Leaching tests for environmental assessment of inorganic substances in wastes, Sweden. Sc. Total Environ., 178, 71–84.
6. Fällman A.M. 1997. Performance and design of the availability test for measurement of potentially leachable amounts from waste materials. Environ. Sc. Tech., 31, 735–744.
7. Egemen E., Yurteri C. 1996. Regulatory leaching tests for fly ash: a case study. Waste Management & Research, 14, 43–50.
8. Wahlström M. 1996. Nordic recommendation for leaching tests for granular waste materials. Sc. Total Environ., 178, 95–102
9. Lewin K. 1996. Leaching tests for waste compliance and characterisation: recent practical experiences. Sc. Total Environ., 178, 85–94
10. Quevauviller Ph., van der Sloot H.A., Ure A., Muntau H., Gomez A., Rauret G. 1996. Conclusions of the workshop: harmonization of leaching/extraction tests for environmental risk assessment. Sc. Total Environ., 178, 133–139.
11. Galas K., Uliasz – Bocheńczyk A. 2005. Źródła i użytkowania popiołów lotnych ze spalania węgla w Polsce. Gospodarka Surowcami Mineralnymi, 21(1), 23–40 [In Polish].
12. Bilitewski B., Hardtle G., Merk K. 2003. Podręcznik gospodarki odpadami, Teoria i praktyka. Wydawnictwo Seidel-Przywecki Sp. z o.o, Warszawa, p. 45 [In Polish].
13. Pachowski J.2002 Rozwój technologii powstawania ubocznych produktów elektrownianych oraz ich charakterystyka i możliwości zastosowań w technologiach budownictwa drogowego. Drogi i Mosty, 1, 59–99 [In Polish].
14. Esakku S., Karthikeyan O.P., Nagendran R. 2008. Heavy metal fractionation and leachability studies on fresh and partially decomposed municipal solid waste. Practice Periodical of Hazardous, Toxic, and Radioactive Waste Management, 12, 127–132.
15. Sophia A.C., Swaminathan, K. 2005. Assessment of the mechanical stability and chemical leachability of immobilized electroplating waste. Chemosphere, 58, 75–82.
16. Janusa M.A., Bourgeois J.C., Heard G.E., Kliebert N.M., Landry A.A. 1998.Effectts of Particle and Contact Time one the Reliability of Toxicity Characteristic Leaching Procedure for Solidified/Stabilized Waste, Microchem. J., 59, 326–332.

17. Gitari W.M., Fatoba O.O., Petrik L.F., Vadapalli V.R.K. 2009. Leaching characteristics of selected South African fly ashes: Effect of pH on the release of major and trace species. J. Environ. Sc. and Health Part A, 44, 206–220.

18. Lu S.G., Chen Y.Y., Shan H.D., Bai S.Q. 2009. Mineralogy and heavy metal leachability of magnetic fractions separated from some Chinese coal fly ashes. J. Hazardous Materials, 169, 246–255.

19. Jing C., Liu S., Korfiatis G.P., Meng X. 2006. Leaching behavior of Cr(III) in stabilized/solidified soil. Chemosphere, 64, 379–385.

20. Dermata S.D., Moon D.H., Menounov N., Meng X., Hires R. 2004. An evaluation of arsenic release from monolithic solids using a modified semi-dynamic leaching test. J. Hazardous Materials, 116, 25–38.

21. Mangialardi T. 2003. Disposal of MSWI fly ash through a combined washing-immobilization process. J. Hazardous Materials, 98, 225–240.

22. Sarode D.B., Jadhaw R.N., Khatik V.A., Ingle S.T., Attarde S.B. 2010. Extraction and leaching of heavy metals from thermal power plant fly ash and its admixtures. Polish J. Environ. Stud., 6, 1325–1330.

23. Fytianos K., Tsaniklidi B. 1998. Leachability of heavy metals in greek fly ash from coal combustion. Environ. Int., 4, 477–486.

24. Hardaway C., Gauthreaux K., Sneddon J., Beck J. 1999. Evaluation of contaminated sed-iments by toxicity characteristic leaching procedure extraction techniques. Microchem. J., 63, 398–404.

25. Praharaj T., Powell M.A., Hart, B.R., Tripathy S. 2002. Leachability of elements from sub – bituminous coal fly ash from India. Environ. Int., 27, 609–615.

26. Sitarz-Palczak E., Kalembkiewicz J., 2012. Study of remediation of soil contamined with heavy metals by coal fly ash, J. Environ.Prot., 3, 1373–1383.

27. Sitarz-Palczak E., Kalembkiewicz J., Zapała l., 2008. A study of the chemical forms or species of manganese found in coal fly ash and soil, Microchem. J., 90, 37–43.

28. The Perkin – Elmer Corporation, Analytical Methods For Atomic Absorption Spectrophotometry, Norwalk Connecticut, USA, 1982.

29. Polska Norma Branżowa BN-81/0623-01. Popioły lotne i żużle z kotłów opalanych węglem kamiennym i brunatnym. Pobieranie i przygotowanie próbek [In Polish].

30. Polska Norma PN-77/G-04528-00.Paliwa stałe. Oznaczanie składu chemicznego popiołu. Przygotowanie próbki popiołu i roztworów do badań [In Polish].

31. Polska Norma PN-ISO-10390. Jakość gleby. Oznaczanie pH [In Polish].

32. Polska Norma PN-95/Z-15005. Oznaczanie zawartości węgla organicznego w kompoście z odpadów miejskich [In Polish].

33. Polska Norma PN-R-04032. Gleba i utwory mineralne – Pobieranie próbek i oznaczanie składu granulometrycznego [In Polish].

34. Łączny J.M. 2002. Model matematyczny oddziaływania składowisk odpadów energetycznych na środowisko. Zakład Narodowy im. Ossolińskich, Wrocław, p. 16 [In Polish].

35. Su D.C., Wong J.W.C. 2003. Chemical speciation and phytoavailability of Zn, Cu, Ni and Cd in soil amended with fly ash-stabilized sewage sludge. Environ. Int., 29, 895–900.

36. Antonkiewicz J. 2014. Effect of fly ashes and sewage sludge on Fe, Mn, Al, Si and Co uptake by grass mixture. J. Ecol. Eng., 3, 6–13.

37. Izquierdo M., Quero X. 2012. Leaching behavior of elements from coal combustion ash; an overvirw. Int. J. Coal Geology, 94, 54–66.

38. Choi S.K., Lee S., Song Y.K., Moon H.S. 2002. Leaching characteristics of selected Korean fly ashes and its implications for the groundwater composition near the ash disposal mound. Fuel, 81, 1083–1090.

39. Reardon E.J., Czank C.A., Warren C.J., Dayal R., Johnston H.M. 1995. Determining controls on element concentrations in fly ash leachate. Waste Manag. Res., 13, 435–450.

40. Steenari B.M., Schelauder S., Lindqvist O. 1999. Chemical and leaching characteristics of ash from combustion of coal, peat and woo in a 12MW CFB – a comparative study. Fuel, 78, 249–258.

41. Cho H., Oh D., Kim K. 2005. A study on removal characteristics of heavy metals from aqueous solutions by fly ash. J Hazardous Materials, 127, 187–195.

42. Lee M.K., Saunders J.A. 2003. Effect of pH on metals precipitation and sorption: Field bioremediation and geochemical modeling approaches. Vadose Zone J., 2, 1770–185.

43. Ho Y.S., Huang C.T., Huang H.W. 2002. Equilibrium sorption isotherm for metal ions on tree fern. Process Biochemistry, 37, 1421–1430.

44. Cuppett J.D., Duncan S.E., Dietrich A.M. 2006. Evaluation of copper speciation and water quality factors that affect aqueous copper tasting response. Chemical Senses, 31, 689–697.

45. Zirino A., Yamamoto S. 1972. A pH-dependent model for the chemical speciation of copper, zinc, cadmium and lead in seawater. Limnology and Oceanography, 17, 661–671.

46. Cárdenas-López C., Camargo G., Giraldo L., Moreno-Piraján J.C. 2007. Design of an adsorbent employing activated carbon fiber to remove lead. Ecletica Quimica, 32, 61–72.

47. Deliyanni E.A., Peleka E.N., Matis K.A., 2007. Removal of zinc ion from water by sorption onto iron-based nanoadsorbent. J. Hazardous Materials, 141, 176–184.

26

CONTENT OF REDUCED GLUTATHIONE FORM AS A BIOMARKER OF OXIDATIVE STRESS IN SPINACH PLANTS GROWING IN SOIL CONTAMINATED WITH ZINC

Monika Arasimowicz[1], Barbara Wiśniowska-Kielian[2], Marcin Niemiec[3]

[1] Department of Agricultural and Environmental Chemistry, University of Agriculture, 21 A. Mickiewicza Ave., 31–120 Kraków, Poland, e-mail: komorowska.monika@interia.pl

[2] Department of Agricultural and Environmental Chemistry, University of Agriculture, 21 A. Mickiewicza Ave., 31–120 Kraków, Poland, e-mail: e-mail: rrkielia@cyf-kr.edu.pl

[3] Department of Agricultural and Environmental Chemistry, University of Agriculture, 21 A. Mickiewicza Ave., 31–120 Kraków, Poland, e-mail: niemiecm@tlen.pl

ABSTRACT

The aim of the paper is the assessment of the efficiency of anti-oxidative system in spinach plants growing in the substratum polluted with zinc. The assessment was conducted on the basis of changes of reduced glutathione (GSH) form concentration in the plant aboveground organs. Spinach plants, 'Matador' c.v., were cultivated in soils contaminated with zinc in two pot experiments conducted in 2010 and 2011. The experimental substratum was light, slightly soil with granulometric composition of sandy silt loam. Zinc in the acetate form, was supplied to the soil in four doses corresponded to this metal critical concentrations in soil with 0, I, II and III degrees of pollution with this element according to IUNG classification. Simultaneously, the control with natural Zn content in soil was maintained. Zn concentrations in spinach ranged from 412.8 to 1722 mg · kg^{-1} d.m. and increased with growing degree of substratum pollution with this element. Over the course of the vegetation period the content of Zn in plants was generally greater. GSH content in spinach grown in both years of experiments fluctuated from 31.70 to 238 μg · g^{-1} f.m. The biggest content of this compound in spinach was stated in the initial phase of plants growth. The plants tolerated only the first two Zn doses supplied to the substratum. Spinach growing in the objects where zinc additions to the soil equalled II and III degree of substratum pollution died shortly after germination. The plants from these objects in the initial growth phase contained significantly less GSH than spinach from the objects with two first degrees (0 and I) of substratum pollution with zinc or from the control. The content of reduced glutathione form in spinach is a good biomarker of oxidative stress caused by zinc presence in plants. Synthesis of a bigger amount of GSH conditions spinach plant resistance to over the norm zinc content in soil. The efficiency of anti-oxidative system in spinach is bigger in the initial phase of this plant growth.

Keywords: environmental stress, glutathione, biomarker of oxidative stress, soil pollution, zinc.

INTRODUCTION

Heavy metals cause oxidative stress in plant organism visible as forming of over the norm amounts of reactive oxygen species (ROS) and free radicals (FR) in cells [2, 13]. The course of this stress in plant cells is affected by the activity of their anti-oxidative system, composed of anti-oxidative enzymes and low molecular, usually nonspecific, antioxidants [8].

One of the nonspecific antioxidants is glutathione, whose molecule is a polymer of three amino acids: glutamic acid, cysteine and glycine [6, 14, 16]. In the presence of ballast trace elements and with an excess of elements crucial for the cell, glutathione fulfils a double function. The compounds decreases oxidative stress: directly by scavenging free radicals and improving the efficiency of the anti-oxidative system through regeneration of other anti-oxidative compounds, but also indirectly by binding these element ions, which leads to their disabling [16]. Moreover, glutathione is a precursor for phytochelatins – specific non-enzymatic protein compounds able to bind heavy metal ions [9, 12].

The paper aimed to assess the disturbances of the metabolism of spinach plants growing in zinc polluted substratum and estimation of anti-oxidative system efficiency in development of spinach plants growing in Zn polluted soil. The assessment was conducted on the basis of changes in GSH content in the aboveground organs of this plant in three subsequent phases of its development: intensive growth, flowering and seed setting.

MATERIAL AND METHODS

In 2010 and 2011 two pot experiments were conducted in a vegetation hall of the University of Agriculture in Krakow, situated at the experimental station in Mydlniki. The substratum for the experiment was taken from experimental plots in the area of this station. Before the experiment outset analyses of the soil substratum were conducted and the following physicochemical properties were determined with methods commonly used in agro-chemical laboratories:

- granulometric composition – Bouyoucose-Casagrande method in Proszynski's modification,
- maximum capillary water capacity,
- pH in suspension in water and in 1 mol · dm^{-3} KCl solution,
- hydrolytic acidity – Kappen's method,
- total nitrogen content – Kjeldahl's distilling method,
- organic carbon content – Tiurin's method,
- contents of available forms of potassium and phosphorus – Egner-Riehm's method,
- total macroelements and trace metals content after sample decomposition in a mixture of nitric and chloric(VII) acids (3:2; *v/v*) following organic matter mineralization in a muffle furnace at 500 °C [10].

Results are presented in Tables 1, 2 and 3.

The experimental soil was light, slightly acid, contained average amount of organic matter and total nitrogen, and had high content of available phosphorus and very high content of available potassium. The content of Fe and Pb corresponded to their average contents in surface levels of light soils, amounting: 6400 mg Fe and 13 mg Pb · kg^{-1}. Soil content of Mn, Zn, Ni, Cr and Cd were little higher than averages, amounting to: 282 mg Mn, 31 mg Zn, 5.9 mg Ni, 11 mg Cr and 0.21 mg Cd · kg^{-1}, and above 2 times higher than average Cu content equals 6.2 mg Cu · kg^{-1} [7]. Trace el-

Table 1. Texture of experimental soil

Fraction diameter [mm]	1.0 – 0.1	0.1 – 0.05	0.05 – 0.02	0.02 – 0.006	0.006 – 0.002	< 0.002
Share of fraction [%]	45	11	23	13	5	3
Agronomic category	sandy silt loam					

Table 2. Basic physicochemical properties of soil used in the experiments

Water capacity [%]	pH		Hh [mmol$^{(+)}$ · kg^{-1}]	Organic C	Total N	P_2O_5	K_2O
	KCl*	H_2O		[g · kg^{-1}]		acc. to Egner-Riehm [mg · kg^{-1}]	
33.34	5.75	6.25	9.70	7.50	1.22	157.1	281.6

Explanation: 1 mol · dm^{-3} KCl solution

Tabela 3. Total content of macroelements and trace elements in experimental soil

Macroelements [g · kg^{-1}]	Ca	Mg	K	Na	P
	1.384	9.913	0.888	0.361	0.443

Trace elements [mg · kg^{-1}]	Fe	Mn	Zn	Cu	Ni	Cr	Pb	Cd
	4921	283.3	39.23	14.35	6.637	11.21	0.930	0.266
Degree of pollution [7]	-	-	0°	0°	0°	0°	0°	0°

ement contents were lower than limit values for natural heavy metals content in surface levels of light soils calculated on the basis their geometric means and geometric standard deviations amounting: 13200 mg Fe, 506 mg Mn, 65 mg Zn, 13.0 mg Ni, 27 mg Cr, 24.5 mg Pb and 0.44 mg Cd · kg^{-1}, and higher than limit for Cd, equals 12.4 mg Cu [7]. Taking into consideration degrees of soil pollution elaborated by the Institute of Soil Science and Plant Cultivation (IUNG) [7] heavy metal contents in soil were natural.

The soil preparation for the experiment comprised its drying in the air, crushing and sifting. Subsequently each pot was filled with 5 kg of soil. A week before plant sowing uniform basic fertilization: 1 g N in the form of ammonium nitrate, 0.25 g P as sodium dihydrogen(V) and 1 g K as potassium chloride, was applied to each pot. Simultaneously, zinc in the form of acetate was added to the soil according to the experiments scheme (Table 4).

Table 4. Scheme of pot experiments

Zinc addition to the substratum	
Degree of contamination acc. to criteria of IUNG [7]	[mg Zn · kg^{-1}]
Control	Natural content
0°	50
I°	100
II°	300
III°	700

Zn was added to the soil in amounts corresponded to critical values of the subsequent degrees of soil pollution with this element as suggested by the Institute of Soil Science and Plant Cultivation [7]. Basic fertilization and zinc supplements were applied as solutions of pure for analysis salts. Spinach, 'Matador' c.v., was cultivated as a test plant.

Reduced form of glutathione (GSH) was determined in compliance with the method described by Gurie [4] with modifications. Glutathione was extracted from plant tissues with a mixture of ethylenediaminetetraacetic acid and trichloroacetic acid (EDTA-TCA). In order to bring pH solution to the value of c.a. 7.0, K-phosphate buffer with pH = 7.0 was used. The content of reduced glutathione was assessed by spectrophotometer using 5,5-dithiobis-2-nitrobenzoic acid (DTNB) – Ellman reagent in Beckman UV/VIS PU 6400 apparatus. The solution extinction was measured at the wavelength $\lambda = 412$ nm. Glutathione concentration in the plant biomass was calculated on the basis of the values read from the standard curve. Absorbance of the plant homogenate, which absorbed a part of the radiation, was measured as a blind sample.

Zinc content in spinach aboveground parts was assessed using ICP-AES method after wet mineralization in a mixture of HNO_3 and H_2O_2 acids (6:1, *v/v*) in a closed system in a Multiwave 3000 microwave oven (Anton Paar). The solution method was selected for the plants and conducted basing on application instructions of the microwave system manufacturer.

Contents of glutathione, ascorbic acid and nickel were assessed in spinach aboveground parts in three phases of development: intensive growth, flowering and seed setting.

The significance of differences in the analysed chemical compound concentrations in the test plants growing in substrata with various degrees of zinc pollution and in successive vegetation phases were assessed using t-Student test at the significance level $\alpha \le 0.05$.

RESULTS AND DISCUSSION

Zinc concentrations in spinach growing on the control objects of experiments in the 2010 and 2011 were from 412.8 to 552.5 mg · kg^{-1} d.m. (Table 5).

The first Zn supplement supplied to the substratum, corresponding to 0 degree of soil pollution (50 mg · kg^{-1}) caused over two-fold, while the second equal to I degree (100 mg · kg^{-1}) brought about a three-fold increase in the content of this element in spinach aboveground parts in comparison with the control. Bigger additions of zinc to the substratum (the II and III degree of pollution) caused an inhibition of germination and growth of spinach plants (Figure 1).

Zn content in plants from individual objects did not change significantly during their growth (Table 5). According to Yanqun et al [18] zinc accumulation in plants depends on their species.

The highest concentrations of glutathione, irrespective of the applied metal, were registered in the first phase of plant growth. With progressing vegetation glutathione content in the aboveground parts was decreasing and in the last investigated phase was twice or three times lower than in the first (Table 6).

Table 5. Zinc contents in biomass of spinach in individual phases of plants development [mg Zn · kg^{-1} d.m.]

Phase of plant development	Zinc addition acc. to degree of substratum contamination				
	control	0° *	I°	II°	III°
2010					
I phase	505.8^{a} **	1013.4^{a}	1341.4^{a}	lack of plant material for analyses	
II phase	531.2^{a}	1152.4^{b}	1346.4^{a}		
III phase	552.5^{a}	1081.1^{ab}	1625.4^{a}		
2011					
I phase	412.8^{a}	1006.7^{ab}	1301.6^{a}	lack of plant material for analyses	
II phase	560.8^{b}	1127.4^{b}	1704.3^{b}		
III phase	511.6^{b}	968.9^{a}	1722.1^{b}		

Explanation: * – concerns Tables 5–7 and Figures 1–3: 0°, I°, II°, III° – Zinc additions corresponded to degree of successive substratum contamination acc. to criteria of IUNG; ** – concerns Tables 5–7: different letters indicate significant differences depending on zinc additions in relation to control object and objects with lower degree of pollution, at $\alpha \leq 0.05$.

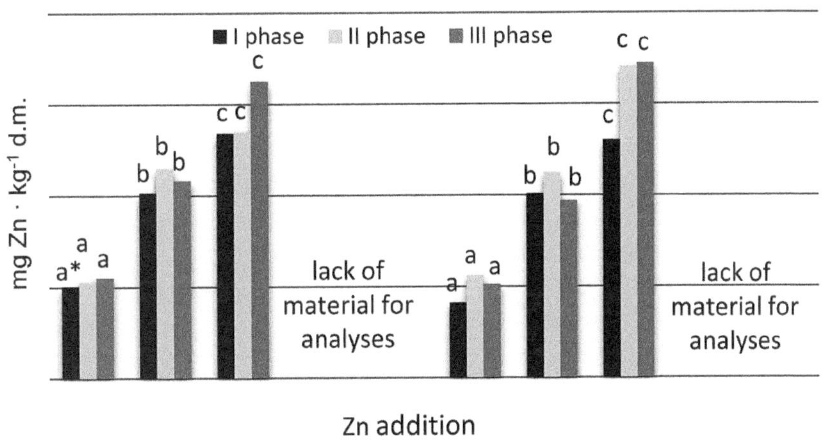

Figure 1. Zinc contents [mg Zn · kg^{-1} d.m.] in aboveground biomass of spinach in individual phases of plants development depending on this metal addition

Explanation: * concerns Figures 1–3: different letters indicate significant differences depending on nickel additions in relation to control object and objects with lower degree of pollution, at $\alpha \leq 0.05$

Table 6. Glutathione content in biomass of spinach in individual phases of plants development [μg GSH · g^{-1} f.m.]

Phase of plant development	Zinc addition acc. to degree of substratum contamination				
	control	0°	I°	II°	III°
2010					
I phase	78.92^{a}	176.83^{a}	180.14^{a}	94.37	59.56
II phase	64.47^{b}	96.83^{b}	190.24^{a}	–	–
III phase	31.70^{c}	59.63^{c}	142.62^{b}	–	–
2011					
I phase	87.33^{a}	132.32^{a}	238.03^{a}	129.67	99.58
II phase	90.76^{a}	130.46^{a}	201.76^{a}	–	–
III phase	59.01^{b}	128.90^{a}	226.81^{a}	–	–

Glutathione concentrations in the aboveground parts of spinach growing in the objects with zinc supplement in the substratum ranged in both years of the experiment from 59.56 to 238.03 μg · g^{-1} f.m. (Figure 2). Guri [4] assessed approximate glutathione content in fresh mass of pea (*Phaseolum vulgaris* L.) a variety sensitive to ozone effect and exposed to this stressor.

In the Authors' own research, in both years of the experiments, glutathione content in spinach increased significantly in response to Zn additions corresponded to 0 and I degree of substratum pollution in relation to the control. After application of Zn in the amounts corresponding to II and III degree of pollution, glutathione concentration in spinach in 2011 decreased markedly in comparison with the previous objects, whereas in 2010 its content declined significantly also in relation to the control. Agarwal [1] in the research on the effect of various UV radiation intensity on antioxidant content in plant tissue determined a notable decrease in glutathione (GSH) content at two-fold increase in the stressor intensity. Two larger zinc additions (300 and 700 mg · kg^{-1} soil d.m.) initially caused inhibition of spinach growth and then the plant die back. Spinach plants which contained lesser quantities of glutathione than the plants from the control and less polluted objects, did not tolerate high zinc concentrations. As a result of high toxicity of the highest applied supplements of this metal, the plants after germination first grew poorly and then died. Presence of reduced form of glutathione and possibility of its quota regeneration in a plant cell is a condition of tolerating high concentrations of trace elements [5]. Heavy metals present in a plant cell cause numerous disturbances and impairments of all live processes. Each organism has its own specific limit of tolerance to individual stressors. When stressor intensity is too high and disturbances due to its presence are too serious, the tolerance mechanisms fail and the cell dies [15]. Low content of reduced glutathione in the plants growing in the substratum with Zn supplements corresponding to II and III degree of pollution was connected with too high concentration of zinc ions in cells. Free metal ions disable functions of enzymes, among others glutathione reductase which is responsible for reducing oxidized glutathione and therefore for recovering its abilities for these ions chelating [11, 17]. Moreover, free ions of toxic elements lead to formation of reactive oxygen species which injure and damage lipids, proteins, carbohydrates and nucleic acids, disturbing the proper metabolism of a living cell [3]. Spinach plants growing in the other objects with Zn supplements, in the 2nd and 3rd development phase, despite smaller amount of produced biomass in comparison with the control plants, tolerated the applied metal supplements. In the 2nd and 3rd phase of growth glutathione content in the plants tolerating Zn quantities added to the soil and growing in the objects with increasing substratum pollution with this metal, was significantly higher than in the control plants.

In the second studied development phase (flowering period) glutathione content in spinach was comparable or smaller than in the analogous objects in the first phase. In the second and third plant development phase glutathione concentrations were markedly higher in the plants growing in substrata with increasing degree of zinc pollution. In the third phase of plant growth, this compound concentration in spinach was generally lower than in the first phase (Table 6). The tendency was more pronounced in 2010.

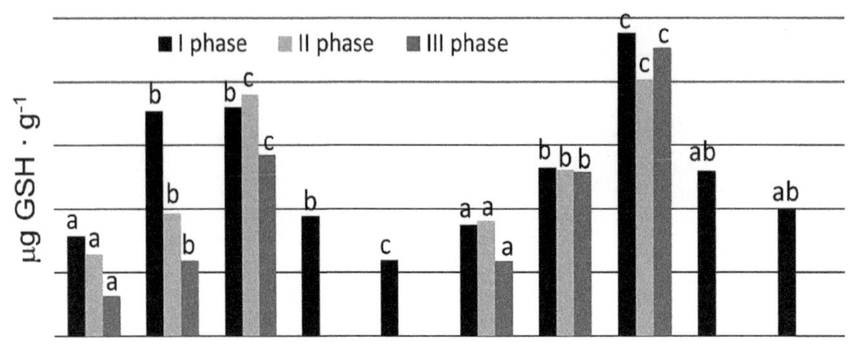

Figure 2. Glutathione content [μg GSH · g^{-1} f.m.] in aboveground parts of spinach in individual phases of development depending on zinc addition

CONCLUSION

The following conclusions may be drawn on the basis of the obtained research results:

1. Spinach plants absorb Zn proportionately to its content in the substratum.
2. Excessive amounts of Zn are tolerated by spinach plants, 'Matador' c.v., greatly owing to the mechanisms connected with glutathione synthesis. There is the tolerance threshold for this metal in plants, above which GSH is no longer synthesised in the amounts sufficient for tolerating the effects of this metal ions presence in cells.
3. The anti-oxidative system in the plant cells functions more efficiently at the initial stage of spinach vegetation, as has been evidenced by much higher glutathione contents in the first phase of growth. At later stages of plant development successively decreasing amounts of this antioxidant are observed.
4. The content of reduced form of glutathione (GSH) in spinach plant biomass is a good indicator (biomarker) of stress caused by Excess of Zn in the substratum.

REFERENCES

1. Agarwal S. 2007. Increased antioxidant activity in Cassia seedlings under UV-B radiation Biol. Plantarum, 51(1): 157–160.
2. Bartosz G. 2003. Druga twarz tlenu. Wolne rodniki w przyrodzie. Wyd. Nauk. PWN, Warszawa.
3. Foyer C.H., Lopez-Delgado H., Dat J.F., Scott I.M. 1997. Hydrogen peroxide- and glutathione- associated mechanisms of acclimatory stress tolerance and signaling. Physiol. Plantarum, 100: 241–254.
4. Guri A. 1983. Variation in glutathione and ascorbic acid content among selected cultivars of *Pftaseolus vulgaris* prior to and after exposure to ozone. Can. J. Plant. Sci., 63: 733–737.
5. Hall J.L. 2002. Cellular mechanism for heavy metal detoxification and tolerance. J. Exp. Bot., 53(366): 1-11.
6. Igamberdiev A., Hill R. 2004. Nitrate, NO and hemoglobin in plant adaptation to hypoxia: an alternative to classic fermentation pathways. J. Exp. Bot., 55: 2473–2482.
7. Kabata-Pendias A., Piotrowska M., Motowicka-Terelak T., Maliszewska-Kordybach B., Filipiak K., Krakowiak A., Pietruch Cz. 1995. Podstawy oceny chemicznego zanieczyszczenia gleb. Metale ciężkie, siarka i WWA. Biblioteka Monitoringu Środowiska, Państwowa Inspekcja Ochrony Środowiska, Warszawa, 34 p.
8. Kopcewicz J., Lewak S. Fizjologia roślin. 2002. Wyd. Nauk. PWN, Warszawa.
9. May M., Leaver V., Montagu M., Inze´ D. 1998. Glutathione homeostasis in plants: implications for environmental sensing and plant development. J. Exp. Bot., 49(321): 649–667.
10. Ostrowska A., Gawliński S., Szczubiałka Z. 1991. Metody analizy i oceny właściwości gleb i roślin. Katalog. Wyd. IOŚ, Warszawa.
11. Pastore A., Federici G., Bertini E., Piemonte F. 2003. Analysis of glutathione: implication in redox and detoxification. Clin. Chim. Acta., 333: 19–39.
12. Rennenberg H., Brunold C. 1994. Significance of glutathione in metabolism in plants under stress. Prog. Bot., 55: 142–156.
13. Scandalios J.G. 1993. Oxygen stress and superoxide dismutases. Plant Physiol., 101: 7–12.
14. Sies H. 1999. Glutathione and its role in cellular function. Free Radic. Biol. Med., 27(9–10): 916–921.
15. Starck Z., Chołuj D., Niemyska B. 1995. Fizjologiczne reakcje roślin na niekorzystne czynniki środowiska. 2 wyd. Wyd. SGGW, Warszawa.
16. Tukendorf A. 1990. Rola kompleksów metaloproteinowych w tolerancji roślin wyższych na toksyczne stężenia metali ciężkich. Rozprawy Wydziału Biologii i Nauk o Ziemi UMCS, Rozprawy Habilitacyjne, Tom 39, 75 p.
17. Winiarska K. 2000. Glutation: niezwykłe funkcje pospolitego peptydu. Post. Biochem., 46: 318–326.
18. Yanqun Z., Yuan L., Schvartzc C., Langlade L., Fan L. 2004. Accumulation of Pb, Cd, Cu and Zn in plants and hyperaccumulator choice in Lanping lead-zinc mine area, China. Environ. Int., 30: 567–576.

ASSESSMENT OF HEAVY METALS CONTENTS IN BOTTOM SEDIMENTS OF BUG RIVER

Elżbieta Skorbiłowicz[1]

[1] Department of Technology in Engineering and Environmental Protection, Faculty of Civil Engineering and Environmental Engineering, Bialystok University of Technology, 45A Wiejska Str., 15-351 Bialystok, Poland, e-mail: e.skorbilowicz@pb.edu.pl

ABSTRACT

The development of industry, agriculture, and transport contributes to an increased environmental pollution by heavy metals. The aim of the study was preliminary assessment of the contents of selected metals (lead, cobalt, copper, chromium, cadmium and nickel) in the sediments of Bug river. The study comprised part of the river flowing through Poland. It was found that the Bug river sediments are not contaminated in respect to the content of tested metals. Based on the analysis of the study results, these metals can be lined up in the following order: Cr > Pb > Cu > Ni > Co > Cd. Statistical analysis showed that copper and chromium occur in Bug river sediments in forms bindings with organic matter in majority of cases. The granulometric analysis of sediments from Bug river revealed the largest percentage of two fractions: 1.0–0.2 mm with average of 47.7 ± 19.77% and 0.2–0.1 mm with average of 20.6 ± 7.7%. These are the dominant fractions with the accumulation of metals in river sediments, which has been confirmed by statistical analysis.

Keywords: Bug river, bottom sediment, heavy metals.

INTRODUCTION

The development of industry, agriculture, and communication contributes to an increased environmental pollution by heavy metals. Heavy metals are stable elements that last in the environment as they are not degraded or destroyed. Therefore, they tend to accumulate in the soil, sediments, and organisms [Yu et al. 2000, Demirezen, Aksoy 2004, Dmochowski et al. 2005, Wiśniowska-Kielian, Niemiec 2005, Sasmaza et al. 2008]. Contamination of rivers due to metals can originate from natural and anthropogenic sources. Heavy metals in the waters are subject to chemical transformations, the adsorption and complexation, then the processes of sedimentation and accumulation in bottom sediments. At low concentrations, most of heavy metals behave as microelements, while in higher concentrations, they are toxic [Nelson et al. 2000]. Analysis of the contents of metals accumulated in bottom sediments of a water reservoir is a valuable material for the description of sources, rate, and ways of metal distribution in the reservoir. Bottom sediments are good indicators of the contamination degree of the river environment [Kucuksezgin et al. 2008, Essien et al. 2009, Skorbiłowicz 2012]. Concentration of metals in sediments strongly exceeds their contents in waters, therefore the chemical analysis of sediments can detect them even when their quantities are low in waters [Bojakowska et al. 2006]. The accumulation of metals in sediments is controlled by a number of environmental factors such as pH, Eh, salinity, type and concentration of organic and inorganic ligands, hydraulic processes in rivers, and sediment grain size (Helios-Rybicka 1991).

The aim of the study was preliminary assessment of the content of selected metals (lead, cobalt, copper, chromium, cadmium and nickel) in bottom sediments of Bug river. The study comprised part of the river flowing through Poland.

MATERIAL AND METHODS

Study area

The Bug river is one of the longest in Poland and also belongs to the largest right tributaries of the Vistula river and left Narew river. The catchment area of the river is situated in the central-eastern Poland, the north-western Ukraine, and south-western Belarus. The Bug river catchment area is 39 420 km^2: 19 284 km^2 in Poland, which represents 49% of the Bug River catchment and 20 136 km^2, i.e. 51% beyond its borders. The average height of the Bug river catchment is 183 meters above sea level and the average river inclination is 0.3‰.

The length of the Bug river is approximately 755 km. The lower section of the river with a length of 207 km (28%) is located on the territory of our country, while 185-kilometer upper section represents 24% of the total catchment area and is situated outside Polish borders. The border zone between Poland, Belarus, and Ukraine stretches through 363 km (48%).

In its lower section, Bug river runs through the lowlands. They consist of sand, here and there covered with illuvial deposits. In the north of Terespol, the river flows through 3–4 meters wide valley located in the surrounding of hills of up to about 80 meters high. The greatest narrowing is found in Melnik surroundings. In a further section, the Bug river valley again expands to about 5 km. In the area of the middle and lower Bug river, arable lands and grasslands dominate. There are also mixed forests, although the lower part is dominated by coniferous afforestation. The main tectonic elements of the ground in Bug river valley is a marginal part cut by numerous faults in a direction perpendicular to its borders. They make this zone is composed of continuous depressions and elevations.

The river cuts across the faults structures on almost the entire length of the valley. Loess is the dominant surface formation on plateau areas. The dammed sediments prevail in the valley coastal zones. The plains are covered with boulder clay and fluvio-glacial formations. Lower terraces are built of sands and gravels covered by high-water flow sediments. Organic sediments, peats, and silts are accumulated in the depressions of terrace surface. In the bottom of the valley, there are also common Aeolian forms – sand dunes and wind-blown sand fields [Dojlido 2003].

The catchment area of Bug river is dominated by arable lands (50.3%) as well as meadows and pastures (17.7%). Forests cover 23.5% of the catchment area. Forests are predominated by pine, fir, spruce, ash, birch, oak, and alder.

There are 25 cities, that differ with economic status, population, and building character, within Bug river catchment. Their economic and industrial growth is a result of localization at the intersection of major rail and road routes, and in the case of Terespol – of the border crossing. Cities with a small number of inhabitants dominate in the river catchment area. Only the three largest cities can be distinguished: Siedlce (72 858 inhabitants), Biała Podlaska (55 424 inhabitants), and Chełm (69 016 inhabitants).

Nine industrial activities can be found in the catchment area. These include tanning, glass, ceramic, machine-metal, light, cement, electrical, food, and tourist industries. The tanning industry is localized in Włodawa, although it is very harmful to the environment. In the vicinity of Wyszków and Chełm, the glass industry facilities operate. In the area from Chełm to Bełżec, pottery plants function. Machine-metal industry can be found in larger cities such as Łuków, Włodawa, Biała Podlaska, Chełm, Międzyrzecz, and Siedlce. One of the biggest Polish cement industry facility is localized in Chełm. Branches of high technology, i.e. electrical industry may be encountered in the north-western part of the catchment, more precisely in Węgrów, Woźniki, and Biała Podlaska.

Top-developed industry is the food industry, which can be located in any locality within discussed catchment. The largest sectors of this industry include such branches as baking, dairy, sugar, meat, fruit and vegetable. The second most developed industry is tourism, which is mainly located in areas with rich and beautiful landscape. The main sectors of the industry are farmhouses, tourist attractions (hiking, cycling, and water tracks), accommodations and dining sites.

Bug river, together with its tributaries, is also the receiver of wastewaters from municipal and industrial sources. The main point sources include sewage inflow from the most populous cities, entering the sewage system and that from urban and industrial areas. The contamination factors are also effluents coming from the Ukraine from mines and industrial plants.

Field and laboratory works

The fieldwork was conducted in 2012 in 11 measurement points located on the Bug river (Drohiczyn, Siemiatycze, Mielnik, Niemirów, Krzyczew, Terespol, Sławatycze, Włodawa, Dorohusk, Horodło, and Kryłów). The object for analyses was composed of bottom sediments, in which contents of lead, cobalt, copper, chromium, cadmium, and nickel, were determined. Bottom sediments were collected in the coastal zone, where suspended material is deposited (Bojakowska and Sokołowska 1992, Bojakowska 2001). Several individual surface samples of bottom sediments (from a depth of 5 cm) from beneath the water, from every selected measurement point. After mixing the test material, representative sample (weighing about 1000 g) was achieved. The samples were then dried in air to the "air-dried" status and stored until laboratory tests [Lis and Pasieczna 1995]. Prior to chemical analyzes, the bottom sediment sample was dried at 40 °C and sieved through a nylon mesh of 0.2 mm. Sediments were mineralized with nitric acid [Frančišković-Bilinski et al, 2005, Kunwar P. Singh et al. 2005, Salminen et al. 2005] in a closed microwave system CEM Mar-5. The concentrations of Pb, Co, Cu, Cr, Cd, and Ni in sediments were determined by means of ASA. The measurements were performed using Varian Spectra AA100 equipment. The correctness of the methodology was verified on a base of the reference material NCSDC 733 analysis. Achieved results for tested metals contents were presented in relation to air-dry sediments and compared with the literature data as well as arithmetic mean of these metals in bottom sediments for Poland (fraction < 0.2 mm) [Lis and Pasieczna 1995] and Europe (fraction < 0.15 mm) (Salminen et al. 2005) and soil of Polish Lowlands area [Pasieczna 2003]. To assess the extent of sediment contamination by heavy metals, the proposed classification of aquatic sediments in Poland on the basis of geochemical criteria [Bojakowska and Sokołowska 1998] and on the basis of threshold values taking into account the harmful effects of pollutants accumulated in sediments towards aquatic organisms [Bojakowska 2001], was applied. For sediments, the granulometric composition analysis was carried out, which consisted in separating the raw air-dry sediment samples to size fractions using an appropriate set of sieves (mesh: 2 mm, 1 mm, 0.2 mm, 0.1 mm, 0.063 mm) mounted to a mechanical shaker. After the screening complete, the individual sieve fractions were weighed, then the weight percentage of particular grain size samples in reference to the total sample weight, was calculated. In addition, the sediments were subject to analysis of organic compounds content by means of annealing (550 °C). The pH of sediments and river water was performed applying potentiometry.

The licensed version of Statistica 10 software was used for statistical processing of results. Arithmetic mean, median, standard deviation, and Spearman correlation coefficient values were calculated. The results were also subject to cluster analysis based on the concept of distance of objects or variables in a multidimensional space.

RESULTS AND DISCUSSION

Granulometric analysis of sediments collected from Bug river showed mainly varied character of size distribution and resembled in majority the features of loose and weak loamy sand. It has been reported the largest percentage of two fractions: 1.0–0.2 mm (average 47.7 ± 19.77%) and 0.2–0.1 mm (average 20.6 ± 7.7%). The smallest percentage characterized fraction 0.1–0.063 of sediments (average 7 mm ± 5.09%). The coarsest grain size (fraction > 2 mm) occurred at the measurement points Niemirow (21.43%) and Wlodawa (13.17%). This was due to the fact that the least favorable conditions for deposition of the smallest of loam and dust parts were present at those points (Table 1).

The organic matter content in analyzed samples of bottom sediments ranged from 0.8% to 19.3%. Samples collected in the localities Mielnik (19.3%) and Drohiczyn (11.2%) were characterized by the highest contents of organic matter. The lowest content, not exceeding 3%, was recorded at the measuring points: Sławatycze, Horodło, and Kryłów (Table 2).

The Bug river water was also subject to pH determination, which ranged from 6.2 to 7.5 pH, while the pH of the bottom sediment in most cases was neutral ranging from 6.3 to 7.3 pH.

Table 2 presents the results of heavy metals contents determinations: lead, cobalt, copper, chromium, cadmium, and nickel, in bottom sediments of Bug river flowing through Poland.

Bojakowska et al. [2006] reported that the lead content in uncontaminated sediments gen-

Table 1. Contents of particle fractions in bottom sediments from Bug river

River	Measurement point	Content of particle fraction [% by weight]					
		>2 mm	2.0–1.0 mm	1.0–0.2 mm	0.2–0.1 mm	0.1–0.063 mm	<0.063 mm
Bug	Drohiczyn	8.23	10.37	37.60	20.80	12.60	10.40
	Siemiatycze	12.97	12.87	22.10	24.07	10.32	17.67
	Mielnik	9.47	21.07	30.20	20.20	12.23	6.83
	Niemirów	21.43	11.5	47.80	14.27	2.17	2.83
	Krzyczew	3.93	3.80	37.17	33.60	9.63	11.87
	Terespol	1.35	4.39	85.03	7.30	1.15	0.78
	Sławatycze	2.30	3.13	52.17	28.13	6.77	7.50
	Włodawa	13.17	11.87	57.43	12.77	2.50	2.26
	Dorohusk	3.23	7.47	60.47	23.63	3.06	2.14
	Horodło	3.40	8.50	24.43	26.87	15.00	21.80
	Kryłów	3.30	7.03	70.60	15.23	1.97	1.87
General statistical data							
Minimum		*1.35*	*3.13*	*22.10*	*7.30*	*1.15*	*0.78*
Maximum		*21.43*	*21.07*	*85.03*	*33.60*	*15.00*	*21.80*
Arithmetic mean		*7.5*	*9.2*	*47.7*	*20.6*	*7.0*	*7.8*
Median		*3.9*	*8.5*	*47.8*	*20.8*	*6.8*	*6.8*
Standard deviation		*6.25*	*5.15*	*19.77*	*7.7*	*5.09*	*6.99*

Table 2. Measured indicators for Bug river

River	Measurement point	River water pH	Sediment pH (pH_{H2O})	Organic matter [%]	Pb	Co	Cu	Cr	Cd	Ni
					Heavy metals contents in bottom sediments [mg·kg^{-1}]					
Bug	Drohiczyn	7.1	6.8	11.2	4.8	3.8	3.6	8.3	0.2	2.9
	Siemiatycze	6.8	6.7	6.1	9.1	6.2	12.1	27.2	0.4	7.8
	Mielnik	6.5	6.6	19.3	4.9	3.2	5.3	10.2	0.3	6.1
	Niemirów	6.2	6.7	5.8	4.6	5.9	6.5	9.8	0.4	2.9
	Krzyczew	6.7	6.8	3.1	2.8	3.6	3.4	7.3	0.5	4.2
	Terespol	7.5	7.3	0.8	6.3	4.5	1.9	5.1	1.1	5.3
	Sławatycze	7.4	6.9	1.8	4.2	5.3	4.6	10.2	0.4	5.1
	Włodawa	7.2	6.3	5.9	5.8	3.7	3.6	13.1	0.3	6.1
	Dorohusk	7.7	6.8	3.1	9.3	4.3	4.5	14.1	0.6	1.9
	Horodło	7.5	6.9	2.2	1.1	3.1	2.3	4.1	0.7	3.1
	Kryłów	6.7	6.8	2.8	6.8	4.3	4.1	6.1	0.5	4.2
General statistical data										
Minimum		*6.2*	*6.3*	*0.8*	*1.1*	*3.1*	*1.9*	*4.1*	*0.2*	*1.9*
Maximum		*7.5*	*7.3*	*19.3*	*9.3*	*6.2*	*12.1*	*27.2*	*1.1*	*7.8*
Arithmetic mean				*5.6*	*5.4*	*4.4*	*4.7*	*10.5*	*0.5*	*4.5*
Median				*3.1*	*4.9*	*4.3*	*4.1*	*9.8*	*0.4*	*4.2*
Standard deviation				*5.37*	*2.44*	*1.04*	*2.76*	*6.35*	*0.24*	*1.76*

erally does not exceed 30 mg·kg^{-1}. The pattern of lead concentration variability in bottom sediments coincides substantially with the geochemical pattern of this element in soils. Geochemical Atlas of Poland indicates that the arithmetic mean for lead is 68 mg·kg^{-1} [Lis and Pasieczna 1995] and for Europe it is 38.6 mg·kg^{-1} [Salmine et al. 2005]. Much less contents were recorded in present studies. The lead content in tested sediments ranged from 1.1 to 9.3 mg·dm^{-1} (average of 5.4 ± 2.44 mg·kg^{-1}). The highest lead concentrations above 9 mg·kg^{-1} were recorded at the measuring points Siemiatycze and Dorokhusk. On this basis, it can be concluded that the Bug river bottom sediments are not contaminated with lead. Sediments from the Odra river collected near Nowa Sól – Kostrzyń contained 24–84 mg·kg^{-1} of lead [Jędrzejczak and Czyrski 1990], whereas sediments from the upper Narew river from 9.2 ± 4.65 to 22.9 ± 15.14 mg·kg^{-1} [Skorbiłowicz 2012].

Cobalt in bottom sediments from non-industrialized areas is usually present at levels up to several mg·kg^{-1}. In general, Co content higher than 10 mg·kg^{-1} is associated with anthropogenic activity [Bojakowska and Sokołowska, 1998]. The study showed variations in the content of this element from 3.1 mg·kg^{-1} to 6.2 mg·kg^{-1}. The average cobalt concentration in bottom sediments of Bug river was 4.4 ± 1.04 mg·kg^{-1}, which was less than the average value for European sediments 11.2 mg·kg^{-1}.

The copper content in bottom sediments is related to the type of macierzysta rocks and its average content reaches only a few mg·kg^{-1}. Fox and Pasieczna [1995] reported that differences in the copper content in aquatic sediments is much higher than in the soils. The average copper concentration in soils of Polish Lowlands is 3 mg·kg^{-1} [Pasieczna 2003], while average content of Cu for European sediments is 22.1 mg·kg^{-1} and for Polish sediments 21 mg·kg^{-1}. Bojakowska and Sokołowska [1998] consider value of 6 mg·kg^{-1} the background of copper in Polish sediments. Copper in bottom sediments collected from Bug river occurred in relatively small contents (1.9–4.6 mg·kg^{-1}) and in majority of studied samples it appeared below geochemical background. The exception were measurement points localized in Siemiatycze (12.1 mg·kg^{-1}), Niemirów (6.5 mg·kg^{-1}) and Mielnik (5.3 mg·kg^{-1}).

The highest contents among the tested metals were recorded in the case of chromium. The amount of chromium in studied bottom sediments ranged from 4.1 to 27.2 mg·kg^{-1} (average 10.5 ± 6.35 mg·kg^{-1}). Variable chromium contents were observed in the sediments along with the river flow. The largest concentration of the element occurred at the measurement point in Siemiatycze (27.2 mg·kg^{-1}), then chromium quantities decreased to the point in Włodawa (13.1 mg·kg^{-1}) and Dorohusk (14.1 mg·kg^{-1}); at subsequent points, chromium concentrations were even lower. The tanning industry using chromium compounds in technological processes, is localized in the Bug river catchment. In part, this may be the cause of increased contents of this element in studied sediments in the case of discharging the wastewaters from tanneries into Bug river water. Considering achieved results in a view of natural geochemical chromium contents provided by Bojakowska and Sokołowska [1998] at the level of 5 mg·kg^{-1}, it must be noted that the amount of chromium has not been exceeded only in 2 examined sediment samples. In contrast, Lis and Pasieczna [1995] argued that natural concentration of chromium in bottom sediments of not contaminated rivers reaches 10 mg·kg^{-1}. In relation to this value, chromium content has been exceeded in five samples of analyzed sediments. It is also worth mentioning that results of Cr content achieved for sediments from Bug river were much lower than the average value for bottom sediments in Europe (92.8 mg Cr·kg^{-1}).

Cadmium contents in the range from 0.2 to 1.1 mg Cd·kg^{-1} (average 0.5 ± 0.24 mg· kg^{-1}) was observed in collected samples of bottom sediments. The geochemical criteria allowed to classify 8 sediment samples to geochemical background, while 2 samples (Dorohusk – 0.6 mg·kg^{-1} and Horodło – 0.7 mg·kg^{-1}) to the first geochemical class [Bojakowska Sokołowska 1998]. Surprisingly high concentrations of cadmium were observed in bottom sediments from Bug river in Terespol (1.1 mg·kg^{-1}) (II geochemical class). The border crossing is localized in Terespol, which is associated with an intense traffic flows. Wang et al. [2006] claimed that among variety of human activities contributing to the natural environment degradation, broadly understood transport is one of the most important. On areas of high development density, products of liquid fuels combustion cannot be quickly and freely moved, which in turn is the cause of very high concentrations of heavy metals in urban atmospheric air [Sutherland 2000]. Increase in heavy metal contamination of aquatic ecosystems localized on areas with intense traffic, is very disturbing trend. These results of cadmium concentrations in bottom sediments from Bug river in Terespol seem to confirm the thesis about the impact of road transport on the content of cadmium in aquatic environment.

Nickel levels in Polish aquatic sediments are similar to those in soils. On Polish Lowland, nickel content does not exceed 10 mg·kg^{-1} [Piaseczna 2003]. The arithmetic mean nickel content in the case of sediments from surface waters for Polish amounts to 11 mg·kg^{-1} and to 35.2 mg·kg^{-1} for Europe. The study upon bottom sediments in Bug river showed that nickel concentrations ranged between 1.9–7.8 mg·kg^{-1} with average of 4.5 ± 1.76 mg·kg^{-1}. The maximum contents were recorded in Siemiatycze (7.8 mg·kg^{-1}) as well as Mielnik and Włodawa (6.1 mg·kg^{-1}).

Statistical analysis of achieved results indicated that copper and chromium contents were significantly positively correlated with the content of organic matter (r_{Cu} = 0.64, r_{Cr} = 0.65). There was also a positive correlation between copper and chromium (r = 0.69 at $p < 0.05$), which may indicate that both elements have similar pollution sources. The cluster analysis (CA) was also performed in relation to the variability of metals concentrations and sediment grain size, water and sediments pH value, and organic matter. Figure 1, being the result of cluster analysis, illustrates two separate main groups and two subgroups. Two fractions are present in main groups. Analysis of particle size distribution of bottom sediments collected from Bug river showed the largest percentage of two fractions: 1.0–0.2 mm (average 47.7%) and 0.2–0.1 mm (average 20.6%). These are the dominant fractions in sediments of Bug river. The accumulation of metals in bottom sediments is affected by sediment grain size. However, the first subgroup consisted of the following components: fraction > 2 mm, 2.0–1.0 mm fraction, organic matter, chromium, fraction 0.1–0.063 mm, and fraction < 0.063 mm, whereas the second subgroup consisted of: sediment and water pH, and metals: lead, cobalt, nickel, copper, and cadmium.

When comparing the results of present studies with those carried out in other rivers [Wardas 2001, Wiśniowska-Kielian Niemiec 2005, Samecka-Cymerman Kempers 2007, Skorbiłowicz 2012], it is evident that heavy metals contents in bottom sediments from Bug river were lower. It has been found that lead, copper, cadmium, and nickel were at a geochemical background level, while cobalt and chromium were within the range of I geochemical class (Table 3).

CONCLUSIONS

1. Granulometric analysis of bottom sediments from Bug river revealed the largest percentage of two fractions: 1.0–0.2 mm (average 47.7 ± 19.77%) and 0.2–0.1 mm (average 20.6 ± 7.7%). These are predominant fractions in Bug river sediments having the largest impact

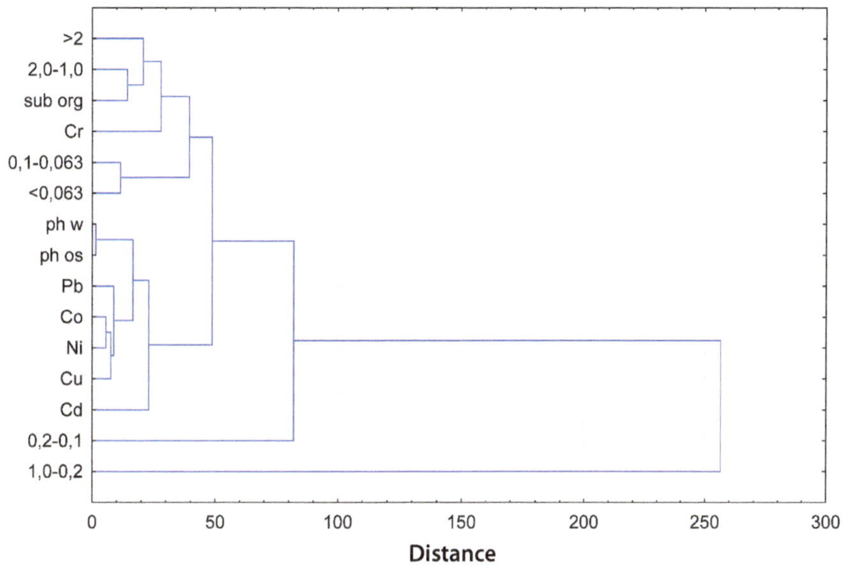

Figure 1. Cluster analysis (CA) by Ward for Bug river

Table 3. Geochemical classification of bottom sediments from Bug river on a base of arithmetic mean value [Bojakowska Sokołowska 1998, Bojakowska 2001]

	Pb	Co	Cu	Cr	Cd	Ni
Bug river	[mg·kg^{-1}]					
	background	I	background	I	background	background

on metals accumulation in bottom sediments, which was confirmed by statistical analysis. Fraction 0.1–0.063 mm (average 7 ± 5.09%) was characterized by the lowest proportion.

2. It was found that bottom sediments from Bug river, due to the contents of lead, cobalt, copper, chromium, cadmium and nickel, were not contaminated. Results of studies upon metals contents allow to line them up in the following sequence: Cr > Pb > Cu > Ni > Co > Cd.
3. Statistical analysis showed that copper and chromium in bottom sediments of Bug river occurred in majority as bound to organic matter.

REFERENCES

1. Bojakowska I., Sokołowska G., 1992. Charakterystyka geochemiczna aluwiów głównych rzek Polski, Przeg. Geolog., 1, 16-20.
2. Bojakowska I. Sokołowska G., 1998. Geochemiczne klasy czystości osadów wodnych, Przeg.Geolog. 46, 1, 49-54.
3. Bojakowska I., 2001. Kryteria oceny zanieczyszczenia osadów wodnych. Przeg. Geolog. 49, 3, 213-218.
4. Bojakowska I., Gliwicz T., Małecka K., 2006. Wyniki geochemicznych badań osadów wodnych Polski w latach 2003–2005, Inspekcja Ochrony Środowiska, Warszawa.
5. Demirezen D., Aksoy A., 2004. Accumulation of heavy metals in Typha angustifolia (L.) and Potamogeton pectinatus (L.) living in Sultan Marsh (Kayseri, Turkey), Chemosphere 56, 685-696.
6. Dmochowski D., 1995. Analiza możliwości zastosowania specjacji metali ciężkich do oceny zanieczyszczenia wód powierzchniowych na terenach zurbanizowanych, (praca doktorska), Politechnika Warszawska, Wydział Inżynierii Środowiska, Warszawa.
7. Dojlido J., 2003. Rzeka Bug – zasoby wodne i przyrodnicze, Instytut meteorologii i gospodarki wodnej, Wyższa Szkoła Ekologii i Zarządzania, Warszawa.
8. Essien J.P., Antai S.P., Olajire A.A., 2009. Distribution, Seasonal Variations and Ecotoxicological Significance of Heavy Metals in Sediments of Cross River Estuary Mangrove Swamp, Water Air Soil Pollut., 197, 91-105.
9. Frančišković-Bilinski S., Bilinski H., Tibljaš D. Andrantitsch G., 2005. Effects of mercury mining regions from NW Dinarides on quality of stream sediments. Fresenius Environmental Bulletin, 14, 10, 913-927.
10. Helios-Rybicka E. 1991. Akumulacja i mobilizacja metali ciężkich w osadach środowiska wodnego: osady datowane jako wskaźnik chronologiczny, Mat. Konf.: Geologiczne aspekty ochrony środowiska, Kraków, 18-23.
11. Jędrzejczak A., Czyrski T., 1990. Zawartość metali ciężkich w wodzie i osadzie dennym Odry na odcinku Nowa Sól-Kostrzyń., Gospodarka Wodna, 12, 280-284.
12. Kucuksezgin F., Uluturhan E., Batki H., 2008. Distribution of heavy metals in water, particulate matterand sediments of Gediz River (Eastern Aegean) Environ Monit Assess 141, 213-225.
13. Kunwar P. Singh,, Amrita Malik, Sarita Sinha, Vinod K. Singh, Ramesh Murthy, 2005. Estimation of source of heavy metal contamination in sediments of Gomti river (India) using principal component analysis, Water, Air, and Soil Pollution, 166, 321-341.
14. Lis J., Pasieczna A., 1995. Atlas geochemiczny Polski w skali 1: 2 500 000, Państw. Inst. Geol., Warszawa, 7.
15. Nelson S.M., Roline R.A., Thullen J.S., Sartoris J.J., Boutwell J.E. 2000. Invertebrate assemblages and traceelement bioaccumulation associated with constructed wetlands. Wetlands, 20, 2, 406-415.
16. Pasieczna A. 2003. Atlas zanieczyszczeń gleb miejskich w Polsce. Atlas of Urban soils qcontamination in Poland. Polish Geological Institute.
17. Salminen R. (chief-editor), Batista M.J., Bidovec M., Demetriades A., De Vivo B., Devos W., Duris M., Gilucis A., Gregorauskiene V., Halamic J., Heitzmann P., Lima A., Jordan G., Klaver G., Klein P., Lis J., Locutura J., Marsina K., Mazreku A., O'connor P.J., Olsson S.Å., Ottesen R.-T., Petersell V., Plant J.A., Reeder S., Salpeteur I., Sandström H., Siewers U., Steenfel T A., Tarvainen T., 2005. Geochemical atlas of Europe. Part 1 – Background Information, Methodology and Maps.
18. Samecka-Cymerman A., Kempers. J., 2007. Heavy Metals in Aquatic Macrophytes from Two Small Rivers Polluted by Urban Agricultural and Textile Industry Sewages SW Poland Arch. Environ. Contam. Toxicol. 53, 198–206.
19. Sasmaza A., Obekb E., Hasarb H., 2008. The accumulation of heavy metals in *Typha latifolia* L. grownin a stream carrying secondary effluent Ecological Engineering, 33, 278-284.
20. Skorbiłowicz E. 2012. Studia nad rozmieszczeniem niektórych metali w środowisku wodnym zlewni górnej Narwi. Rozprawy Naukowe Nr 222, Białystok, p. 212.
21. Sutherland R., 2000. Bed sediment – associated trace metals in an urban stream Oahu. Hawai. Environ. Geol., 39, p. 6.

22. Wang S., Gao Z., Lan D., Zheng A., Li G., 2008. Concentration distribution and assessment of several heavy metals in sediments of west-four Pearl River Estuary, Environ Geol., 55, 963-975.

23. Wang K., Huang T. Li, Mi H., Luo H., Tai J. 2006. Emissions of fuel metals content from a diesel vehicle engine. Atmospheric Environment, 37, p. 4637.

24. Wardas M., 2001. Badania metali ciężkich w osadach rzecznych w rejonie Polski południowej, Problemy analityczne badań osadów dennych, Sympozjum Naukowe Komitetu Chemii Analitycznej PAN, Komisja Analizy Wody, Radom-Jedlnia 2001, Materiały pokonferencyjne pod redakcją Jerzego Siepaka, Poznań, 63-74.

25. Wiśniowska-Kielian B., Niemiec M., 2005. Heavy metals contents in sediments of selected tributaries of river Dunajec, J. Elementom, 10, 2, 435-443.

26. Yu K.C., Chang C.Y., Tsai L.J., Ho S.T., 2000. Multivariate analysis on heavy metal binding fractions of river sediments in southern Taiwan. Water Sci. Technol. 42, 7–8, 193-199.

28

ECOLOGICAL ASSESSMENT OF THE SAN RIVER WATER QUALITY ON THE AREA OF THE SAN VALLEY LANDSCAPE PARK

Teresa Noga[1], Jadwiga Stanek-Tarkowska[2], Anita Pajączek[2], Natalia Kochman[3], Łukasz Peszek[3]

[1] Department of Biological Foundations of Agriculture and Environmental Education, Faculty of Biology and Agriculture, University of Rzeszów, Ćwiklińskiej 2, 35–601 Rzeszów, Poland, e-mail: teresa.noga@interia.pl

[2] Department of Soil Studies, Environmental Chemistry and Hydrology, Faculty of Biology and Agriculture, University of Rzeszów, Ćwiklińskiej 2, 35–601 Rzeszów, Poland, e-mail: jagodastanek@wp.pl; chuanita66@poczta.fm

[3] International Doctoral Studies in Natural Sciences, Faculty of Biology and Agriculture, University of Rzeszów, Ćwiklińskiej 2, 35–601 Rzeszów, Poland, e-mail: lukaspeszek@gmail.com; kochman_natalia@wp.pl

ABSTRACT

Water of the San River was characterized by good and high ecological status and high diatom richness. During four studied seasons at two sites 353 diatom taxa were identified in total, from which the most numerous were *Achnathidium pyrenaicum* (Hust.) Kobayasi and *Achnanthidium minutissimum* (Kütz.) Czarnecki var. *minutissimum*. Water of the San River, was characterized by high and good water quality (II and I class) in all seasons and at both studied sites.

Keywords: Diatoms, ecological status, diatomaceous indices IPS,GDI,TDI.

INTRODUCTION

Diatoms are a specialized group of organisms, occurring in most aquatic ecosystems or constantly moist ones. Therefore, they are always available to collect. In unfavorable conditions (including drying, freezing) they can survive in a vegetative form. Species diversity is important in assessing the environmental status of various aquatic ecosystems and is expressed in the lists of species, their autecology, number of individuals, etc. In assessment of purity of rivers it is important to distinc sensitive species, tolerant, resistant, with quantitative shares. Diatoms occurring in different biotopes can be accurately determined and counted and their communities can be compared in different types of environments, thus, they are an excellent material for the study of species diversity [Siemińska 1964, Rakowska 2001]. Monitoring studies with the use of diatoms to assess the quality of water are conducted in many European countries as well as in Poland [Prygiel, Coste 1993, Kelly, Whitton 1995, Kelly et al. 1995, Kwandrans et al. 1999, Eloranta 1999, Kawecka, Kwandrans 2000, Bogaczewicz-Adamczak et al. 2001, Rakowska 2001, Kwandrans 2002, Prygiel 2002, Bogaczewicz-Adamczak, Dziengo 2003, Kelly 2003, Kelly et al. 2008, Żelazowski et al. 2004, Dumnicka et al. 2006, Rakowska, Szczepocka 2011]. For this purpose a computer software OMNIDIA [Lecointe et al. 1993] was prepared, which includes taxonomic and ecological databases, together with indicative values and sensitivity degrees of individual diatom taxa.

Regular studies which aim to investigate the diversity of diatoms in the flowing water, together with an assessment of water quality have been carried out in the Podkarpacie Province only since 2007 [Noga, Siry 2010, Tambor, Noga 2011, Noga 2012, Pajączek et al. 2012, Żelazna-Wieczorek 2012, Noga et al. 2013a,b,c,d]. Previously, individual studies were carried out only in the upper reach of the San River in connection with the massive growth of *Didymosphenia geminata* below the reservoirs of "Solina" and "Myczkowce" dams [Kawecka, Sanecki 2003].

The aim of this work is to recognize the diatom species diversity and the ecological water quality of the upper section of the area of the San River on the territory of the San Valley Landscape Park using three diatomaceous indices (IPS, GDI, TDI) calculated with the computer software OMNIDIA.

STUDY AREA

The San River with a 443.4 km length is the largest and has one of the richest of water among the Carpathian rivers.

The upper San basin is one of the most valuable area in Poland and includes more than 90-km stretch of the river from the source to the reservoir at Solina dam. It is a highland river for the first 50 kilometers, in the further part it is a mountain river. The San River Valley is characterized by a shallow and wide riverbed, steep shores and therefore the Carpathian flysch, which is typical for this region, can be observed. Numerous rapids are in the riverbed especially on the section between Stuposiany and Rajskie [Kukuła 2002, Sieradzki 2002, Szlak wodny „Błękitny San" 2005].

The study sites were located in Procisne near the bridge on the San River on the national road No. 896 and in Sękowiec also near the bridge over the San River (Figure 1). Detailed description of the positions is presented in Table 1.

MATERIAL AND METHODS

The material for this study was collected four times from October 2010 to September 2011 from all available habitats (stones, slit and aquatic macrophytes) in the San River on the San Valley Landscape Park area. pH, conductivity and water temperature were measured directly in the field. Samples of water for chemical analysis were taken during the last study season (September 2011).

The collected material was preserved in a 4% solution of formalin. In order to obtain pure valves of diatoms a part of the obtained material was subjected to maceration in a mixture of sulphuric acid and potassium dichromate in a ratio 3:1, and next it was rinsed in a centrifuge (at 2500 rpm). The diatoms were mounted in permanent diatom slides with synthetic resin – PLEURAX (refractive index 1.75). The material was prepared according to the methods applied by Kawecka [2012].

Diatoms were identified using the Nikon 80i light microscope, according to keys: Krammer, Lange-Bertalot [1986, 1988, 1991a,b], Lange-Bertalot [2001], Krammer [2000, 2002, 2003], Hofmann et al. [2011].

Diatoms were identified and counted using DIC (Differential Interference Contrast) under a Nikon ECLIPSE 80i microscope. The ultrastructure was examined under a Hitachi S-4700 electron microscope in the laboratory of Field Emis-

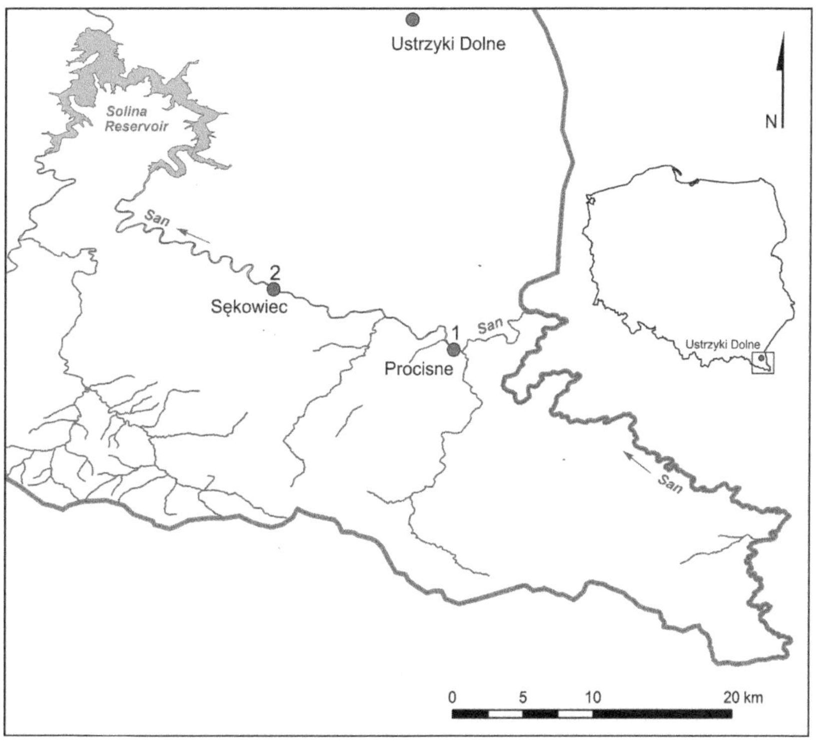

Figure 1. Location of the study sites

sion Microscopy, Scanning Electron Microscopy and Microanalysis (Institute of Geological Sciences of the Jagiellonian University in Cracow). Chemical analysis of water was carried out in the Departmental Laboratory of Analysis of Wholesomeness of Environment and Materials of Agricultural Origin, by means of a liquid chromatography apparatus PeakNet Dionex 2001–2006, version 6.80.

Species composition in the samples was determined through counting specimens on randomly selected fields of view under a light microscope. The number of valves counted was 400. Species with a content above 5% in a given diatom assemblage were defined as abundant.

Ecological diatoms classification was presented according to the Van Dam et al. [1994] list. The following indicators were used: pH, saprobity and trophic state.

Biological water quality assessment was performed using OMNIDIA software [Lecointe et al. 1993], version 4.2. This also contains ecological and taxonomic data [Prygiel, Coste 1993]. Results of the analyses were presented applying the chosen diatomic indices, for which a range of ecological classes of water quality and the ecological status conforming to them was outlined by mutual agreement [Dumnicka et al. 2006]:

Water Quality Clas*	Ecological status	IPS	GDI	TDI	Trophic status
I	high	> 17	> 17	<35	oligotrophic
II	good	15–17	14–17	35–50	oligo-mesotrophic
III	moderate	12–15	11–14	50–60	mesotrophic
IV	poor	8–12	8–11	60–75	eutrophic
V	bad	<8	<8	>75	hypertrophic

* According to the Decree of the Minister of the Environment from 9 Nov. 2011 (Dz. U. No 257, pos. 1545).

Indices of organic pollution: the SPI – Specific Pollution Sensitivity Index [CEMAGREF 1982] and the GDI – Generic Diatom Index based on genera are scaled from 1 to 20 (when water quality increases there is an increase in indicator value). The TDI – Trophic Diatom Index is scaled from 1 to 100 (the higher the value, the bigger trophy of water). The percentage of participation of species characteristic for organic pollution (PT) must be taken into account in interpretation of the TDI index. There is a possibility of organic pollution if PT values are above

Table 1. Characteristic of the investigated sites, designated on the San River, including the dominant, endangered and rare species

Site	PROCISNE	SĘKOWIEC
Width [m]	about 18	about 20
Depth [m]	0.2–0.5	0.2–0.7
Insolation	Medium	Low
Bottom	Stony, large boulders in the center of riverbed	Stony, rock outcrops of Carpatian flysh in riverbed
Dominant species [>5%]	*Achnanthidium pyrenaicum, A. minutissimum* var. *minutissimum, A. catenatum, Encyonopsis minuta, E. subminuta*	*Achnanthidium pyrenaicum, A. subatomus, A. minutissimum* var. *minutissimum, Denticula tenuis, Fragilaria perminuta, Encyonopsis subminuta*
Endangered and rare species (accoding to Siemińska et al. 2006)		
Endangered [E]	*Encyonema gauemanni, Fallacia lenzii, Neidium alpinum, Pinnularia subruperstris, Sellaphora pseudopupula*	*Fallacia lenzii, Pinnularia schonfelderii, P. subruperstris, P. viridiformis, Sellaphora pseudopupula*
Vulnerable [V]	*Fragilaria tenera, Gomphonema sarcophagus, Navicula splendicula, Pinnularia microstauron, Psammothidium laueunburgianum, P. subatomoides, Sellaphora bacillum*	*Fragilaria nanana, F. tenera, Gomphonema sarcophagus, Navicula splendicula, Psammothidim laueunburgianum, P. subatomoides, Sellaphora bacillum*
Rare [R]	*Achnanthes coarctata, Amphora cimbrica, Amphipleura pellucida, C. lancettula, C. molaris, C. schumaniana, Cymbella helvetica, C. lanceolata, C. proxima, Diploneis pseudoovalis, Fallacia subhamulata, Fragilaria brevistriata, Geissleria deccusis, Hippodonta neglecta, Luticola acidocinata, Navicula cariocincta, N. upsaliensis, Nitzschia flexa, Parlibellus protractoides, Stauroneis thermicola, Surirella brebissonii* var. *brebissonii*	*Amphipleura pellucida, Amphora cimbrica, Caloneis lancettula, C. molaris, Cocconeis disculus, Cymbella helvetica, C. lanceolata, Fallacia subhamulata, Geissleria deccusis, Luticola acidocinata, Navicula cariocincta, N. oligotraphenta, N. oppugnata, N. upsaliensis, Stauroneis thermicola, Surirella brebissonii* var. *brebissonii*

20% [CEMAGREF 1982, Coste, Ayphassorho 1991, Kelly, Whitton 1995].

On the basis of the Red List of Algae for Polish species by Siemińska et al. [2006], various degrees of threat were distinguished (Ex – Extinct or probably extinct, E – Endangered, V – Vulnerable, R – Rare, I – Indeterminate).

RESULTS

The San River water was characterized by alkaline reaction (6.8–8.2) and moderate electrolyte content. High water quality (I water quality class) on the area of the San Valley Landscape Park was determined by chemical parameters (Table 1).

In total 353 diatom taxa were found at two sampling sites (293 taxa in Procisne and 272 in Sękowiec Village), among which eight taxa were considered as dominant. *Achnathidium pyrenaicum* (Hust.) Kobayasi and *A. minutissimum* (Kütz.) Czarnecki var. *minutissimum* were reached 50% of abundance in the assemblage, while the other dominant were developed less frequently (5–10%) (Table 1, Plate 1).

Ecological preferences of diatom with respect to pH reaction, saprobity and trophy were presented according to Van Dam et al. [1994] list.

Based on pH reaction, two groups of diatom taxa were dominated in the San River – alkaliphilous and neutral (Figure 2). Based on saprobity, β-mesosaprobous (water quality class II) taxa was the most frequent. Oligosaprobous (class I and II) and α-mesosaprobous (class III) taxa were also frequent (Figure 3). Oligotraphentic and mesotraphentic taxa were the most numerous in all sampling seasons (Figure 4).

All diatomaceous indices values (IPS, GDI, TDI) in the San River at both sampling sites were similar and indicated good, and high water quality (class II and I), what corresponded to good and high ecological status in all sampling seasons. A small share of taxa resistant to organic pollutants was also found (PT = 3.2–9.8%) (Table 2).

Endangered and rare diatom taxa constituted about 10% of the total number of taxa. Seven diatom taxa were found in the category of endangered, eight in the category of vulnerable, and twenty-four in the category of rare (Table 1, Plate 2–3).

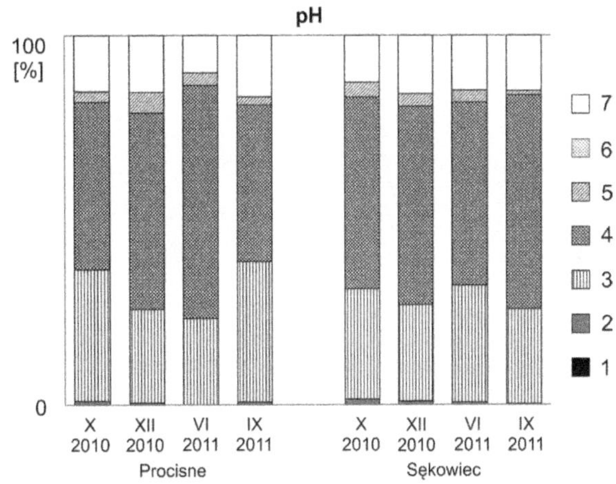

Figure 2. Percentage participation of diatoms at studied sites in the San River, according to Van Dam et al. [1994] classification system: pH range: 1 – acidobiontic, 2 – acidophilous, 3 – neutral, 4 – alkaliphilous, 5 – alkalibiontic, 6 – indifferent, no apparent optimum, 7 – unknown

Table 2. The chemical parameter values measured in the San River in years 2010–2011 including Shannon-Wiener diversity index (H') and IPS, GDI and TDI indices

Site	PROCISNE				SĘKOWIEC			
Date	X.2010	XII.2010	VI.2011	IX.2011	X.2010	XII.2010	VI.2011	IX.2011
pH	7.3	6.9	7.2	8.2	7.3	7.1	7.3	8,0
Conductivity [μS/cm]	162	171	287	165	210	193	272	295
Cl^- [mg/l]	–	–	–	3.47	–	–	–	2.11
NO_2–N [mg/l]	–	–	–	0.27	–	–	–	<0.1
NO_3–N [mg/l]	–	–	–	1.10	–	–	–	0.99
SO_4–S [mg/l]	–	–	–	23.78	–	–	–	23.86
H'	4.92	5.82	4.37	5.28	5.59	5.60	4.22	4.34
IPS	17.1	16.5	17.6	16.0	16.2	17.1	18.1	17.2
GDI	15.5	15.2	16.3	14.8	14.7	15.2	16.2	15.7
TDI	35.7	45.9	41.4	39.5	42.2	40.7	35.0	34.4
%PT	5.2	4.7	3.8	9.2	8.0	6.2	3.2	5.3

Ecological status IPS, GDI, TDI	high	good
	(unshaded)	(shaded)

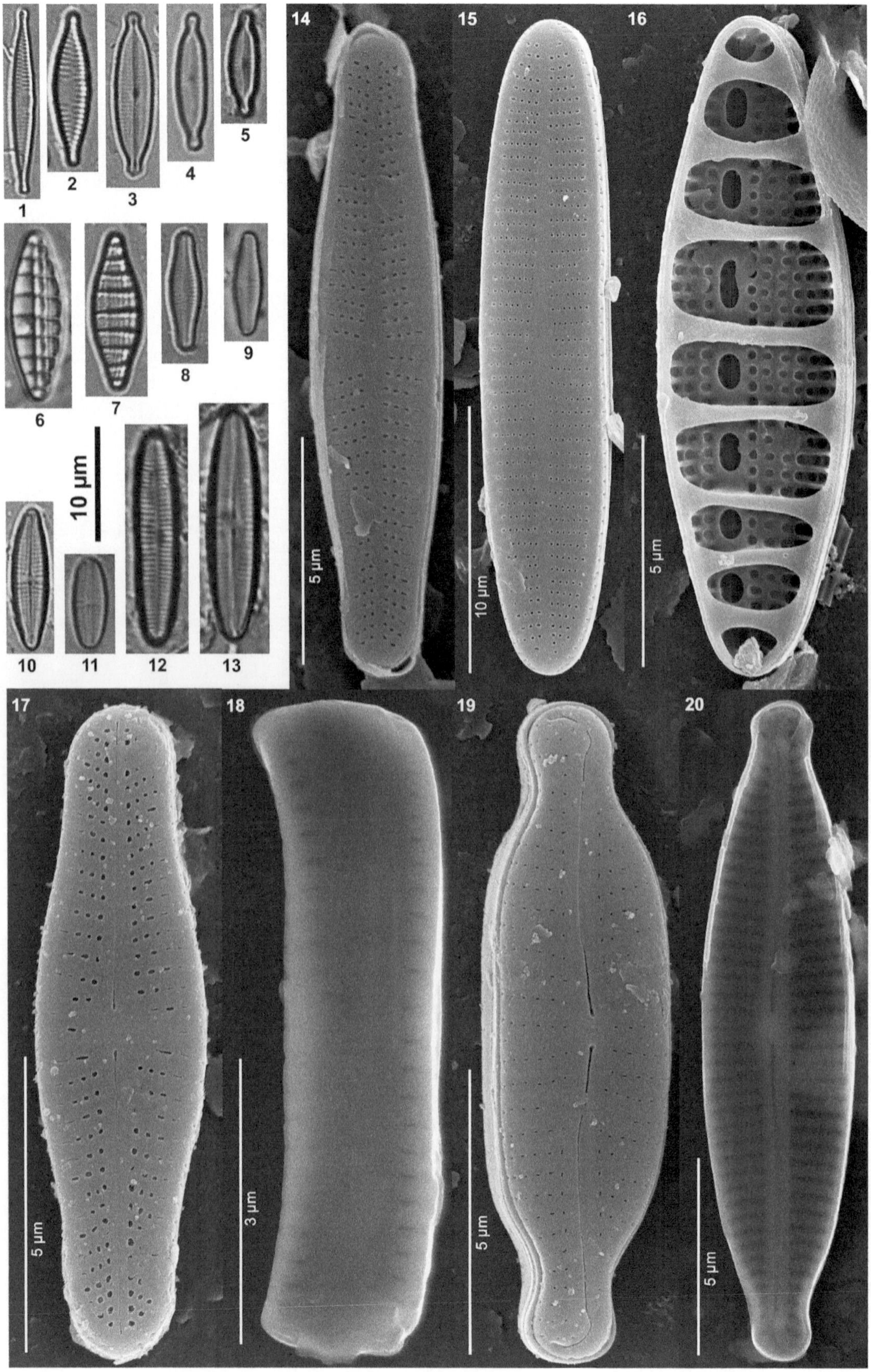

Plate I. Dominant taxa at studied sites

1-2 – *Fragilaria perminuta* (Grunow) Lange-Bert., 3, 20 – *Encyonopsis subminuta* Krammer & Reichardt, 4-5, 19 – *E. minuta* Krammer & Reichardt, 6-7, 16 – *Denticula tenuis* Kütz., 8, 17-18 – A*chnanthidium catenatum* (Bily & Marvan) Lange-Bert., 9, 14 – *A. minutissimum* (Kütz.) Czarnecki var. *minutissimum,* 10 – *A. pyrenaicum* (Hust.) Kobayasi, 11-13, 15 – A. *subatomus* (Hust.) Lange-Bert.

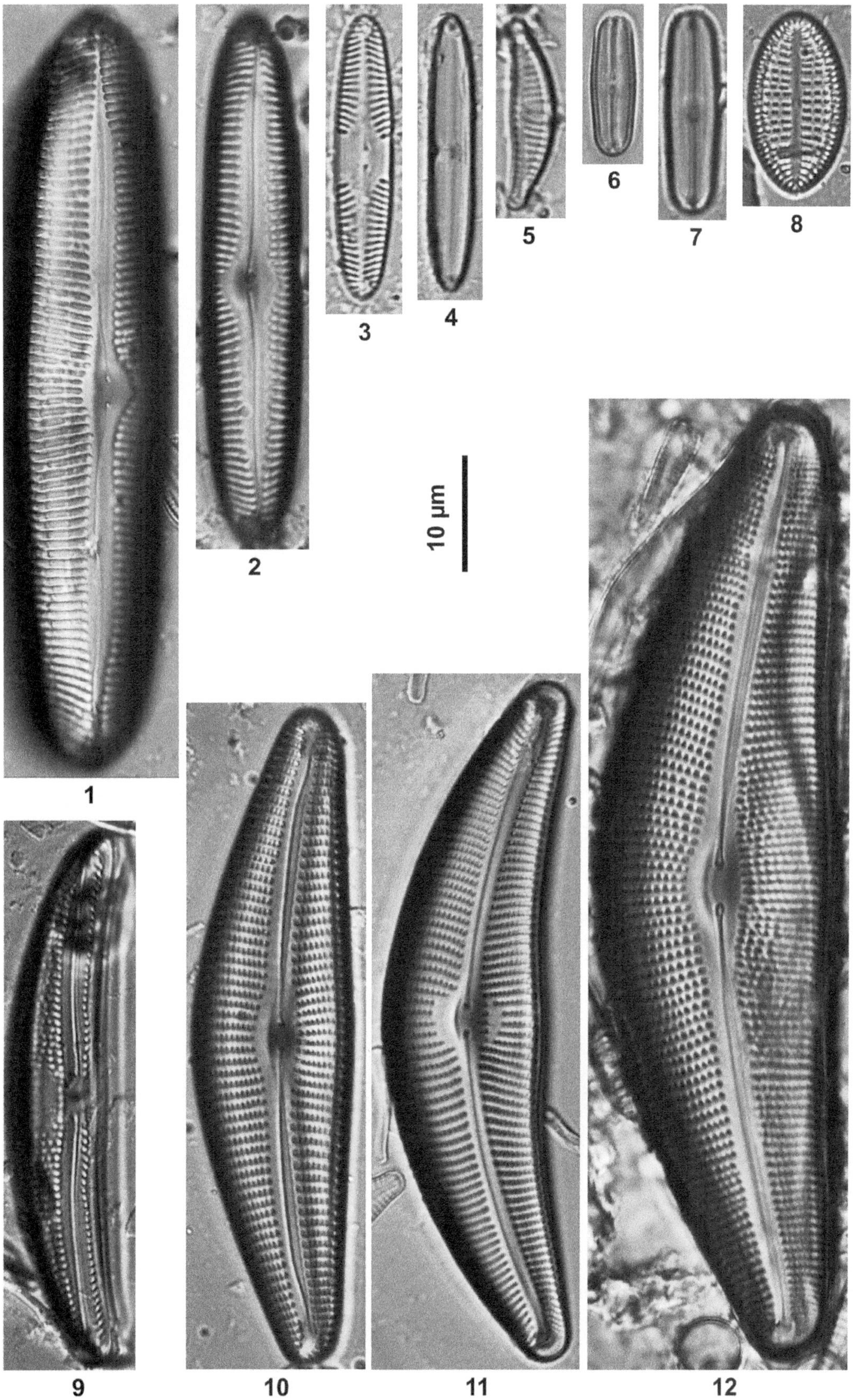

Plate II. Taxa from Red list of the algae in Poland (according to Siemińska et al. 2006)
1 – P*innularia viridiformis* Krammer, 2 – *P. subrupestris* Krammer, 3 – *P. schoenfelderi* Krammer, 4 – *Neidium alpinum* Hust., 5 – *Encyonema gauemannii* (Meister) Krammer, 6 – *Fallacia lenzii* (Hust.) Lange-Bert., 7 – *F. subhamulata* (Grunow) D.G. Mann, 8 – *Cocconeis disculus* (Schumann) Cleve, 9 – *Amphora cimbrica* Østrup, 10 – *Cymbella helvetica* Kütz., 11 – *Cymbella proxima* Reimer, 12 – *Cymbella lanceolata* (Agardh) Kirchner

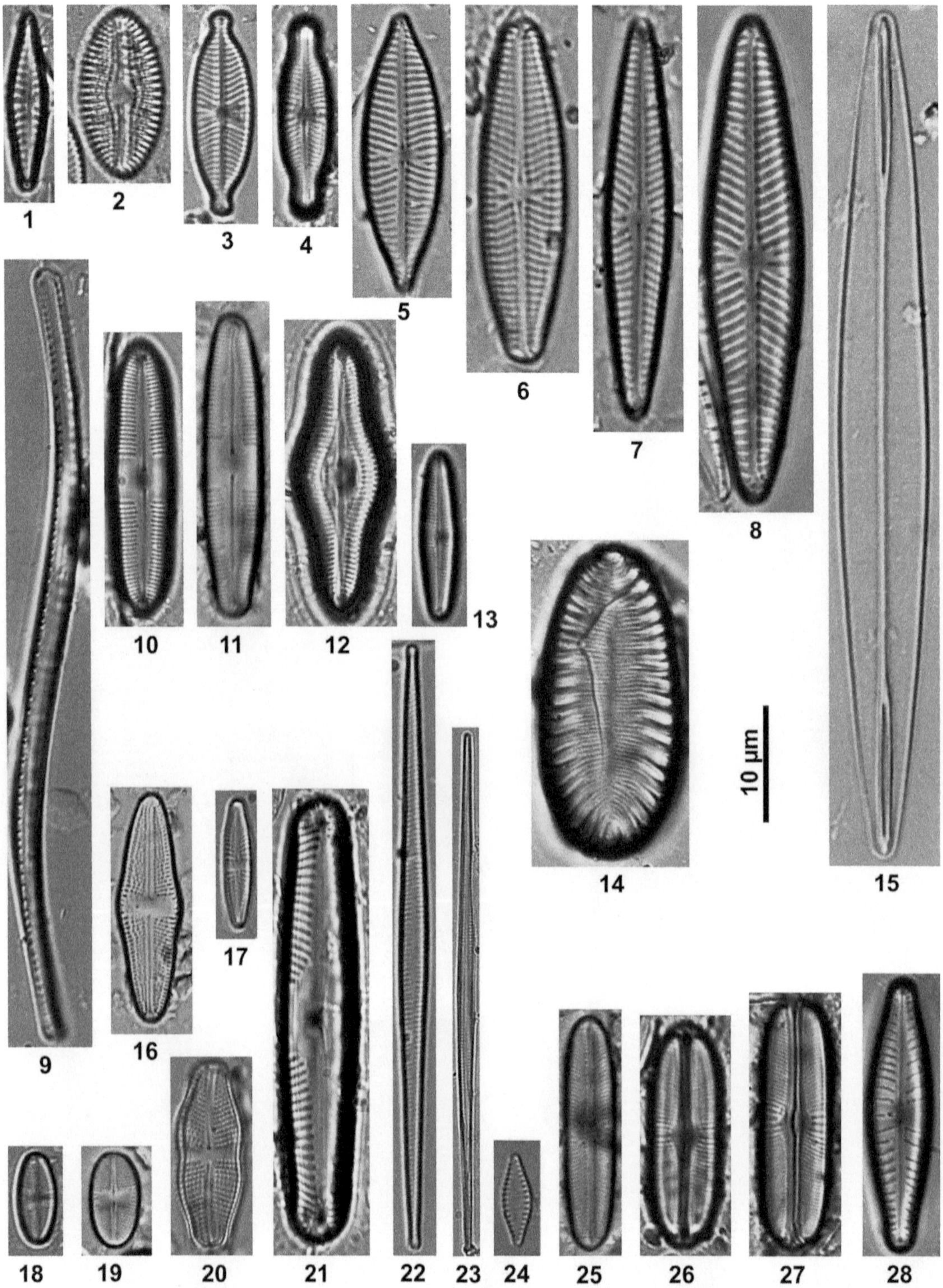

Plate III. Taxa from Red list of the algae in Poland (according to Siemińska et al. 2006)
1 – *Hippodonta neglecta* Lange-Bert., Metzeltin & Witkowski, 2 – *Diploneis pseudoovalis* Hust., 3 – *Geissleria deccusis* (Østrup) Lange-Bert. & Metzeltin, 4 – *Parlibellus protractoides* (Grunow) Witkowski, Lange-Bert. & Metzeltin, 5 – *Navicula upsaliensis* (Grunow) Paragallo, 6 – *N. splendicula* Land, 7 – *N. cariocincta* Lange-Bert., 8 – *N. oppugnata* Hust., 9 – *Nitzschia flexa* Schumann, 10-11 – *Caloneis molaris* (Grunow) Krammer, 12 – *C. schumaniana* (Grunow) Cleve, 13 – *C. lancettula* (Schulz) Lange-Bert. & Witkowski, 14 – *Surirella brebissonii* Krammer & Lange-Bert. var. *brebissonii*, 15 – *Amphipleura pellucida* (Kütz.) Kütz., 16 – *Luticola acidoclinata* Lange-Bert., 17 – *Stauroneis thermicola* (Petersen) Lund, 18 – *Psammothidium lauenburgianaum* (Hust.) Bukhtiyarova & Round, 19 – *P. subatomoides* (Hust.) Bukhtiyarova & Round, 20 – *Achnanthes coarctata* (Bréb.) Grunow, 21 – *Pinnularia microstauron* (Ehrenb.) Cleve, 22 – *Fragilaria tenera* (W. Smith) Lange-Bert., 23 – *F. nanana* Lange-Bert., 24 – *F. brevistriata* Grunow, 25 – *Sellaphora pseudopupula* (Krasske) Lange-Bert., 26-27 – *S. bacillum* (Ehrenb.) D.G. Mann, 28 – *Gomphonema sarcophagus* Gregory

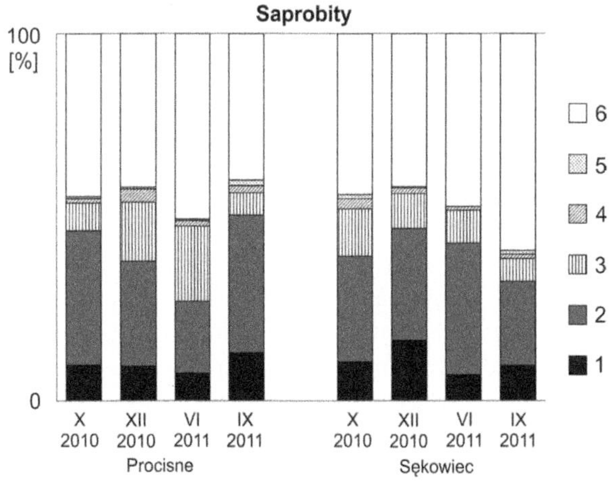

Figure 3. Percentage participation of diatoms at studied sites in the San River, according to Van Dam et al. [1994] classification system: saprobity range: 1 – oligosaprobous, 2 – β-mesosaprobous, 3 – α-mesosaprobous, 4 – α-meso-polysaprobous, 5 – polysaprobous, 6 – unknown

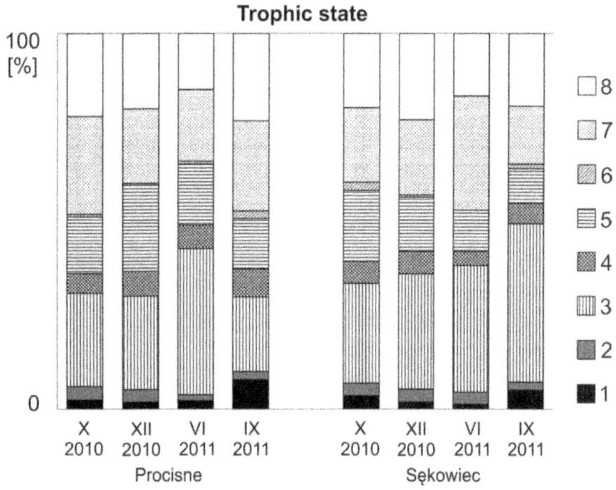

Figure 4. Percentage participation of diatoms at studied sites in the San River, according to Van Dam et al. [1994] classification system: trophic state: 1 – oligotraphentic, 2 – oligo-mesotraphentic, 3 – mesotraphentic, 4 – meso-eutraphentic, 5 – eutraphentic, 6 – hypereutraphentic, 7 – oligo- to eutraphentic, 8 – unknown

DISCUSSION

The spring areas of the San River are mountainous terrain of Bieszczady with sparse buildings for tourism and recreation. A significant percentage of the catchment area is covered by legal protection due to high natural and landscape values as in case of Bieszczady National Park and eight Landscape Parks, including the San Valley Lanscape Park. The San river has the highest quality of water in upper parts of its course. This is connected with the fact that source sections of the river flowing through the forest areas which are under inconsiderable influence of human pressure [WIOŚ 2004].

The water quality studies of the San River conducted in 2010–2011 in the area of the San Valley Landscape Park using benthic diatoms as bio-indicators indicated a good and very good water quality. This was confirmed by the chemical analysis, which classified waters of the San in Sękowiec and Procisne Villages in the first class of water quality. The waters of the San were characterized by very high species richness of diatoms - at two sites designated 353 taxa. Żelazna-Wieczorek [2012] gave a similar number of diatom species from the springs and the upper sections of streams which were tributaries of the San, but in the material collected from 17 positions. Other rivers and streams in the Podkarpacie region are also characterized by high species richness, but research on them was always conducted on a larger number of sites [Noga 2012, Pajączek et al. 2012, Noga et al. 2013a, c, d].

The most numerous were two species from the genus *Achnanthidium* in all study seasons. Numerous populations were created by *A. pyrenaicum*, especially on the site in Sękowiec (even> 50% share in assemblage). Optimum occurrence was in oligo- to mesotrophic water rich in calcium with medium to high content of electrolytes [Krammer, Lange-Bertalot 1991b]. According to Hofmann [1994], it was tolerant species, which prefer alkaline waters with pH>7. It was very abundant in the territory of the Podkarpacie Province in upper course of Wisłok River [Noga 2012, Noga et al. 2013b). High pH values in the San River may be related to geological structure of terrain. All the Western and Eastern Carpathians, that is where the upper valley of the San is located, are composed of sedimentary rock such as sandstone, mudstone, marlston, conglomerates and shales that make up the so-called Carpathian flysch [Kondracki 2001, Sieradzki 2002]. Therefore *Achnanthidium pyrenaicum* may find here favorable conditions to grow.

The species that grow very abundantly on both study sites were also *Achnanthidium minutissimum* var. *minutissimum,* cosmopolitan species having a wide ecological spectrum. It occurs in a wide range of pH (4.3–9.2) in all types of water bodies with the conditions from oligo- to eutrophic. It is an indicator of water rich in oxygen. Numerous populations develop i.a. in high mountain streams [Krammer, Lange-Bertalot 1991b, Van Dam ety al. 1994, Hofmann et al. 2011, Kawecka 2012]. It is also one of the most common and most frequently noted diatoms in the Podkarpacie region especially in the upper courses of rivers and streams [Noga 2012, Pajączek et al. 2012, Noga et al. 2013a, b, c, d].

The alkaline character of water is reflected also in ecological preferences of the other species of diatoms developing in the San River. In all seasons, on both sites predominated alkalophiles (pH> 7) and neutral (pH = 7) taxa, as in most of the watercourses in Podkarpacie [Van Dam et al. 1994, Pajączek et al. 2012, Noga et al. 2013b,c].

Calculation on diatomaceous indices with the OMNIDIA software [Lecointe et al., 1993] showed that San River on the area of the San Valley Landscape Park carried waters of good quality. The IPS and GDI indices values reflecting the degree of organic and non-organic contaminants and trophic index TDI, indicated good or very good ecological status (second and first class of water quality). A small proportion (> 10%) of diatoms tolerant to organic pollution – % PT was found on both sites, which is why it can be said that the San is not exposed to the risk of eutrophication [CEMAGREF 1982, Coste, Ayphassorho 1991, Kelly, Whitton 1995].

The study showed that the San River has the best ecological potential in relation to other studied watercourses of the Podkarpacie region [Noga et al. 2013a, b, c].

San River is the second largest watercourse in the south-eastern Poland. It is the main source of water supply in the Podkarpacie region. The studies carried out in the San Valley Landscape Park showed that the river is clean in the upper reaches with large diatom species richness with a high share of rare and endangered taxa. On both sites, in the four study seasons indicates 35 rare and endangered taxa [Siemińska et al. 2006], including 7 in the category of endangered (E): *Encyonema gauemanni* (Meister) Krammer, *Fallacia lenzii* (Hust.) Lange-Bert, *Neidium alpinum* Hust., *Pinnularia subruperstris* Krammer, *P. schonfelderii* Krammer, *P. viridiformis* Krammer, *Sellaphora pseudopupula* (Krasske) Lange-Bert were found on both sites, in the four study seasons. Therefore, further monitoring of these waters, especially in the spring areas is necessary.

REFERENCES

1. Bogaczewicz-Adamczak B., Kłosińska D., Zgrundo A. 2001. Diatoms as indicators of water pollution in the coastal zone of the Gulf of Gdańsk (southern Baltic Sea). Oceanol. Hydrobiol. St. 30(3-4), 59–75.
2. CEMAGREF 1982. Etude des méthodes biologiques quantitative d'appréciation de la qualité des eaux – Rapport Division Qualité des Eaux LyonAgence financiére de Bassin Rhone – Méditerranée – Corse, Pierre – Bénite. pp. 218.
3. Coste M., Ayphassorho H. 1991. Étude de la qualité des eaux du Bassin Artois–Picardie a l'aide des communautés de diatomées benthiques (Application des indices diatomiques) – Raport Cemagref. Bordeaux – Agence de l'Eau Artois–Picardie, Douai. pp. 277
4. Dumnicka E., Jelonek M., Kwandrans J., Wojtal A., Żurek R. 2006. Ichtiofauna i status ekologiczny wód Wisły, Raby, Dunajca i Wisłoki. Institute of Nature Conservation, Polish Academy of Sciences, Kraków, pp. 220.
5. Eloranta P. 1999. Application of diatom indices in Finnish Rivers. [In:] J. Prygiel, B.A. Whitton, J. Bukowska (eds), Use of Algae for Monitoring Rivers III. Agence de I' Eau Artois – Picardie, 138–144.
6. Hofmann G., Werum M., Lange-Bertalot H. 2011. Diatomeen im Süßwasser – Benthos von Mitteleuropa. Bestimmungsflora Kieselalgen für die ökologische Praxis. Über 700 der häufigsten Arten und ihre Ökologie. [In:] H. Lange-Bertalot (ed.). A.R.G. Gantner Verlag K.G. pp. 908.
7. Kawecka B. 2012. Diatom diversity in streams of the Tatra National Park (Poland) as indicator of environmental conditions. W. Szafer Institute of Botany, Polish Academy of Sciences, Kraków.
8. Kawecka B., Kwandrans J. 2000. Diversity of flora and fauna in running waters of the Province of Cracow (southern Poland) in relation to water guality. Acta Hydrobiol. 42(3/4), 145–173.
9. Kawecka B., Sanecki J. 2003. *Didymosphenia geminata* in running waters of southern Poland – Symptoms of change in water quality? Hydrobiol. 495, 193–201.
10. Kawecka B., Kwandrans J., Szyjkowski A. 1999. Use of algae for monitoring rivers in Poland – Situation and development. [In:] J. Prygiel, B. A. Whitton, J. Bukowska (eds), Use of algae for monitoring rivers III. Agence de l'Eau Artois-Picardie, 57–65.
11. Kelly M.G. 2003. Short term dynamics of diatoms in an upland stream and implications for monitoring eutrophication. Environmental Pollution 125, 117–122.
12. Kelly M.G., Whitton B.A. 1995. The Trophic Diatom Index: a new index for monitoring eutrophication in rivers. J. Appl. Phycol. 7, 433–444.
13. Kelly M.G., Penny C.J., Whitton B.A. 1995. Comparative performance of benthic diatom indices used to assess river water quality. Hydrobiol. 302, 179–188.
14. Kelly M.G., Juggins S., Guthrie R., Pritchard S., Jamieson J., Rippey B., Hirst H., Yallop M. 2008.

Assessment of ecological status in U.K. rivers using diatoms. Freshwater Biol. 53, 403–422.

15. Kondracki J. 2001. Geografia regionalna Polski. Państwowe Wydawnictwo Naukowe, Warszawa, pp. 441.
16. Krammer K. 2000. The genus *Pinnularia.* [In:] H. Lange-Bertalot (ed.). Diatoms of Europe. A.R.G. Ganter Verlag K.G., Vaduz 1, 1–703.
17. Krammer K. 2002. *Cymbella.* [In:] H. Lange-Bertalot (ed.). Diatoms of Europe. A.R.G., Gantner Verlag K.G., Rugell 3, 1–584.
18. Krammer K. 2003. *Cymbopleura, Delicata, Navicymbula, Gomphocymbellopsis, Afrocymbella.* [In:] H. Lange-Bertalot (ed.). Diatoms of Europe. A.R.G., Gantner Verlag K.G., Rugell 4, 1–530.
19. Krammer K., Lange-Bertalot H. 1986. Bacillariophyceae. 1. Naviculaceae. [In:] H. Ettl, J. Gerloff, H. Heyning & D. Mollenhauer (eds). Süsswasserflora von Mitteleuropa. G. Fischer Verlag, Stuttgart – New York 2(1), 1–876.
20. Krammer K., Lange-Bertalot H. 1988. Bacillariophyceae. 2. Bacillariaceae, Epithemiaceae, Surirellaceae. [In:] H. Ettl, J. Gerloff, H. Heyning & D. Mollenhauer (eds). Süsswasserflora von Mitteleuropa G. Fischer Verlag, Stuttgart – New York 2(2), 1–596.
21. Krammer K., Lange-Bertalot H. 1991a. Bacillariophyceae. 3. Centrales, Fragilariaceae, Eunotiaceae. [In:] H. Ettl, J. Gerloff, H. Heyning & D. Mollenhauer (eds). Süsswasserflora von Mitteleuropa. G. Fischer Verlag, Stuttgart – Jena 2(3), 1–576.
22. Krammer K., Lange-Bertalot H. 1991b. Bacillariophyceae. 4. Achnanthaceae, Kritische Ergänzungen zu *Navicula* (Lineolate) und *Gomphonema*, Gesamtliteraturverzeichnis. [In:] H. Ettl, J. Gerloff, H. Heyning & D. Mollenhauer (eds). Süsswasserflora von Mitteleuropa. G. Fischer Verlag, Stuttgart – Jena 2(4), 1–437.
23. Kukuła K. 2002. Dorzecze Górnego Sanu i jego walory. [In:] F. Oberc (ed.), Zeszyty Archiwum Ziemi Sanockiej. Fundacja „Archiwum Ziemi Sanockiej", Sanok – Zahutyń 2, 30–35.
24. Kwandrans J., Eloranta P., Kawecka B., Wojtan K. 1999. Use of benthic diatom communities to evaluate water quality in rivers of southern Poland. [In:] J. Prygiel, B.A. Whitton, J. Bukowska (eds). Use of algae for monitoring rivers III. Agence de l'Eau Artois-Picardie, 154–156.
25. Kwandrans J. 2002. Upper Vistula River: Response of aquatic communities on pollution and impoundment. IX. Benthic diatom communities. Polish J. Ecol. 50(2), 223–236.
26. Lecointe C., Coste M., Prygiel J. 1993. OMNIDIA: software for taxonomy, calculation of diatom indices and inventories management. Hydrobiol. 269/270, 509–513.
27. Noga T. 2012. Diversity of diatom communities in the Wisłok River (SE Poland). [In:] K. Wołowski, I. Kaczmarska, J. M. Ehrman & A. Z. Wojtal (eds), Phycological Reports: Current advances in algal taxonomy and its applications: phylogenetic, ecological and applied perspective. Institute of Botany Polish Academy of Sciences, Krakow, 109–128.
28. Noga T., Siry K. 2010. Różnorodność flory okrzemek w potoku Łubienka (Pogórze Dynowskie, Polska SE). Zeszyty naukowe PTIE i PTG 12, 75–86.
29. Noga T., Stanek-Tarkowska J., Peszek Ł., Pajączek A., Kowalska S. 2013a. Use of diatoms to assess water quality of anthropogenically modified Matysówka stream. Journal of Ecological Engineering 14(2), 1–11.
30. Noga T., Stanek-Tarkowska J., Pajączek A., Peszek Ł., Kochman N. 2013b. Ecological characterization of diatom communities in the Wisłok River with application of their indicatory role to the evaluation of water quality. Journal of Ecological Engineering 14(4),18–27.
31. Noga T., Stanek-Tarkowska J., Pajączek A., Peszek Ł., Kochman N., Woźniak K. 2013c. Application of diatoms to assess the quality of the waters of the Baryczka stream, left-side tributary of the River San. Journal of Ecological Engineering 14(2), 8–23.
32. Noga T., Stanek-Tarkowska J., Pajączek A., Peszek Ł., Kochman **N.,** Kozak E., Kędziora Ł., Wąsacz P. 2013d. Wstępne rozpoznanie okrzemek Bacillariophyceae Jeziorek Duszatyńskich (Bieszczady Zachodnie). Roczniki Bieszczadzkie 21, 127–146.
33. Pajączek A., Musiałek M., Pelczar J. & Noga T. 2012. Diversity of diatoms in the Mleczka River, Morwawa River and Różanka Stream (tributaries of the Wisłok River, SE Poland), with particular reference to threatened species. [In:] K. Wołowski, I. Kaczmarska, J. M. Ehrman & A. Z. Wojtal (eds), Phycological Reports: Current advances in algal taxonomy and its applications: phylogenetic, ecological and applied perspective. Institute of Botany Polish Academy of Sciences, Krakow, 129–152.
34. Prygiel J. 2002. Mangement of the diatom monitoring network in France. J. Appl. Phycol. 14, 19–26.
35. Prygiel J., Coste M. 1993. The assessment of water quality in the Artois–Picardie water basin (France) by the use of diatom indices. Hydrobiol., 269/270, 343–349.
36. Rakowska B. 2001. Studium różnorodności okrzemek ekosystemów wodnych Polski niżowej. Wydawnictwo Uniwersytetu Łódzkiego, Łódź.
37. Rakowska B., Szczepocka E. 2011. Demonstration of the Bzura River restoration using diatom indices. Biologia 66(3), 411–417.
38. Siemińska J. 1964. Bacillariophyceae – Okrzemki, [In:] K. Starmach (ed.), Flora słodkowodna Polski, 6. PWN, Warszawa.

39. Siemińska J., Bąk M., Dziedzic J., Gąbka M., Gregorowicz P., Mrozińska T., Pełechaty M., Owsiany P. M., Pliński M., Witkowski A. 2006. Red list of the algae in Poland. [In:] Z. Mirek, K. Zarzycki, W. Wojewoda, Z. Szeląg (eds). Red list of plants and fungi in Poland W. Szafer Institute of Botany, Polish Academy of Sciences, Kraków, 37–52.

40. Sieradzki S. Geologia Doliny Sanu. [In:] F. Oberc (ed.), Zeszyty Archiwum Ziemi Sanockiej. Fundacja „Archiwum Ziemi Sanockiej", Sanok – Zahutyń 2, 30–35.

41. Szczepocka E., Szulc B. 2009. The use of benthic diatoms in estimating water quality of variously polluted rivers. Oceanol. Hydrobiol. St. 38(1), 17–26.

42. Szlak wodny „Błękitny San" – przewodnik. 2005. Podkarpacka Regionalna Organizacja Turystyczna, Rzeszów.

43. Tambor A., Noga T. 2011. Różnorodność flory okrzemek w rzece Lubcza i potoku Lubenia (Podgórze Rzeszowskie, Polska SE). Rocznik Przemyski 47(3), 105–118.

44. Van Dam H., Martens A., Sinkeldam J. 1994. A coded checklist and ecological indicator values of freshwater diatoms from the Netherlands. Netherlands J. Aquatic Ecol. 28, 117–133.

45. WIOŚ Rzeszów. 2004. Raport o stanie środowiska w Województwie Podkarpackim (The state of the environment, report in Subcarphatian Voivodeship).

46. Żelazna-Wieczorek J. 2012. Okrzemki *Bacillariophyta* źródeł i odcinków źródłowych potoków w górnym odcinku rzeki San. Roczniki Bieszczadzkie 20, 220–229.

47. Żelazowski E., Magiera M., Kawecka B., Kwandrans J., Kotowicz J. 2004. Use of algae for monitoring rivers in Poland – in the light of a new law for environmental protection. Oceanol. Hydrobiol. St. 33(4), 27–39.

EVALUATION OF LANDFILL POLLUTION WITH SPECIAL EMPHASIS ON HEAVY METALS

Magdalena Vaverková[1], Dana Adamcová[1]

[1] Department of Applied and Landscape Ecology, Faculty of Agronomy, Mendel University in Brno, Zemědělská 1, 613 00 Brno, Czech Republic, e-mail: magda.vaverkova@uake.cz; dana.adamcova@mendelu.cz

ABSTRACT

Monitoring leachate, groundwater and surface water aiming to determinate the concentration of heavy metals (Hg, Zn, Ni, Cr^{6+},Cd, Pb) at the municipal solid waste landfill Štěpánovice took place in the years 2002–2010. The values of heavy metals concentration oscillated as follows: Zn (0.05 – 0.37 μg/dm^3), Cr^{+6} (0.01 – 3.3 μg/dm^3), Hg (0.0001 – 0.001 (μg/dm^3), Ni (0.001 – 0.19 μg/dm^3, Cd (0.001 – 0.007 μg/dm^3) and Pb (0.002 – 0.176 μg/dm^3). The measured data show that the concentrations of heavy metals met the limits provided by law. Under the current landfill operation mode, the results of measurements do not indicate any negative impact on the quality of surface water, groundwater or leachate water.

Keywords: heavy metals, leachate, surface water, groundwater, landfill, waste.

INTRODUCTION

For a long time, landfilling was the most common waste management option. In many European countries, not to mention the rest of the world, this is still the case [Laner et al. 2012]. In several respects landfilling is the opposite of sustainability. Not only is it a waste of resources but landfills as such constitute a health hazard and an environmental burden [Ettler et al. 2008, Öman and Junestedt 2008, Kalčikova et al. 2011]. Organic waste biodegradation in landfills creates greenhouse gas called landfill gas. Water that enters the landfill forms leachate which can carry pollutants to the surroundings. Environmental problems related to leachate include groundwater pollution and oxygen depletion and ecotoxicity in surface waters [Kjeldsen et al. 2002, Pablos et al. 2011, Modin 2012].

Landfilling and leachate generation

Leachate is the liquid percolated through the waste layers within a landfill cell [Fauziah et al. 2013]. Landfill leachate might contain, among many other constituents, heavy metals in considerable concentrations. The heavy metals may constitute an environmental problem, if the leachate migrates into surface or groundwater, or a treatment issue where the leachate is collected and treated prior to discharge [Baun and Christensen 2004].

The leachate generation is highly dependent on soil types, waste composition, degree of compaction of the waste disposed, amount of landfill received, evapotranspiration and also the type and the age of landfill. Leachate generation begins as soon as the disposal site receives its first batch of waste, where the liquid sourced from the moisture that exists in the waste, water from degradation of putrescible waste and rain. This will continue until the disposal sites cease its operation and no longer receive additional waste. At this instant, leachate will be produced mainly from the degradation of the waste within the waste cells. Due to this factor, characteristics of leachate will vary from time to time. Normally, the concentration will be the highest during the first 3–8 years when biodegradation is occurring very rapidly [Fauziah et al. 2013]. Due to the pollution potential to the surrounding environment, such as local surface water and groundwater, landfill leachate must be treated to meet the local standard for discharge into the receiving water [Kumar and Alappat 2005, Zhang et al. 2013].

Landfilling of waste at sanitary landfills is an important part of the global waste management scheme, and it is crucial to assess the environmental impact of this. In this paper, leachate from Štěpánovice landfill has been size charge fractionated in field and selected metals analysed. The results acquired enable a better understanding of the bioavailability and biological impact of metals discharged via landfill leachate.

In this paper we describe investigations of the fate of heavy metals for leachate from Štěpánovice landfill and surrounding waters.

MATERIAL AND METHODS

The landfill site

Leachate used in this study was collected from municipal solid waste landfill (MSW) located in Czech Republic, 1 km north of Štěpánovice commune and 1 km south of Dehtín commune. It has been operating since summer 1996 (GPS coordinates of the landfill – 49°26'15.934"N, 13°16'55.352"E). The landfill receives waste from Klatovy with satellite towns and villages for a total of about 37 725 inhabitants, producing in all approx. 20 000 Mg/year of waste to be Landfilled annually. The climate is typically inland, with annual rainfall of 580 mm $year^{-1}$.

The landfill (Figure 1) is formed by three sub-landfills: landfill A (closed in 2003, area 8750 m^2); landfill B (working from 2003, area 26 000 m^2); landfill C (that will work after closing part B).

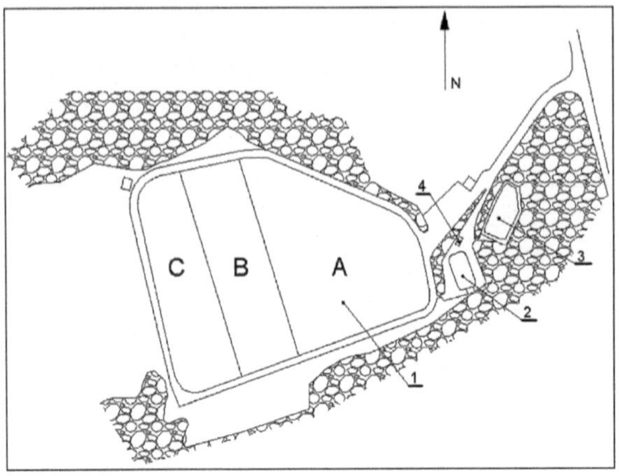

1 – landfill, 2 – detention receiver of leachate, 3 – rainwater reservoir, 4 – entrance gate

Figure 1. Map of three sub-landfills [Kotovicová et al., 2011]

The total volume of both (A, B) parts of the landfill is 289 000 m^3. Planned service life of the facility is up to year 2018.

The activities carried out in this landfill are: landfilling of municipal waste, biogas measurement and monitoring, leachate management. In the landfill, there are deposited municipal waste apart from fluid waste, faecal matter, hazardous substances, radioactive and toxic waste. In 1996–2011 approximately 25% of disposed waste was of organic origin. The rest was glass, paper, plastics, metals, textiles and inorganic remains like construction waste etc.

Sampling

Leachate monitoring was carried out by means of samples taken from leachate pond (Figures 1 and 2). The samples were collected with a sampler from the surface of the leachate pond. Analyses of the leachate samples were carried out twice a year. The assessed parameters of the leachate in the research included the following heavy metals: Hg, Zn, Ni, Cr^{6+}, Cd, Pb.

In order to determine the parameters of heavy metals in groundwater and surface water, monitoring wells and sampling points were used (Figure 2). The samples were collected from the wells in a dynamic state, after short-time drainage (approx. triple exchange of water in the well, or to stable state). Sampling methods were in accordance with the Czech National Standard ČSN ISO 5667, part 11 (groundwater). Surface water samples were collected from the surface of a nameless stream – left tributary stream of Točnického. All samples were collected into airtight glass or plastic containers based on the analyzed component and immediately transported to accredited laboratory for analysis.

Sampling points and monitoring wells for leachate, groundwater and surface water were defined in the area of the landfill. Their distribution and frequency were established by an accredited laboratory. The allocation of sampling points and monitoring wells for the assessment of leachate, groundwater and surface water at the landfill MSW Štěpánovice is shown in Figure 2.

Samples of leachate, groundwater and surface water were collected within time period of 9 years (April 2002 – August 2010). The analyses of the samples were carried out by the company SOM, s.r.o., Mníšek pod Brdy and Health Institute seated in Pilsen, the Centre of laboratories Klatovy (accreditation No. 150/2007), Czech Republic.

Figure 2. Scheme of sampling points and monitoring wells at MSW landfill Štěpánovice

RESULTS

Monitoring of landfill leachate, surface and groundwater quality

The landfill is subjected to monitoring of surface water, groundwater and leachate at regular intervals. Samples of groundwater and surface water are taken from five sampling points, which are marked as follows (designation – characterization): A – old borehole (below the landfill), B – new borehole (above the landfill), C – bedrock, D – brook below the landfill (distant less than 2 m from the former place) and E – brook above the landfill. Samples of leachate (F) are taken from the detention receiver of leachate. Thus, there are altogether six sampling points. The research was divided into three parts. The specifications of particular parts of the research are described in Table 1.

Table 1. Specification of particular parts of the research

Research	Specialization	Parameters [μg/dm^3]	Period	Sampling point
I	Leachate water	Cd	4/2005–4/2006	F
		Cr^{6+}	5/2003–4/2006	
		Hg	5/2003–8/2010	
		Ni	5/2003–10/2004	
		Pb	4/2005–3/2010	
		Zn	5/2003–4/2006	
II	Groundwater	Cd	4/2005–3/2010	A, B, C
		Cr^{6+}	4/2002–3/2010	
		Hg	4/2002–8/2010	
		Ni	4/2002–10/2004	
		Pb	4/2005–3/2010	
		Zn	4/2002–3/2010	
III	Surface water	Cd	4/2005–3/2010	D, E
		Cr^{6+}	4/2002–3/2010	
		Hg	4/2002–8/2010	
		Ni	4/2002–10/2004	
		Pb	4/2005–3/2010	
		Zn	4/2002–3/2010	

The obtained values were assessed pursuant to the criteria set forth in the Methodological Guide of the Ministry of Environment of the Czech Republic – „Soil and Groundwater Contamination Criteria" (1996) and according to the Czech National Standard ČSN 75 7221 „Classification of Surface Water Quality".

All samples were monitored for parameters required by valid legislation as well as for the occurrence of heavy metals: Zn, Cr^{6+}, Hg, Ni, Cd and Pb. Figures 3–8 illustrate the courses of heavy metal concentrations (μg/dm^3) from all six sampling points in the period 2002–2010 and the limits for particular water categories according to the legislation.

The course of the measured values and the limits for Zn (μg/dm^3) in the reporting period for all sampling points are shown in Figure 3. The value of Zn concentration ranged from 0.05 to 0.37 μg/dm^3. The highest values were obtained at sampling point F. The limits for Zn stipulated by legislation were not exceeded at any of the six sampling points. The limits for particular types (categories) of water were not exceeded for this parameter.

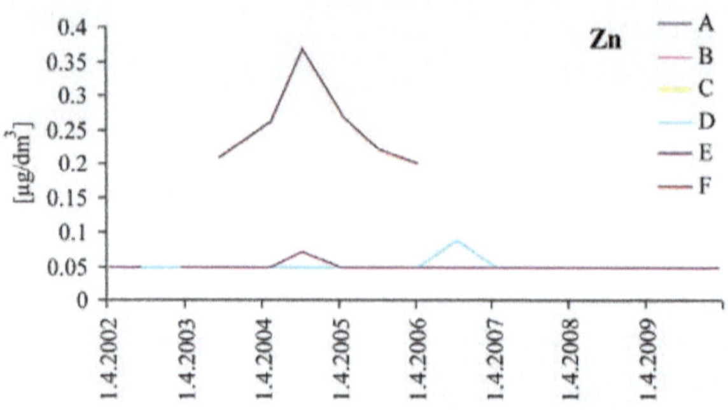

Zn	Limit [μg/dm^3]
Groundwater	150
Surface water	15

Figure 3. Concentrations of heavy metal (Zn) from the sampling points in the period 2002–2010 [Vaverková and Adamcová, 2014]

The course of the measured values and limits for Cr^{+6} (μg/dm3) in the reporting period for all sampling points is shown in Figure 4. The value of Cr^{+6} concentration ranged from 0.01 to 3.3 μg/dm^3. The highest values were observed at sampling point C. At five sampling points (A, B, D, E, F) the values of Cr^{+6} did not exceed the limits stated by legislation. The limits for particular types (categories) of water were not exceeded for this parameter. Only in one sample from 8 April 2008 at sampling point C (bedrock – the groundwater) the value defined by legislation (Methodological Guide of the Ministry of Environment of the Czech Republic – „Soil and Groundwater Contamination Criteria" [1996], 1 μg/dm^3 was exceeded; the measured value reached 3.3 μg/dm^3. Considering the long term monitoring (9 years), it can be stated that the average value for Cr^{+6} (0.177 μg/dm^3) at sampling point C did not exceed the required limit.

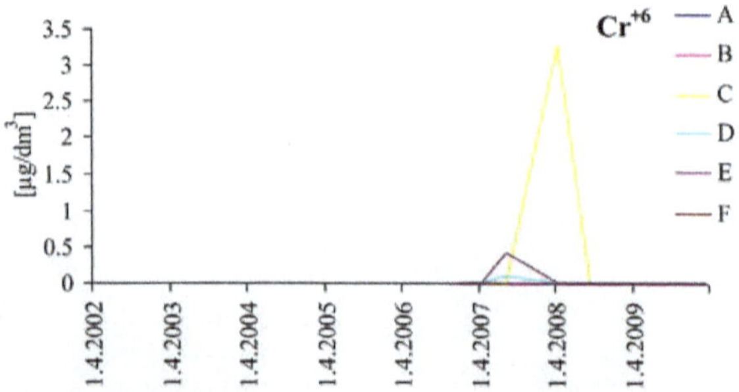

Cr^{+6}	Limit [μg/dm^3]
Groundwater	1
Surface water	5

Figure 4. Concentrations of heavy metal (Cr^{+6}) from the sampling points for the period 2002–2010 [Vaverková and Adamcová, 2014]

Another monitored heavy metal was Hg. The course of measured values and limits for Hg (μg/dm^3) in the reporting period for all sampling points is illustrated in Figure 5. The value of Hg concentration oscillated between 0.0001 and 0.001 μg/dm^3. The limits for Hg stipulated by legislation were not exceeded at any of the six sampling points. The limits for particular types (categories) of water were not exceeded for this parameter.

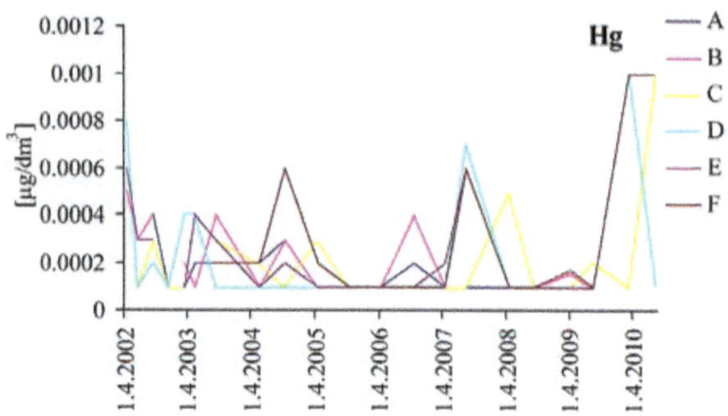

Hg	Limit [μg/dm^3]
Groundwater	0.1
Surface water	0.05

Figure 5. Concentrations of heavy metal (Hg) from the sampling points for the period 2002–2010 [Vaverková and Adamcová, 2014]

Besides Zn, Cr^{+6} a Hg the heavy metal Ni was examined. The course of measured values and limits for Ni (μg/dm³) in the reporting period at all sampling points is shown in Figure 6. The value of Ni concentration ranged from 0.001 to 0.19 μg/dm³. The highest figures were recorded at sampling point F. The limits for Ni stipulated by legislation were not exceeded at any of the six sampling points. The limits for particular types (categories) of water were not exceeded for this parameter.

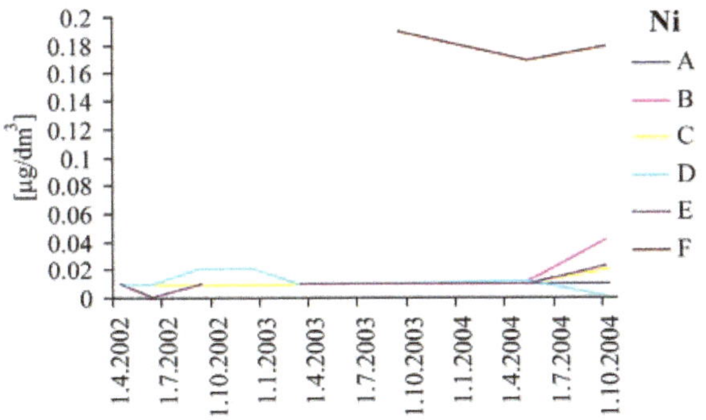

Ni	Limit [μg/dm³]
Groundwater	20
Surface water	5

Figure 6. Concentrations of heavy metal (Ni) from the sampling points for the period 2002–2004 [Vaverková and Adamcová, 2014]

The course of the measured values and limits for Cd (μg/dm³) in the reporting period for all sampling points is shown in Figure 7. The value of Cd concentration ranged from 0.001 to 0.007 μg/dm³. The highest values were observed at sampling point C. The limits for Cd stipulated by legislation were not exceeded at any of the six sampling points. The limits for particular types (categories) of water were not exceeded for this parameter.

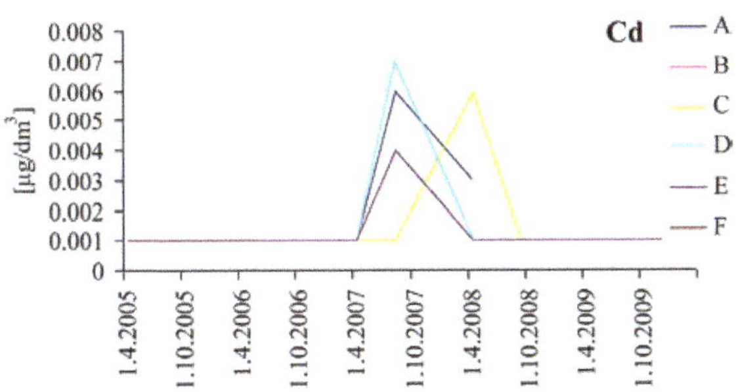

Cd	Limit [μg/dm³]
Groundwater	1.5
Surface water	0.1

Figure 7. Concentrations of heavy metal (Cd) from the sampling points for the period 2005–2010 [Vaverková and Adamcová, 2014]

The last examined heavy metal was Pb. The course of measured values including the limits for Pb (μg/dm³) in the reporting period for all sampling points is demonstrated in Figure 8. The value of Pb concentration ranged between 0.002 and 0.176 μg/dm³. The highest figures were recorded at sampling point F. The limits for Pb stipulated by legislation were not exceeded at any of the six sampling points. The limits for particular types (categories) of water were not exceeded for this parameter.

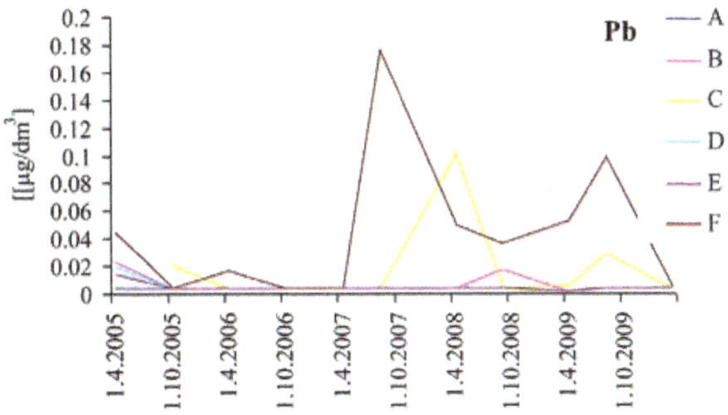

Pb	Limit [μg/dm³]
Groundwater	20
Surface water	3

Figure 8. Concentrations of heavy metal (Pb) from the sampling points for the period 2005–2010 [Vaverková and Adamcová, 2014]

CONCLUSION

Monitoring of leachate, groundwater and surface water aiming at determining the concentration of heavy metals (Hg, Zn, Ni, Cr^{6+},Cd, Pb) at the MSW landfill Štěpánovice took place in the years 2002–2010. The values of heavy metals concentration oscillated as follows: Zn (0.05 – 0.37 μg/dm³), Cr^{+6} (0.01 – 3.3 μg/dm³), Hg (0.0001 – 0.001 (μg/dm³), Ni (0.001 – 0.19 μg/dm³, Cd (0.001 – 0.007 μg/dm³) a Pb (0.002 – 0.176 μg/dm³).

The measured data show that the concentrations of heavy metals meet the limits provided by law. Under the current landfill operation mode, the results of measurements do not indicate any negative impact on the quality of surface water, groundwater or leachate water.

This conclusion can be additionally supported by the results obtained within biological monitoring. Biological monitoring with the use of bioindicators has been taking place in the surroundings of Štěpánovice landfill since 2007 (still ongoing).

In the rainwater reservoir (Figure 1) the occurrence of *Triturus vulgaris* species was recorded. It is necessary to mention that its occurrence may reflect correct operation of the landfill considering the sensitivity of *Triturus vulgaris* to the environment. This species is legally protected and in accordance with the Regulation No. 395/1992 Coll. belongs to highly endangered species. Its presence in the rainwater reservoir water proves clarity of this water and shows that no contamination of water by leachate from the landfill or from leachate pond takes place.

REFERENCES

1. Baun D.L., Christensen T.Ch. 2004. Speciation of Heavy Metals in Landfill Leachate: A Review. Waste Management & Research, 22, 3-23.
2. Ettler V., Mihaljevic M., Matura M., Skalova M., Sebek O., Bezdicka P. 2008. Temporal variation of trace elements in waters polluted by municipal solid waste landfill leachate. Bulletin of Environmental Contamination and Toxicology, 80, 274-279.
3. Fauziah S. H., Nor Izzati M., Agamuthu P. 2013. Toxicity on Anabas Testudineus: a case study of sanitary landfill leachate. Procedia Environmental Sciences, 18, 14-19.
4. Kalčíková G., Vávrová M., Zagorc-Končan J., Žgajnar Gotvajn A. 2011. Seasonal variations in municipal landfill leachate quality. Management of Environmental Quality, 22, 612-619.
5. Kjeldsen P., Barlaz M.A., Rooker A.P., Baun A., Ledin A., Christensen T.H. 2002. Present and Long-Term Composition of MSW Landfill Leachate: A Review. Critical Reviews in Environmental Science and Technology, 32, 297-336.
6. Kumar D., Alappat B.J.2005. Analysis of leachate pollution index and formulation of sub-leachate pollution indices. Waste Management & Research, 22, 230-239.
7. Laner D., Crest M., Scharff H., Morris J.W.F., Barlaz M.A. 2012. A review of approaches for the long-term management of municipal solid waste landfills. Waste Management, 32, 498-512.
8. Modin H. 2012. Modern landfill leachates – quality and treatment. Doctoral Thesis. Water Resources Engineering. Department of Building & Environmental Technology, Lund University.
9. Öman C.B., Junestedt C. 2008. Chemical characterization of landfill leachates – 400 parameters and compounds. Waste Management, 28, 1876-1891.
10. Pablos M.V., Martini F., Fernandez C., Babin M.M., Herraez I., Miranda J., Martinez J., Carbonell G., San-Segundo L., Garcia-Hortiguela P., Tarazona J.V. 2011. Correlation between physicochemical and ecotoxicological approaches to estimate landfill leachates toxicity. Waste Management, 31, 1841-1847.
11. Vaverková M., Adamcová D. 2014. Can vegetation indicate municipal solid waste landfill impact on the environment?. Polish Journal of Environmental Studies. In Press.
12. Zhang Q.Q., Tian B.H., Zhang X., Ghulam A., Fang Ch.R., He R. 2013. Investigation on characteristics of leachate and concentrated leachate in three landfill leachate treatment plants. Waste Management, 33, 2277-2286.

PINE NEEDLES (*PINUS SYLVESTRIS* L.) AS BIOINDICATORS IN THE ASSESSMENT OF URBAN ENVIRONMENTAL CONTAMINATION WITH HEAVY METALS

Agnieszka Parzych[1], Jerzy Jonczak[2]

[1] Environmental Chemistry Research Unit, Institute of Biology and Environmental Protection, Pomeranian University in Slupsk, Arciszewskiego St. 22b, 76-200 Słupsk, Poland, e-mail: parzycha1@op.pl

[2] Department of Geoecology and Geoinformation, Pomeranian University in Słupsk, Partyzantów St. 27, 76-200 Słupsk, Poland, e-mail: jonczak@apsl.edu.pl

ABSTRACT

The studies on Zn, Pb and Mn content in the needles of *Pinus silvestris* and the surface layers of soil (0.0–0.2 m and 0.4–0.6 m) were conducted in 2010. The research stations covered residential areas, municipal parks, neighborhoods of busy streets, industrial establishments, waste dump sites and sewage treatment plants. The tests showed variability of the analyzed elements in the needles of *P.silvestris* and in municipal soils, depending on concentration of urban and industrial activity, road traffic and the depth of the examined layer of soil. The content of heavy metals in soil varied. Zn was characterized by the highest variability, depending on the depth (50.3–52.4%), Mn slightly lower (38.5–42.2%), and Pb, the lowest (29.1–33.2%). It was evidenced that the values of the heavy metals enrichment factors of the needles are closely connected with concentration of the examined metals in the soil. Along with the rise of Zn, Mn and Pb content in the soil, the values of enrichment factors decreased. The strongest negative correlations were found in the case of Zn, both in 1-year old needles ($r = -0.82$, $p<0.05$), as well as in 2-years' old ones ($r = -0.83$, $p<0.05$), slightly weaker in the cases of lead and manganese.

Keywords: polluted area, soil, needles, pinus sylvestris, accumulation, enrichment factor.

INTRODUCTION

Municipal areas are ecosystems of the prevailing human impact, in which buildings and transport networks dominate. In waters, soils and plants of such ecosystems, increase of concentrations of various substances of anthropogenic origin, including heavy metals, is usually observed [Buszewski et al. 2000, Yang et al. 2001, Grzebisz et al. 2002, Dąbkowska-Naskręt and Różański 2009]. Emissions from industrial establishments and transportation means constitute main sources of heavy metal pollution of the environment [Lis and Pasieczna 1995, Krauss et al. 2000]. Heavy metals, as constituents of dust (PM10 i PM2.5) and aerosols, can be transported in the air at substantial distances, leading to pollution of ecosystems situated at longer distances from the source of emission [Brozek and Zarembski 2011]. Heavy metals deposited at the surface of the soil are fixed by the solid phase of the soil, absorbed by living organisms or they undergo migration along with the filtering water. Intensity of such processes depends on the complex of physical and chemical properties of the soils. Reaction plays a very important role and decides about mobility of heavy metals in the soil as well as on their bioavailability and toxicity in reference to living organisms [Tyler 1992, Kabata-Pendias and Pendias 1999, Karczewska 2002]. Toxic impact of heavy metals on living organisms appears after the threshold concentration of bioavailable forms characteristic for each of them has been exceeded [Lis and Pasieczna 1995, Gambuś 1997, Selim and Sparks 2001,

Martinez and Motto 2000, McAlister et al. 2006]. Bioavailability of most metals increases along with the decrease of pH. In neutral and alkaline reaction conditions, labile forms are immobilized by creation of relatively durable complex connection with organic matter of the soils. In municipal ecosystems, alkalization of soils being the consequence of deposit of alkaline dust, has positive impact on this process. Absorption of metals in the soil, however, leads in many cases to their excessive pollution. There is also a risk that in case of change of conditions in the soils, the volume of metals absorbed by the soil becomes bioavailable and gets to the food chain undergoing bioaccumulation. At adequately high concentration, functions of ecosystems can be disturbed causing threat to living organisms [Kabata-Pendias and Pendias 1999, Gruca-Królikowska and Wacławek 2006]. Their contamination hazard increases with subsequent chains of the trophic chain, at the top of which, there is the man. Studies of pollution of municipal agglomerations found in the literature, in most instances refer to soils. Due to toxicity of heavy metals, systematic control of the status of pollution of soils [Czarnowska and Gworek 1991, Madrid et al. 2002, Imperatoa 2003, Pasieczna 2003] as well as needles of *Pinus silvestris*, especially at urban areas [Dmuchowski and Bytnerowicz 1995, Czarnowska 1997, Dmuchowski et al. 2011] is especially important.

The paper aimed at: (i) determination of their level Zn, Mn and Pb in the soil of the city of Słupsk in the layers: 0.0–0.2 m and 0.4–0.6 m,(ii) determination of the content of the tested metals in 1-year old and 2-year-old needles of Scots pine, (iii) specification of accumulation coefficients of the above mentioned metals in the needles of *Pinus silvestris* L. and (iv) evaluation of the level of heavy metals pollution of soils and in the needles of pine trees at the area of the city of Słupsk.

MATERIAL AND METHODS

Soil and plant sampling

The samples for laboratory tests were taken in September 2010 from 25 research stations all over the city of Słupsk (Central Pomerania, Northern Poland), Figure 1. Research stations were distributed around the city including the residential buildings areas, municipal parks as well as vicinity of busy streets, industrial establishments, surrounding area of waste dumps and sewage treatment plants. Soil samples were taken at two depths: 0–0.2 m and 0.4–0.6 m [Lis and Pasieczna1995]. Needles samples (1-year old and 2-year-old) were taken from the seventh whorl. Collective samples of 10–20 g were prepared on a basis of the collected plant matter, taking into consideration the age of the needles (separately 1-year old needles and

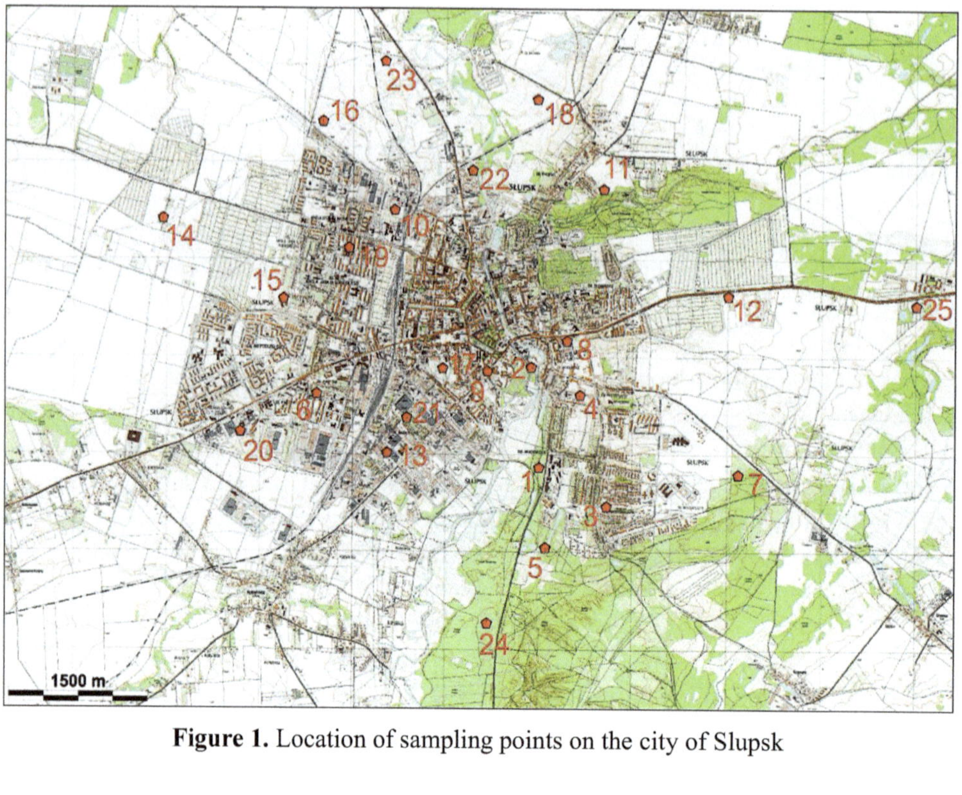

Figure 1. Location of sampling points on the city of Slupsk

2-years' old needles) according to the recommendations ICP Manual Forest [Rautio et al. 2010].

Soil and plant analysis

Soil samples were dried at the temperature of 65 °C and sieved through a sieve with the mesh diameter of 2 mm. Subsequently, the following were determined: active acidity (pH, H_2O) and exchangeable acidity (pH, KCl) by potentiometric method, organic matter content by the method of loss on ignition at the temperature of 550 °C, granulometric content by the sieve method, the content of Zn, Mn and Pb by AAS method (Perkin Elmer Aanalyst 300) in a solution after mineralization in a mixture of the concentrated HNO_3 and 30% H_2O_2. The analyses were performed in the oxy-acetylene flame. The tests were carried out following the original standards (Merck KGaA, 1 g/1000 ml). After bringing to the laboratory, the pine needles were washed in deionized water, dried at the temperature of 65 °C and homogenized in a laboratory grinder. The content of Zn, Mn and Pb was determined by the same method as in the case of soils.

Data processing and statistical analysis

In order to compare concentration of the examined heavy metals in the needles of Scots pine and in the soil, average, minimum and maximum values, standard deviations and coefficients of variation were calculated (CV). Coefficients of correlation of Spearman rank as well as coefficients of enrichment (*Enrichment factor, EF*) of the needles of Scots pine in heavy metals were calculated (Zn, Mn, Pb):

$$EF_{(Zn)} = C_{n(Zn)} : C_{s(Zn)},$$

where: $EF_{(Zn)}$ – enrichment factor of zinc,
$C_{n(Zn)}$ – zinc content in needles,
$C_{s(Zn)}$ – zinc content in soil.

RESULTS AND DISCUSSION

The content of Zn, Mn and Pb in soils

Anthropogenic soils of the city of Słupsk are made of sandy deposits in which coarse- grained (4.98–50.42%) and medium-grained sand (8.28–32.90%) dominate. Participation of framework fraction in the layer 0.0–0.2 m was 1.4–27.8%, and in layer 0.4–0.6 m 1.4–22.0%. The dust content did not exceed 6.9%.

The organic matter content in the tested soil was small and remained, depending on the station, at the level from 0.9 to 7.9% at the layer 0.0–0.2 m and from 0.8 to 4.6% at the layer 0.4–0.6 m, showing variation at the level of 44–46% within the area of 25 stations (Table 1). The soils were characterized by the reaction close to neutral, with values from 6.7 to 7.0 at the layer 0.0-0.2 m and from 6.7 to 7.2 (0.4–0.6 m). High pH values of the tested soils are likely to be a consequence of the deposition of alkalic dust of anthropogenic origin, mostly from coal combustion [Lis and Pasieczna 1995, Morel 1997]. According to Brożek and Zarembski [2011] average concentration of dust hanging over the area of the city of Słupsk in 2010 was: PM10 = 27–28 $\mu g{\cdot}m^{-3}$ and PM2.5 = 19 $\mu g{\cdot}m^{-3}$. Influence of urbanization on the reaction of soils was discovered both in the central part of the city and at its outskirts.

The heavy metals content in the soil of the city of Słupsk varied, depending on the depth and location of the point of test sampling. The highest quantity of Zn, Mn and Pb was found in the surface layer of the soil (0.0–0.2 m). According to Pasieczna [2003] heavy metal accumulation increases with the number of inhabitants of the cities and is more evident at the surface level. In case of zinc, the values from 12.3 to 93.6 $mg{\cdot}kg^{-1}$ were determined in the layer 0.0–0.2 m (average 45.3 $mg{\cdot}kg^{-1}$) and from 5.8 to 90.2 $mg{\cdot}kg^{-1}$ (average 35.7 $mg{\cdot}kg^{-1}$) at the depth of 0.4–0.6 m (Table 2).

Table 1. Selected properties of the soils

Parameter	Organic matter [%]		pH (H_2O)		pH (KCl)	
Depth [m]	0.0–0.2	0.4–0.6	0.0–0.2	0.4–0.6	0.0–0.2	0.4–0.6
Minimum	0.9	0.8	6.7	6.7	6.4	5.7
Maximum	7.9	4.6	7.0	7.2	7.4	6.1
Average ± SD	3.1 ± 1.4	2.5 ± 1.1	6.8 ± 0.2	6.9 ± 0.3	6.4 ± 0.5	5.9 ± 0.2
Median	2.9	2.4	6.8	6.7	6.6	5.8
CV, %	46	44	2.6	4.3	7.8	3.6

SD – standard deviation, CV – coefficient of variation.

Table 2. Heavy metals contents ($mg \cdot kg^{-1}$) in the soils

Parameter	Zn		Mn		Pb	
	0–0.2 m	0.4–0.6 m	0–0.2 m	0.4–0.6 m	0–0.2 m	0.4–0.6 m
Minimum	12.3	5.8	68.3	47.0	10.8	15.9
Maximum	93.6	90.2	343.5	313.4	65.1	55.6
Average ± SD	45.3 ± 22.8	35.7 ± 18.7	138.8 ± 53.5	129.4 ± 54.6	35.8 ± 11.9	31.9 ± 9.3
Median	45.8	33.3	136.8	115.8	38.6	31.4
CV, %	50.3	52.4	38.5	42.2	33.2	29.1

In surface areas, at 9 stations, the highest value from 40 to 50 $mg \cdot kg^{-1}$ Zn was discovered (Figure 2). In the deeper layer, the dominating values were from 30 to 40 $mg \cdot kg^{-1}$ Zn. Concentrations of Zn in soil samples exceeded the geochemical background values for a tested area, but were within the acceptable limits for municipal areas, not exceeding toxic values [The Minister of environment 2002]. The largest quantities of Zn were discovered at the stations located in the neighborhood of busy streets. According to Ottesen et al. [1999] about 1.5–2.0% of zinc comes from abrasion of car tyres. In other cities of Poland the level of Zn in anthropogenic soils was: 44.8–893.7 $mg \cdot kg^{-1}$ in Opole [Kusza et al. 2009], 9.0–400.0 $mg \cdot kg^{-1}$ in Poznań [Grzebisz et al. 2002], about 19.9 $mg \cdot kg^{-1}$ in Toruń [Buszewski et al. 2000]. The average volume of zinc in sandy soils of Poland varies usually from 7 to 150 $mg \cdot kg^{-1}$ [Kabata-Pendias and Pendias 1999].

Concentration of manganese in the surface layer of the soil (0.0–0.2 m) was from 68.3 to 343.5 $mg \cdot kg^{-1}$ (average 138.8 $mg \cdot kg^{-1}$), and in the layer 0.4–0.6 m from 47.0 to 313.4 $mg \cdot kg^{-1}$ (average 129.4 $mg \cdot kg^{-1}$) (Table 2). At 18 from among 25 stations, concentration of Mn was found at the level of 100–150 $mg \cdot kg^{-1}$ (0.0–0.2 m) and at 12 stations in the level of 0.4–0.6 m (Figure 2). The obtained results for the soils of the city of Słupsk exceed the values of geochemical background of manganese, but do not constitute threat to living organisms. The highest concentration of that metal was found at the stations in the central part of the town. According to Kabata-Pendias and Pendias [1999], the occurrence of manganese in the soil depends on its content in the parent rock as well as on the character of soil-forming processes, and average content for various kinds and types of soils is usually from 100 to 1300 $mg \cdot kg^{-1}$.

The content of Pb in the layer 0.0–0.2 m was between 10.8 and 65.1 $mg \cdot kg^{-1}$ (average 35.8%), and in the 0.4–0.6 m layer between 15.9 and 55.6 $mg \cdot kg^{-1}$ (average 31.5 $mg \cdot kg^{-1}$) (Table 2). In 8 from among 25 test stations, the content of Pb was at the level of 40–50 $mg \cdot kg^{-1}$ in the surface layer (0.0–0.2 m), and in 10 – at the level of 30–40 $mg \cdot kg^{-1}$ in the layer (0.4–0.6 m). The largest concentration of lead was found in the vicinity of the streets and in the central part of the town. No excess of acceptable values of Pb was found. Much higher variation of Pb content was found in the soils of Opole from 14.6 to 1378 $mg \cdot kg^{-1}$ [Kusza et al. 2009], Białystok 50.6–145.6 $mg \cdot kg^{-1}$ [Czubaszek and Bartoszuk 2011] or Poznań 5.4-280 $mg \cdot kg^{-1}$ [Grzebisz et al. 2002].

The content of the analyzed heavy metals in the soil varied to different extent. Zn (50.3–52.4 %) was the most changeable depending on the depth, Mn was slightly less so (38.5–42.2%), and Pb the least changeable (29.1–33.2%). Similar diversification of elements was observed in the soils of other agglomerations in Poland [Grzebisz et al. 2002]. Highly vital from statistical point of view, positive values of correlation coefficients between concentration of Zn, Mn and Pb in the layers of soil: 0.0–0.2 m and 0.4–0.6 m, reflect small diversification of pollutants of anthropogenic origin depositing independent of the depth of the samples of taken soil (Figure 3).

The content of Zn, Mn and Pb in the needles of *Pinus Sylvestris*

Heavy metal content in the needles of the *Pinus silvestris* within the area of Słupsk varied, depending on the metal, age of needles and the place of sampling (Table 3, Figure 4). From among the tested metals, manganese appeared in the highest quantity, zinc in slightly lower, and lead the lowest (Mn>Zn>Pb). At all examined stations, higher contents of the above mentioned metals were found in 2-years' old needles.

Concentration of zinc remained at the level from 45.8 to 85.2 $mg \cdot kg^{-1}$ in 1-year old needles and from 44.3 to 88.8 $mg \cdot kg^{-1}$ in 2-years; old needles. The highest Zn content in the needles

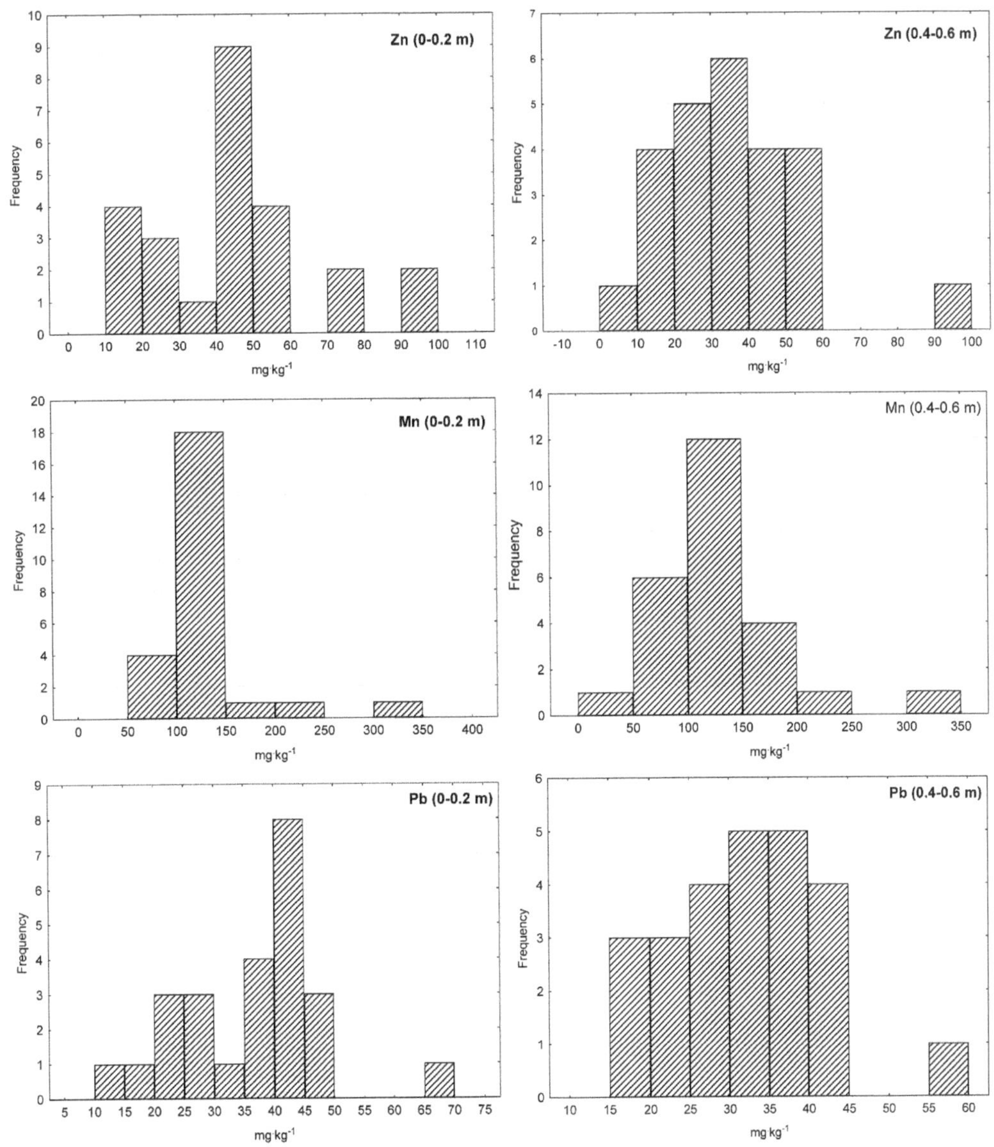

Figure 2. Histograms of heavy metals (Zn, Mn, Pb) concentrations in the soils of Słupsk

Table 3. Heavy metals contents (mg·kg^{-1}) in the needles of Scots pine in Słupsk

Parameter	Zn		Mn		Pb	
	1-needles	2-needles	1-needles	2-needles	1-needles	2-needles
Minimum	45.8	44.3	239.5	351.0	6.7	7.8
Maximum	85.2	88.8	579.0	789.0	22.1	23.9
Average ± SD	59.4 ± 9.9	66.8 ±12.1	372.7 ± 113.1	532.0 ± 137.4	12.2 ± 3.9	13.3 ± 4.1
Median	56.5	64.4	322.0	512.7	12.2	13.0
CV, %	16.7	18.1	30.3	25.8	31.9	30.8

was collected at the stations in the central part of the town and in so called *Słupsk Economic Zone* in which production facilities are concentrated. None of the research stations showed toxic levels of Zn (>89 mg·kg^{-1}) in needles. The average zinc content in overground parts of the plants, which are not under influence of pollutants, usually remains at the level of 10–70 mg·kg^{-1}. Concentration in the leaves of most of plants between 15–30 mg·kg^{-1} [Pasieczna 2003] is enough to cover physiological needs of most plants. Similar zinc levels were found in the needles of *Pinus silvestris* over the area of Stalowa Wola: from 31 to 61 mg·kg^{-1} (1-years; old needles) and from 45–99 mg·kg^{-1} (2-year-old needles) [Samecka-Cymerman et al. 2006].

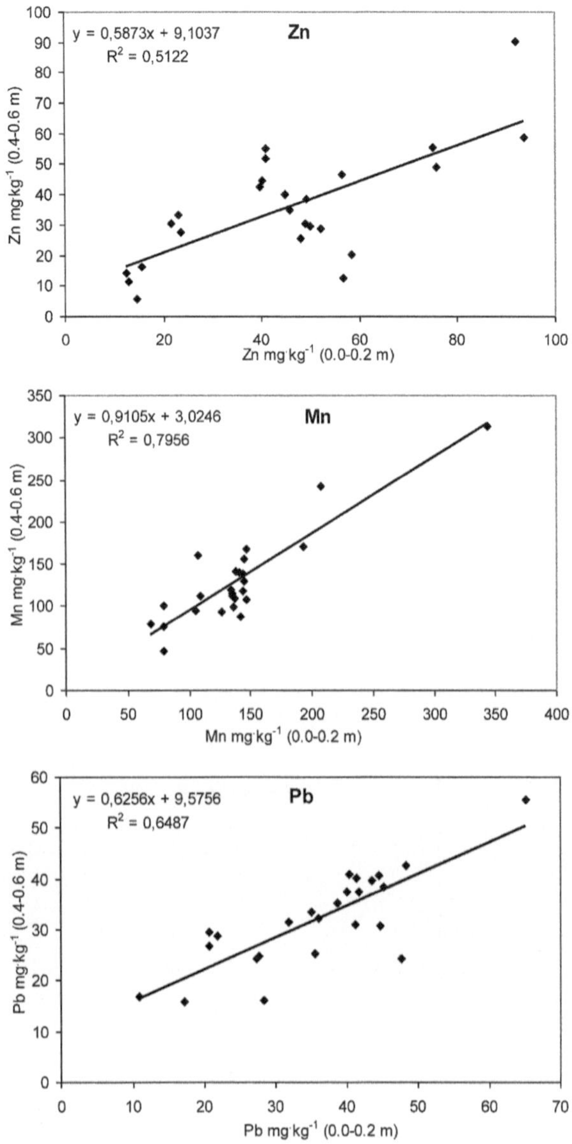

Figure 3. Relation between concentration of heavy metals in soil (0–0.2 m) and (0.4–0.6 m), $p<0.05$, $n=50$

Concentration of manganese in the needles of *Pinus silvestris* remained at the level from 239.5 to 579 mg·kg^{-1} in 1-year old needles and from 351.0 to 789.0 mg·kg^{-1} in 2-year old needles (Table 3, Figure 4). According to various authors, concentration of Mn in plants at the area outside the influence of pollutants is most often: 1540–3952 mg·kg^{-1} [Kozanecka et al. 2002], and in municipal agglomerations such values may be substantially higher. In case of concentration of Mn in needles, increased values of (> 400 mg·kg^{-1}) were found at 9 research stations in 1-year old needles and at 19 stations in 2-years' old needles. The lowest Mn content was found at the research stations found at the outskirts of cities.

The lead content in the needles of *Pinus silvestris* remained at the level from 6.7 to 22.1 mg·kg^{-1} in 1-year old needles and from 7.8 to 23.9 mg·kg^{-1} in 2-years' old needles (Table 3, Figure 4).

Increased content of Pb (>9 mg·kg^{-1}) was found in the needles originating from the research stations located in the central part of the city as well as in those collected in the vicinity of busy streets. In other cities of Poland, increased concentrations of that element in needles were found as well: 5.8–11.5 mg·kg^{-1} in Toruń [Buszewski et al. 2000] and 50.6–145.6 mg·kg^{-1} in Białystok [Czubaszek and Bartoszuk 2011].

The content of the analyzed heavy metals in the needles of *Ps* varied at the area of Słupsk to different extent. The most changeable, depending on the age of the needles was Pb (30.8–31.9%), slightly less changeable was Mn (30.3–25.8%), Zn was the least changeable (16.1–18.1%). Zn, Mn and Pb content in 1-year old needles and 2-years' old needles reflected statistically vital positive values of correlation coefficients (Figure 5).

The highest value of the enrichment factor was discovered in the case of manganese (EF = 1.4–9.5), slightly lower in the case of zinc (EF = 0.7–4.9), and the lowest in the case of Pb (EF = 0.2–0.6) (Figure 6). The high value of EF for manganese may result from the highest assimilability of that element by plants in soils of slightly alkaline reaction. It was also discovered that the value of heavy metals enrichment factors for the needles of *Pinus silvestris* are strictly connected with the concentration of the examined metals in the soil (Figure 6). Along with the increase of Zn, Mn and Pb content in the soil, the values of enrichment factor decreased. The strongest negative correlations were found in the case of Zn both in 1-year old needles ($r = -0.82$, $p<0.05$) as well as in 2-years'old needles ($r = -0.83$, $p<0.05$) and slightly lower in the case of lead ($r = -0.70$ and $r = -0{,}71$, $p<0.05$) and manganese ($r = -0.60$ i $r = -0.69$, $p<0.05$) (Figure 6, Table 4). Similar dependencies resulted from the research done by Gambuś [1997] and Królak [2003]. Substantially lower values of EF were found in the needles of

Table 4. The correlation coefficient (r) of heavy metals in relation: soil/1-needles, soil/2-needles, soil/EF-1-needles, soil/EF-2-needles. ($n=50$, $p<0.05$, $r_{crit.}=0.30$)

r in relation:	Zn	Mn	Pb
soil/1-needles	0.47	0.11	0.12
soil/2-needles	0.28	-0.04	0.12
soil/EF-1-needles	-0.82	-0.60	-0.70
soil/EF-2-needles	-0.83	-0.69	-0.71

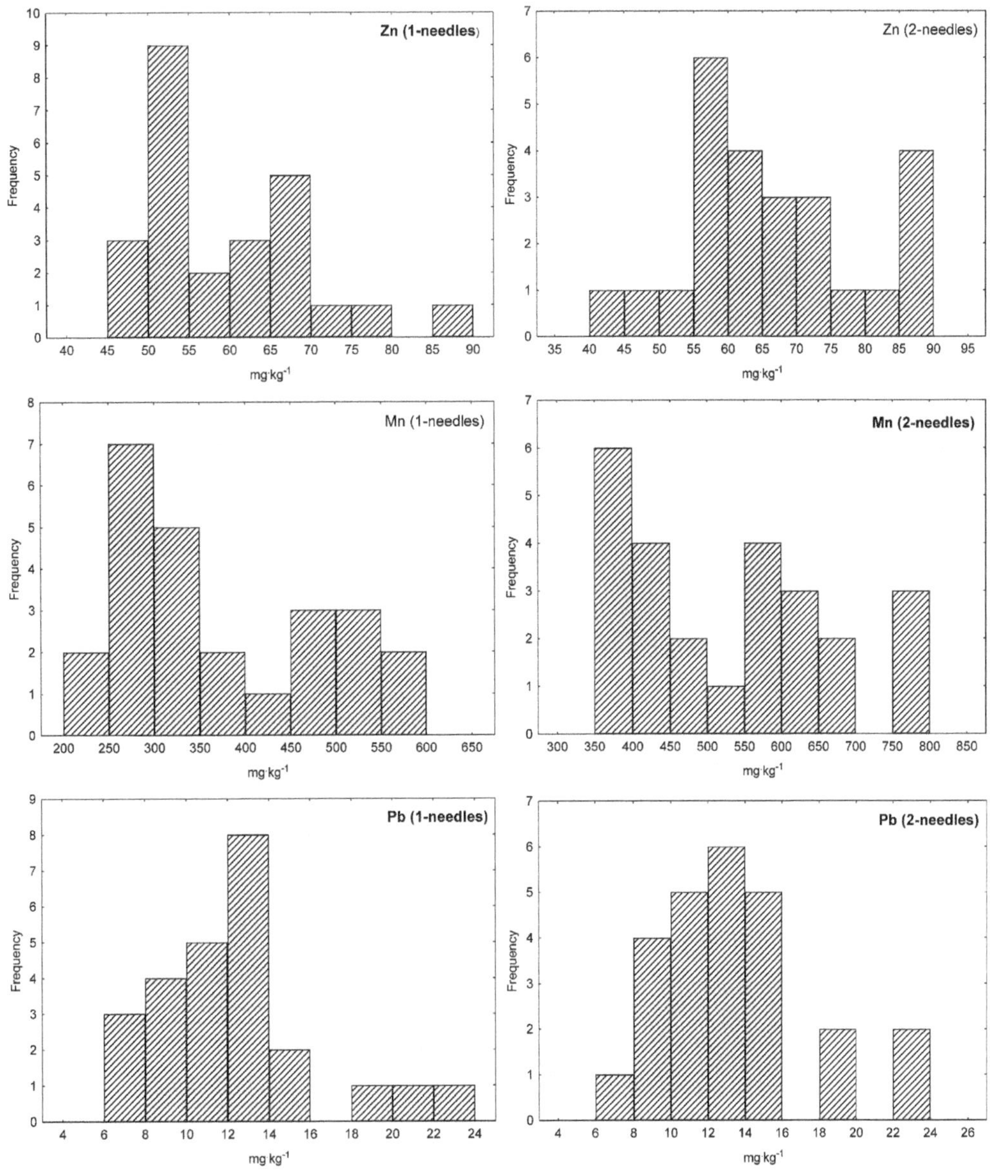

Figure 4. Histograms of heavy metals concentrations in needles of Scots pine in Słupsk

Scots pine at the territories at farther distance from anthropogenic sources of pollution [Parzych and Jonczak 2013]. Obtained *EF* values show that Mn showed highest cumulative properties of the tested metals. According to Kłos [2009] the values of enrichment factors *EF*<10 indicate the alluvial soil, as source of origin of metals. The obtained values of enrichment factors *EF* should be treated only for orientation purposes since they were determined on a basis of total concentrations of metals in the soil, where the plants accumulate only bio - available form of such contamination.

From among the examined heavy metals, only in the case of zinc, a vital statistically, positive correlation was found between the content of Zn in the soil and its concentration in 1-years old needles (Table 4). No vital statistical correlations between the content of Mn and Pb in the soil and the concentration of these metals in the needles of *Pinus silvestris* were found, which can suggest that these elements have been absorbed from the atmosphere. According to Kabata-Pendias [2001] from 73 to 95% Pb in plants is of atmospheric origin, and has not been absorbed from the soil. There are many factors which influence the availability of heavy metals for plants. One of them is soil reaction [Takáć et al. 2009]. In alkaline environment, most heavy metals are not available for plants. Activation of zinc takes place at pH<6.2, lead with pH <5.2 [Martinez and Motto 2000], and manganese in acid environment with pH<5.5 or alkaline with (pH~8), [Pasieczna 2003]. The lack of the above mentioned relationship may be also an effect of examination of surface layers of

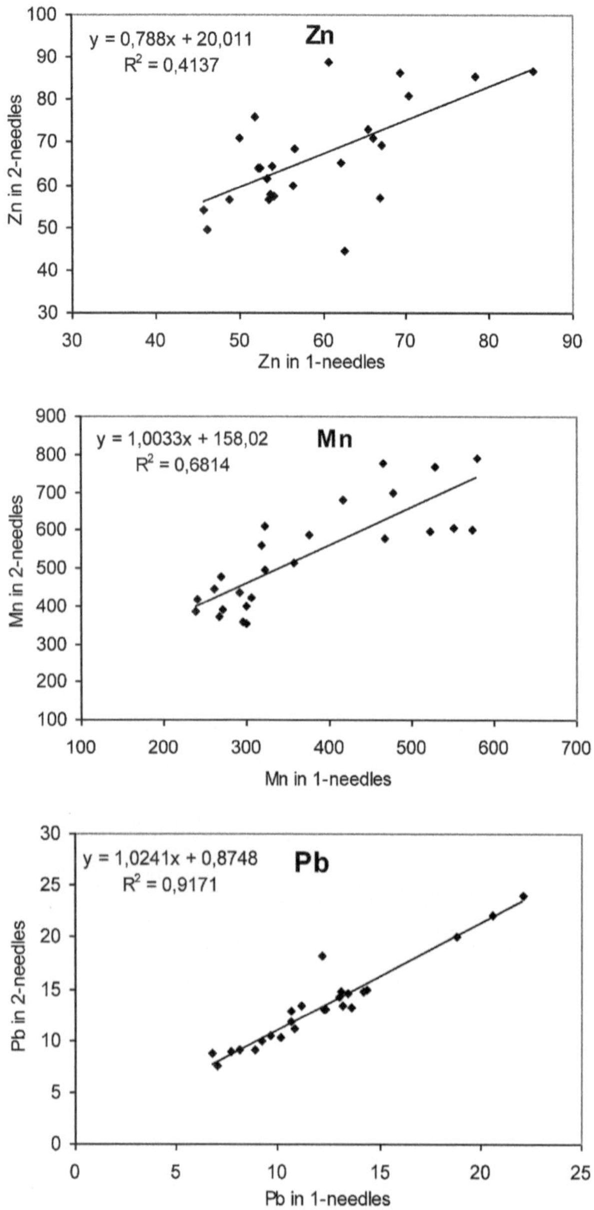

Figure 5. Relation between concentration of heavy metals in 1-needles and 2-needles of Scots pine, $p<0.05$, n=50

soil, which can in turn have impact on concentrations of heavy metals in plants, especially perennial ones.

CONCLUSIONS

The tests reflected variability of the analyzed elements in municipal soils and in the needles of *Pinus sylvestris* L., depending on concentrated urban – industrial activity and transport, as well as from the depth of the tested layer of the soil. Contamination of the soils of Słupsk with Zn, Mn and Pb is small. Their content exceeds the values of geochemical background of these elements but is within the limits of the concentrations acceptable for municipal areas. Manganese showed the highest concentration in the surface layers of the soil, while lead showed the lowest. Heavy metals content in the needles of *Pinus silvestris* also indicates small pollution of the city of Słupsk with the tested elements under consideration. The highest content of Zn, Mn and Pb in the soil and the needles of *Pinus silvestris* was found in the central part of the city, in the *Słupsk Economic Zone* as well as in vicinity of busy streets. The lowest concentrations in the soil and in the needles were found in the outskirts of the city. No vital statistical relations were found between the concentration of Mn and Pb in the soil, and the concentration of these metals in the needles of *Pinus silvestris*, which can reflect substantial assimilation of these elements from atmosphere. The largest values of the enrichment factors were found in the case of Mn, slightly lower in the case of zinc and the lowest in the case of Pb. The values of enrichment coefficients of the needles with heavy metals were strictly connected with concentration of examined metals in the soil. Along with the increase of Zn, Mn and Pb in the soil the values of enrichment factors decreased.

REFERENCES

1. Brożek A., Zarembski A. 2011. The annual assessment of air quality in the region Pomeranian, Report for the year 2010, WIOŚ, Gdańsk (in Polish).
2. Buszewski B., Jastrzębska A., Kowalkowski T., Górna-Binkul A. 2000. Monitoring of selected heavy metals uptake by plants and soils in the area of Toruń, Poland. Pol. J. Environ. Stud. 9, 6, 511-515.
3. Czarnowska K. 1997. The level some heavy metals in soils and leaves of trees in the city of Łódź. Rocz. Glebozn. 48, 3/4, 49-61 (in Polish).
4. Czubaszek R., Bartoszuk. 2011. Content of selected heavy metals in silos in accordance with its distance from the street and land use. Civil and Environ. Engineering, Bud. Inż. Środ. 2, 27-34 (in Polish).
5. Gworek B. 1991. Pollution level zinc, lead and copper Warsaw soils. Rocz. Glebozn. 42, 1/2, 49-56 (in Polish).
6. Dąbkowska–Naskręt H., Różański S. 2009. Forms of Pb and Zn in urbansoils of Bydgoszcz agglomeration, Ochrona Środowiska i Zasobów Naturalnych 41, 489-496 (in Polish).

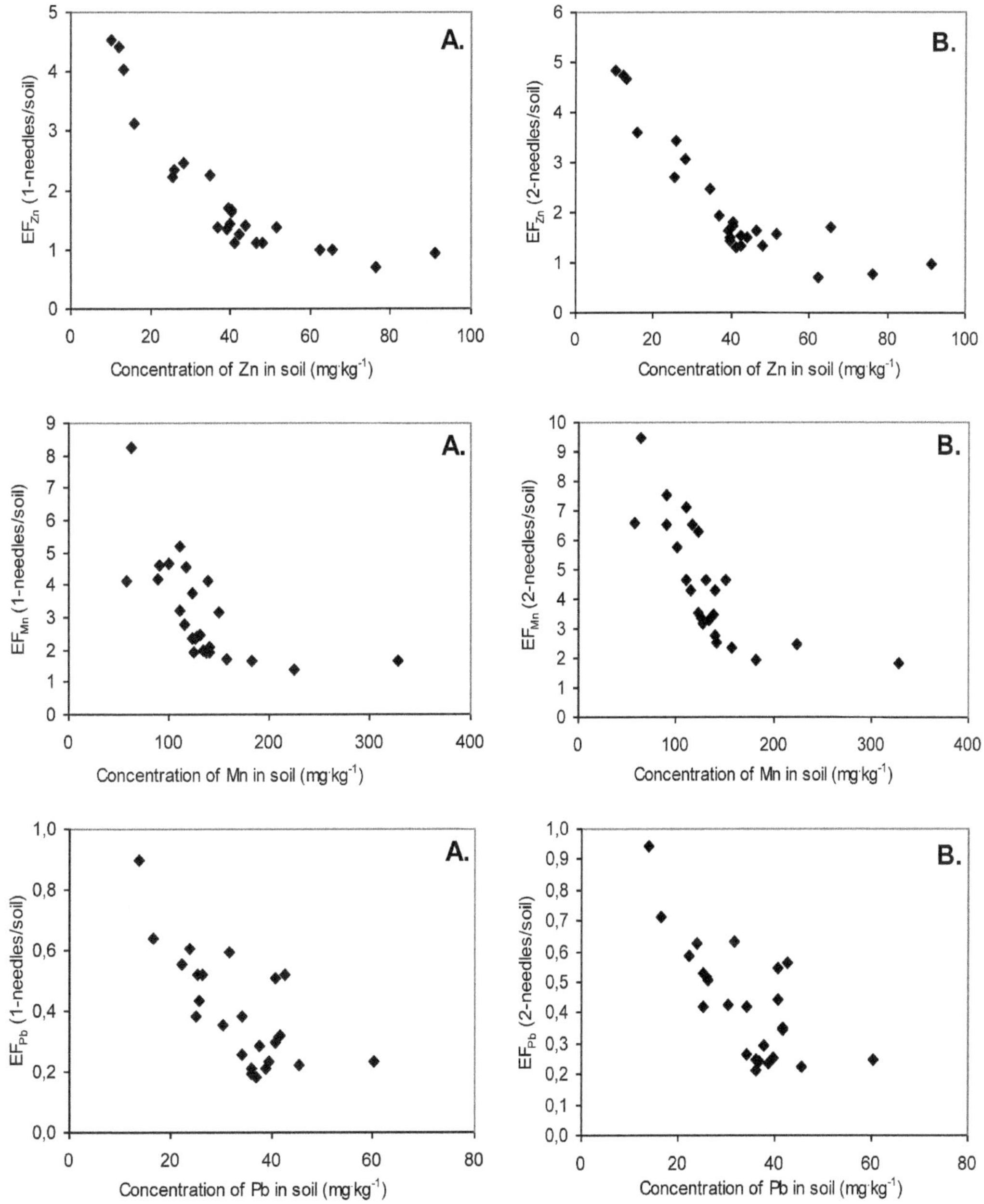

Figure 6. Relation between concentration of heavy metals in soil and enrichment factor (*EF*) in relation: A – concentration heavy metals in 1-needles, B – concentration of heavy metals in 2-needles

7. Dmuchowski W., Bytnerowicz A. 1995. Biomonitoring of the environmental pollution in Poland using chemical analysis of Scott pine (Pinus sylvestris L.) needles, Environ. Pollut. 87, 87-104.
8. Dmuchowski W., Gozdowski D., Baczewska A.H., Brągszewska P. 2011. The comparison bioindication methods in the assessment of environmental pollution with heavy metals, Ochrona Środowiska i Zasobów Naturalnych 50, 170-179.
9. Gambuś F. 1997. Intake of heavy metals by different plant species. Part II. Accumulation of heavy metals by plants. Acta Agraria, Series Agraria, 35, 31, 1997 (in Polish).
10. Gruca-Królikowka S., Wacławek W. 2006. Metals in the environment. Part. II. Influence of heavy metals in plants, Chemistry · Didactics · Ecology · Metrology 11, 1-2, 41-56, 2006 (in Polish).
11. Grzebsz W., Cieśla L., Komisarek J., Potarzycki J., 2002. Geochemical assessment of heavy metals pollution of urban soils. Pol. J. Environ. Stud. 11, 5, 493-499.
12. Imperatoa M., Adamob P., Naimoa D., Arienzob M., Stenzionea D., Violanteb P. 2003. Spatial distribution of heavy metals in urban soils of Naples city (Italy). Environ. Pollut. 124, 247-256.
13. Kabata-Pendias A. 2001. Trace elements in soil and plants. CRC Press, Londyn.
14. Kabata-Pendias A., Pendias H., 1993. Biogeo-

chemistry of trace elements, PWN, Warszawa. [in Polish].

15. Karczewska A. 2002. Heavy metals in soils polluted emissions from copper smelters – form and solubility. Zesz. Nauk. AR Wrocław 432, Dissertations 184, Wrocław [in Polish].

16. Kłos A. 2009. The use of enrichment factor (EF) for the interpretation of results Biomonitoring test. Chemia-Dydaktyka-Ekologia-Metrologia 14, 1-2, 49-55 [in Polish].

17. Kozanecka T., Chojnicki J., Kwasowski W. 2002. Content of heavy metals in plant from pollution-free regions. Pol. J. Environ. Stud. 11, 4, 395-399.

18. Krauss H., Wilcke W., Zech W. 2000. Reactivity and bioavailability of PAHs and PCBs Urban soils of Beyreuth. Proceedings First International Conference Soils of Urban, Industrial, Traffic and Mining Areas. (W. Burghardt and C. Dornauf Eds.) Essen 12-18, July 657-661.

19. Królak E. 2003. Accumulation of Zn, Cu, Pb and Cd by dandelion (*Taraxacum officinale* Web.) in environments with various degrees of metallic contamination. Pol. J. Environ. Stud. 12, 6, 713-721.

20. Kusza G., Ciesielczuk T., Gołuchowska B., 2009. Heavy metal content in soils from the adjacent areas of cement plants in the city of Opole. Ochrona Środowiska i Zasobów Naturalnych. 40: 70-75 [in Polish].

21. Lis J., Pasieczna A. 1995. Geochemical Atlas of Polish, PIG, Warszawa [in Polish].

22. Madrid L., Diaz-Barrientose E., Madrid F. 2002. Distribution of heavy metal contents of urban soils in park sof Seville. Chemosphere 49, 10, 1301-1308.

23. Martines C.E., Motto H.L. 2000. Solubility of lead, zinc and copper added to mineral soils. Environ. Pollut. 170: 153-158, 2000.

24. McALISTER J.J., SMITH B.J., TÖRÖK A. 2006. Element partitioning and potential mobility within surface dusts on buildings in a polluted Urban environment. Budapest. Atmospheric Environment 40, 6780-6790.

25. Morel J.L. 1997. Bioavailability of trace elements to terrestrial plants. Soil Ecotoxicology 141-176.

26. Ottesen R.T., Alexander J., Reidun de Lange. 1999. Ground pollution in Bergen – consequences for health and the environment. NGU Rapport 1999. Geol. Survey of Norway/Trondheim.

27. Parzych A., Jonczak 2013. Content of heavy metals in needles of Scots pine (*Pinus sylvestris* L.) in selected of pine forest in the Słovinski National Park, Arch. Environ. Prot. 1, 39, 41-51.

28. Pasieczna A. 2003. Atlas contamination of urban soils in Poland, PIG, Warszawa [in Polish].

29. Rautio P., Fürst A., Stefan K., Raitio H., Bartels U. 2010. Sampling and analysis of needles and leaves. 19 pp. Manual Part XII. [In]: Manual and methods and criteria for harmonized sampling, assessment, monitoring and analysis of the effects to air pollution on forests. UNECE ICP Forest Programme Coordinating Centre, Hamburg.

30. Samecka-Cymerman A., Kosior G., Kempers A.J. 2006. Comparison of the moss *Pleurozium schreberi* with needles and bark of *Pine sylvestris* as biomonitors of pollution by industry in Stalowa Wola (southeast Poland), Ecotoxicology and Environmental, Safety 65, 108-117.

31. Selim M, Sparks D.L. 2001. Heavy Metals Release in Soils. CRC Press, Boca Raton, FL.

32. Takáć P., Szabová T., B Kozáková L., Benková M. 2009. Heavy metals and their bioavailability from soils in the long-term polluted Central Spis region of SR, Plant, Soil and Environment 55, 4, 167-172.

33. The Minister of Environment on 9 September 2002 in on standards for soil quality and soil quality standards [Dz.U. z 2002 r. Nr 165, poz.1359] [in Polish].

34. Tyler G. 1992. Critical concentrations of heavy metals in the mor horizon of Swedish forests. Swedish Environmental Protection Agency, Report 4078. Solna, p. 38.

35. Yang Y., Peterson E., Campbell C. 2001. Accumulation of heavy metals in urban soils and impacts on microorganism. Huan Jing Ke Xue 22, 3, 44-48.

31

MERCURY AND ALUMINIUM IN ARABLE SOILS IN THE AREA OF INFLUENCE OF BROWN COAL MINE AND POWER PLANT BASED ON THE EXAMPLE OF THE BOGATYNIA REGION (SOUTHWESTERN POLAND)

Ewa Kucharczak-Moryl[1], Andrzej Moryl[2], Jolanta Dąbrowska[3], Romuald Żmuda[4]

[1] Department of Biochemistry, Pharmacology and Toxicology, Wroclaw University of Environmental and Life Sciences, 31 Norwida Str., 50-375 Wroclaw, Poland, e-mail: ewa.kucharczak@up.wroc.pl

[2] Institute of Environmental Engineering, Wrocław University of Environmental and Life Sciences, 24 Grunwaldzki Sq., 50-363 Wrocław, Poland, e-mail: andrzej.moryl@up.wroc.pl

[3] Institute of Environmental Engineering, Wrocław University of Environmental and Life Sciences, 24 Grunwaldzki Sq., 50-363 Wrocław, Poland, e-mail: jolanta.dabrowska@up.wroc.pl

[4] Institute of Environmental Protection and Development, Wrocław University of Environmental and Life Sciences, 24 Grunwaldzki Sq., 50-363 Wrocław, Poland, e-mail: romuald.zmuda@up.wroc.pl

ABSTRACT

This study presents research on the content of mercury and aluminium in arable soils within the range of influence of the Coal Mine and Power Plant Turów (Bogatynia region). The results were compared for reference with the region of Zgorzelec, which is situated outside the range of emission and with the results of tests conducted by other authors. The tests did not show a statistically significant increase in the mercury level in arable land from the range of influence of the Coal Mine and Power Plant Turów, neither in comparison to soils from Zgorzelec region. It is noticeable that the maximum level of mercury in the analysed area was nearly twice as high as in the values obtained in soils from industrial area in Opole and 25 times lower than the maximum mercury level found in the soils of Jelenia Góra District. On the other hand, tests of aluminium content showed a significant, nearly 3 times higher level of this metal in the range of direct influence of the Coal Mine and Power Plant Turów in comparison with the Zgorzelec region. The noted values were higher than in soils from industrial area in Opole.

Keywords: soil, mercury, aluminium, contaminated soils, industrial contamination.

INTRODUCTION

Diffuse pollution is one of the main threats for soil. This type of contamination is characteristic for strongly industrialized areas. One of the major sources of this type of contamination is opencast and underground coal mining, the connected usage of the mined material in energy production and the storage of industrial waste. Regions where large quantities of coal are mined and processed are located throughout the world, mainly in China, the USA, Australia, Indonesia and South Africa. In Europe the problem exists, among others, in Germany and Poland (Tables 1 and 2). These results in the fact that mining and energy production areas are more exposed to soil contamination, in comparison to other regions. In spite of the promotion of renewable energy sources, the interest in the usage of coal as the main energetic resource is not decreasing [6]. It is believed that coal will continue to play the main role in the global structure of energy sources as the power consumption continues to increase. Moreover, in spite of the application of new technologies limiting the emission of harmful substances, the problem of contamination resulting from the mining and processing of coal and its impact on the environment will continue to worsen.

The problem presented in this study is significant also due to the obligations resulting from

Table 1. Coal in electricity generation in 2012 according to the statistics by World Coal Association [30]

Country	Coal in electricity generation (%)
Mongolia	98
South Africa	94
Poland	86
PR China	81
Australia	69
India	68
Israel	59
Indonesia	44
USA	43
Germany	43
UK	29
Japan	27

Table. 2. Top 10 coal producers in 2012 according to the statistics by World Coal Association [30]

Country	Coal production (Mt)
PR China	3549
USA	935
India	595
Indonesia	443
Australia	421
Russia	359
South Africa	259
Germany	197
Poland	144
Kazachstan	126

Poland's membership in the EU and the implementation of the IPCC (Integrated Pollution Prevention and Control) Directive, including the creation of *European Pollutant Release and Transfer Register (E-PRTR) and EMEP* (European Monitoring and Evaluation Programme). Mercury is considered as a priority harmful substance, which was included in the Water Framework Directive.

One of the areas prone to contamination connected with the mining and processing of brown coal in energy production in Poland is the Bogatynia region. Sources of gaseous and dust pollutants are two main facilities on the Polish side, i.e. Coal Mine and Power Plant Turów, but also some plants located on the territory of the Czech Republic, and, until recently, on the German side of the border. It is estimated that cross-border pollutants inflowing to Poland account for as much as 60% of the contamination. Integrated environmental activities that were started in mid-1990s led to a decrease in the emission of air pollutants from power plants located in this region by approximately 60-70%, which has undoubtedly contributed to the improvement of the condition of the natural environment [22].

The aim of the conducted study was to determine the total content of mercury and aluminium in samples of arable soils collected within the influence range of the Coal Mine and Power Plant Turów, to compare the obtained results with Zgorzelec region, located outside the range of emitters, with results obtained by other authors for this region, the geochemical background of Poland and with results obtained for areas that were subject to similar types of anthropogenic pressure in other countries. The research hypothesis assumed that there is a connection between the emission of pollutants from the Coal Mine and Power Plant Turów and the content of mercury and aluminium in the Bogatynia region and that the obtained results should be significantly higher than the values obtained in the Zgorzelec region and the geochemical background of Poland.

MATERIAL AND METHODS

The study was conducted in the years 2008-2011 as part of the evaluation of content of metals in abiotic and biotic elements of the environment in Southwestern Poland. Soil samples originated from two regions:

- Bogatynia (localities: Bogatynia, Działoszyn, Bratków, Ręczyn, Wyszków, Wolanów), encompassing the area of influence of the Coal Mine and Power Plant Turów.
- Zgorzelec (localities: Jerzmanki, Łagów, Jagodzin, Gronów, Sławnikowice), situated outside the range of direct influence of emitters.

A total of 64 soil samples, collected at the depth of 0.0 to 0.3 m b.g.s. (below ground surface) and from 0.3 to 0.6 m b.g.s. were analysed (Figure 1).

The AAS method with amalgamation of cold mercury vapour with the use of the mercury analyser MA 2000 was used to determine the content of total mercury was determined directly in air dry mass of soil. In order to determine the content of total aluminium in soil 0.5 g of weighed quantity was subject to wet mineralisation in aqua regia with use of hypotension microwave system. In the obtained mineralisate, the content of total aluminium was determined with the ICP-AES method, with use of plasma spectrometer Varian-

Figure 1. Location of sampling areas

Liberty 220. The obtained results were subject to statistical analysis, including the calculation of arithmetical mean and standard deviation. The t-student test was used for the purposes of statistical analysis. The value $p \leq 0.05$ was assumed as statistically significant difference. The analyses took into account the physical and chemical properties of samples: the content of organic matter and grain size distribution, which are the key factors influencing the content of heavy metals in soils.

RESULTS AND DISCUSSION

Basic characteristic of soil samples

Soils from both regions are characterised by an acidic pH (Table 4). The pH values measured in 1M KCl for all analysed samples fell within the pH range from 4.70 to pH 6.22. while measured in H_2O fell into the range from pH 3.42 to pH 6.42. In both analysed regions the content of organic matter was higher at the depth from 0.0 to 0.3 m and it fell within the range from 2.83 to 7.60 %, which is beneficial for the neutralisation of heavy metals. At the depth from 0.3 to 0.6 m the average content of organic matter varied from 2.52 to 4.62 % (Table 3). Soils from the Bogatynia region are medium and heavy silty soils (silt loam) with a high content of silt fraction, whereas soils in the proximity of Zgorzelec include very light, light and medium soils (sand, loamy sand, sandy loam), in which a high content of sand fraction was found [9, 15, 21].

Mercury

Mercury belongs to the group of chemical elements that are characterised by particularly strong chemical and biological activity and variability of the forms of existence [5]. It is one of the strongest environmental poisons and its toxicity depends on the form.

In Poland there are no deposits of mercury ore. The main sources of this element in raw materials mined in our country are brown coal, copper ore and hard coal (Table 4). The average mercury content in Polish brown coal is 332 ppb. In the Turów deposit maximum values reaching 950 ppb were noted. Dutch research showed a similar

Table 3. Results of the analysis of physical and chemical properties and grain size distribution of soil samples [9]

Research area	Depth (m b.g.s.)	Parameters	pH		Organic matter (%)	Fraction content (%)		
			H_2O	1M KCl		A	B	C
Bogatynia Region	0.0–0.3	range	5.38–6.32	4.73–6.22	3.63–7.60	23–29	56–68	8–12
		mean	5.80	5.64	5.76	25.67	63.00	9.67
	0.3–0.6	range	5.58–6.42	4.81–6.05	2.52–4.62	15–26	58–72	9–13
		mean	6.02	5.46	3.79	21.00	65.00	11.33
Zgorzelec Region	0.0–0.3	range	5.51–5.96	4.70–5.32	2.83–6.71	52–98	2–41	0–7
		mean	5.63	5.02	5.30	78.80	16.40	3.60
	0.3–0.6	range	3.42–6.01	4.88–5.42	2.54–4.18	51–98	1–39	0–9
		mean	5.65	5.15	3.53	81.80	13.60	3.20

Explanation: A - Sand fraction 2.0-0.05 mm; B - Silt fraction 0.05-0.002 mm; C - Clay fraction <0.002 mm.

Table 4. Sources of mercury in raw materials mined in Poland [12]

Source	Participation in mercury yield [%]
Brown coal	48.4
Copper ore	29.2
Hard coal	19.4
Zinc ore	1.6
Clay raw materials	0.8
Carbonate raw materials	0.6
Total	100

mercury content in brown coal imported from Poland – 350 ppb. Compared to the values obtained in brown coal originating from other countries these are very high: Germany – 160 ppb, the USA – 140 ppb. The lowest content was noted in coal from Columbia and Russia, i.e. 60 ppb [12].

According to the data of PSE S.A. [23], the total installed capacity of hard coal power plants in Poland in the year 2011 was 20 152 MW, and of brown coal power plants – 9630 MW. The coal mine Turów mined 861 tonnes of coal from the beginning of its operation to 2010, while 881 million tonnes were mined in the Bełchatów mine, 553 million tonnes in Konin and 187 million in Adamów. Brown coal-based power plants account for 26% and hard coal power plants account for 54% of the installed capacity in the National Energetic System. The analysis of the composition of brown coal from the Turów coal mine presented hereinabove, together with the amount of coal combusted in the power plant Turów, combined with the additional exposure to cross-border contamination points to a potential threat of contamination of soils in this region with mercury.

Mercury is released to the environment mainly during the combustion of coal, which naturally contains some quantities of this element. Mercury is usually found in coal in combination with sulphur (65–70%), and the other 30–35% are connected with the mineral ash and organic fraction of coal [4]. Global mercury emission from natural and anthropogenic sources ranges from 4400 to 7500 tonnes per year, where the anthropogenic emission accounts for as much as 3500 tonnes per year. In Europe the main source of contamination, accounting for approximately 69% is the combustion of fuels, and 6% is caused by metallurgy of non-ferrous metals. The emission of mercury into the air in quite high in Poland, and the highest amounts of mercury (over 60%) are released as a result of combustion of hard coal and brown coal [19, 24]. Mercury bound in the Earth's crust or in fuels does not pose a significant threat. Only when released it becomes highly mobile and remains in the environment permanently. It is one of the elements characterised by the highest accumulation coefficient, i.e. the relation between the quantity of metal obtained from natural resources to the amount for which demand exists [5]. Natural mercury content in soils varies within the range from 0.05 to 0.3 mg Hg kg^{-1} and the acceptable value in industrial areas, minerals and communication areas (lands in the C group) is 30 mg Hg kg^{-1} dry mass in superficial layers of soil [24, 26]. This element accumulates mainly in the upper layer of soils, and its concentration decreases with the increase in depth [27]. In soil solution it is found in a form of ions, where it is subject to methylation processes with the participation of microorganisms, and the emerging methylated derivatives become more available for plants. High concentrations of mercury were noted in clayey soils (where the sorption of this metal is the strongest) and in peat soils. The pH of soils is an important factor influencing its distribution. In soils with a pH from slightly acidic to alkaline it is strongly tied to large- and small-particle humus substances which leads to its immobilisation, limits the migration of contaminants to the root system of plants and thus into the circulation of matter [1]. In acidic soils it is tied mainly to organic matter and the lowering of the pH value results in an increased availability of organic mercury compounds to plants and to the release of its toxic forms, even if originally they existed in a non-available form. Factors influencing the transformation of this metal in soil also include the content of organic matter, soil structure, mineral composition or the solubility of soil solution [14, 24]. Study conducted in the area of the Bogatynia commune in 2002 showed the presence of mercury in arable soils in the amount from 0.01 to 0.17 mg Hg kg^{-1} whereas in the proximity of Działoszyn these amounts reached even 0.85 mg Hg kg^{-1} [22].

In own analyses (Table 5), in samples of soil collected from the depth from 0.0 to 0.3 m, originating from the area directly influenced by the Power Plant and Coal Mine the maximum mercury level amounted to 0.1028 mg Hg kg^{-1} and it was nearly twice as high as that obtained by Kusza et al. [10] in soils from industrial area in Opole, although 25 times lower than the maximum content of this metal found in the soils of Jeleniogórski District, adjacent to the region of Zgorzelec and

Table 5. Content of mercury in soils

Research area	Parameter	Mercury (mg Hg kg^{-1}) Depth (m b.g.s.)	
		0.00–0.30	0.30–0.60
Bogatynia Region	range	0.042–0.102	0.043–0.073
	mean ± SD	0.067±0.025	0.062±0.013
Zgorzelec Region	range	0.033–0.078	0.018–0.061
	mean ± SD	0.059±0.017	0.040±0.017

Explanation: SD – standard deviation.

Bogatynia [8]. Similar mercury contents for the analysed area were obtained by Pasieczna [20]. The average mercury content in world soils is 0.06 mg Hg kg^{-1} [13], in the research presented by Li [11] the mercury content varied from 0.24 for agricultural soils to 0.35 mg Hg kg^{-1} for soils in China, 3.13–3.82 mg Hg kg^{-1} for mining areas in China and Iran and 52.9 mg Hg kg^{-1} for contaminated areas in the proximity of a mercury mine in Spain. In the context of the presented mercury content in the soils of agricultural and industrial areas, the values obtained in the Bogatynia region are low, they fall into the range determined by Polish geochemical background and average mercury content in world soils.

Aluminium

Aluminium is a basic component of soils. It is a part of the composition of most minerals, where it is usually tied to hardly soluble, non-toxic aluminium silicates or creates amorphous hydroxides, characterised by a high capacity to absorb trace metals, among others. Only 0.1% of aluminium contained in soils has a soluble or exchangeable form, and the content is strictly linked to the type of soil and to its physical and chemical properties [25]. Particularly high levels of this metal are found in soils with a high content of the fraction below 0.02 mm, as well as these characterised by a lower content of organic matter. This dependence was noted during the analyses of hydrogenic soils and peat-muck soils originating from river valleys in the region of Grójec and the upper course of Liwiec river [7, 18]. The highest aluminium content, i.e. 3.38 g Al kg^{-1}was found in muck soils, the second highest – in the peat layer (2.71 g Al kg^{-1}), and the lowest, i.e. 1.5 g Al kg^{-1} in the mineral bed. Moreover, it was observed that in the deeper situated layers the quantities of this metal are significantly lower, which is also connected with lower organic matter content. The factor that strongly determines the increase in the activity of aluminium and its compounds is the pH of soils. As aluminium is easily soluble in acidic environment, the most significant processes occur in soils with a pH ranging from 4.0 to 4.5 [2]. In particularly acidic soils, with a pH ranging from 3.0 to 4.0 they may additionally replace other, useful cations from the absorption complex, which results in the decrease in the content of nutrients necessary for plant development in soils. High aluminium content, through its antagonistic activity, also contributes to the decrease in the content of calcium and magnesium in soil [17]. The increased content of this metal is particularly dangerous in the surface-adjacent layers of soils, as it causes both poor rooting of plants and a weakening of the development of the root system. High content of available forms of aluminium in arable soils leads to an increase of the quantities of this metal in plants, in particular in their underground parts. This leads to disturbances in the transport and collection of nutrients (such as phosphorus, potassium, calcium and natrium) from soils, along with a simultaneous increase in the collection of iron and manganese [2]. The increase in the activity of aluminium is also caused by anthropogenic factors, connected with human activity, which have a significant influence on soil acidification process [16]. Although one of the major sources of this metal is the chemical industry, some quantities of it are also released during the combustion of coal and in the metallurgical industry. Inorganic forms of aluminium are released with particular intensity in regions exposed to the deposition of strong, inorganic acids, in the proximity of mines, deposits of pyrite waste and means of transport emitting nitrogen and sulphur oxides. As a result, alkaline cations are replaced by hydrogen and aluminium cations, what causes a further increase in soil acidity [28]. Additionally, the acidification of the environment by acid rains increases the activity of aluminium,

Table 6. Content of aluminium in soils

Research area	Parameter	Aluminium (g Al kg^{-1}) Depth (m b.g.s.)	
		0.00–0.30	0.30–0.60
Bogatynia Region	range	19.32–32.42	21.80–30.20
	mean ± SD	24.56±5.65*	25.50±3.50*
Zgorzelec Region	range	3.07–16.02	3.46–17.80
	mean ± SD	7.71±5.16*	8.56±5.88*

Explanation: SD – standard deviation; * statistically significant differences at $p \leq 0.05$

stimulates the process of dissolution of its compounds and of silty minerals. The analyses of aluminium content in soils collected at the depth from 0 to 30 cm in Opole showed that the soils contained from 3.44 g Al kg^{-1} to 14.8 g Al kg^{-1} and that its mobility was strongly connected with the pH of soils [3]. These values are lower than those obtained for the range of influence of the Coal Mine and Power Plant Turów that are discussed in this study. One of the regions where soils are exposed to the emission of numerous toxic metals, similarly to the Zgorzelec and Bogatynia region is the Silesian Beskids [31], where the influence of contamination originating both from the Czech Republic and Upper Silesia is noticeable. The aluminium content in soil determined by the authors fell into the range from 23.2 to 50.4 g Al kg^{-1} and they were higher than those obtained as a result of our analyses of soils in the Zgorzelec and Bogatynia regions. Certainly the local and cross-border contamination overlapping in these regions and the physical and chemical properties of the soil itself may significantly influence the adsorption of this metal. The conducted study of aluminium content (Table 6) showed a significant, nearly 3 times higher level of this metal in the range of direct influence of the Coal Mine and Power Plant Turów in comparison with the Zgorzelec region, regardless of the depth from which the samples were collected.

The mean content of aluminium in soil samples collected from the depth from 0.0 to 0.3 m in the Bogatynia region was 24.56 g Al kg$^{-1.}$ whereas in the Zgorzelec region – 7.71 g Al kg$^{-1.}$ and in samples collected from the depth from 0.3 to 0.6 m, respectively: 25.50 and 8.56 g Al kg^{-1}. For soils in the industrialized area of Hamburg the value 6.0 g Al kg^{-1} was obtained, and in Bangkok 13.8 g Al kg^{-1} [29]. Slightly higher aluminium content in samples of soil collected from the depth from 0.30 to 0.60 m in both analysed regions confirm the relation between accumulation and depth, as well as the results of tests obtained by other authors. In the context of the presented literature survey the aluminium content in the analysed area may be considered as increased.

CONCLUSIONS

The authors did not note a significantly higher mercury level in arable soils in the range of influence of the Coal Mine and Power Plant Turów, in comparison with soils from the Zgorzelec region, neither for samples collected from the depth from 0.0 to 0.3 m nor for those collected from the depth from 0.3 to 0.6 m, in spite of the disadvantageous, acidic pH of soils and higher content of small grain fractions in the soil. The obtained results do not differ from the geochemical background in Poland and they are similar to the results for the analysed area presented in literature. No influence of industrial contamination on the increase in mercury content in arable lands in the discussed area was found in the tests.

The significantly higher level of aluminium content in the range of direct influence of the Coal Mine and Power Plant Turów in comparison with the Zgorzelec region, regardless of the depth from which the samples were collected, is connected with the emission of contaminants and with the physical and chemical properties of soils – mainly the acidic pH caused by the emission of nitrogen and sulphur oxides. The aluminium content in soils in the Bogatynia region was higher than in soils from industrial area in Opole and lower than in soils originating from the Silesian Beskids.

REFERENCES

1. Bielicka A., Ryłko E., Bojanowska I. 2009. Zawartość pierwiastków metalicznych w glebach i warzywach z ogrodów działkowych Gdańska i okolic. Ochr. Śr. Zasobów Nat., 40: 209-216.

2. Brzeziński M. 2004. Wpływ zakwaszenia gleby na zawartość glinu w roślinach. Ann. UMCS, Sect.E, 3. 59: 1313-1317.
3. Ciesielczuk T., Kusza G., Kowalska-Góralska M., Senze M. 2011. Aluminium and selenium content in soils of industrial area in Opole (southern Poland). Archiv. Environ. Prot., 1, 37 (1): 25-32.
4. Glodek A., Pacyna J.M. 2009. Mercury emission from coal – fired Power plants in Poland. Atmos. Environ., 43: 5668-5673.
5. Gworek B., Rateńska J. 2009. Migracja rtęci w układzie powietrze – gleba – roślina. Ochr. Śr. Zasobów Nat., 41: 614-623.
6. Juda-Rezler K., Kowalczyk D. 2013. Size distribution and trace elements contents of coal fly ash from pulverized boilers. Pol. J. Environ. Stud., 1 (22): 25-40.
7. Kalembasa D., Pakuła K., Becher M. 2008. Profile differences of Fe, Al and Mn in the peat-muck soils in the upper Liwiec river valley. Acta.Sci. Pol.- Agric., 8(2): 3-8.
8. Kaszubkiewicz J., Kawałko D. 2008. Zawartość wybranych metali ciężkich w glebach i roślinach na terenie powiatu jeleniogórskiego. Ochr. Śr. Zasobów Nat., 40: 177-189.
9. Kucharczak E, Moryl A. 2011.Wpływ Elektrowni i Kopalni „Turów" na zawartość wybranych metali ciężkich w glebach uprawnych. Ochr. Śr. Zasobów Nat., 49: 178-185.
10. Kusza G., Ciesielczuk T. 2007. Wpływ wybranych zakładów przemysłowych na wzrost zawartości metali ciężkich w glebach terenów przyległych. Ochr. Śr. Zasobów Nat., 31: 110-114.
11. Li Z., Ma Z., Kuijp J., Yuan Z., Huang L. 2014. A review of soil heavy metal pollution from mines in China: Pollution and health risk assessment. Sci. Total Environ., 468-469: 843-853.
12. Lorenz U., Grudziński Z. 2007. Zawartość rtęci jako potencjalny czynnik ograniczający wartość użytkową węgla kamiennego i brunatnego. Górnictwo i Geoinżynieria., 31 (3/1): 335-350.
13. Luo X.S., Yu S., Zhu Y.G., Li X.D. 2012. Trace metal contamination in urban soils of China. Sci. Total Environ., 421-422: 17-30.
14. Malczyk P., Długosz J. 2009. Zmienność przestrzenna całkowitej zawartości rtęci w poziomie powierzchniowym gleb wybranego obszaru równiny sępopolskiej. Ochr. Śr. Zasobów Nat., 40: 39-48.
15. Marcinek J., Komisarek J., Bednarek R., Mocek A., Skiba S., Wiatrowska K. 2011. Systematyka gleb Polski. Rocz. Glebozn., 62(3): 5-142.
16. Marcos F.,Lugo E.A. 2002. Aqueous aluminium species in forest soils affected by fluoride emissions from an aluminium smelter in NW Spain. Res. Rep., 35: 110-121.
17. Meriño-Gergichevich C., Alberdi M., Ivanov A.G., Reyes-Diaz M. 2010. Al^{3+}-Ca^{3+}interaction in plants in acid soils: Al-phytotoxicity response to Calcareous amendments. J. Soil. Sci. Plant Nutr., 10: 217-243.
18. Mocek A., Spychalski W. 2007. Zawartość pierwiastków śladowych w glebach hydrogenicznych doliny grójeckiej. Ochr. Śr. Zasobów Nat., 31: 52-56.
19. Pacyna E.G., Pacyna J.M., Sundseth K., Munthe J., Kindbom K., Wilson S., Steenhuisen F., Maxson P. 2010. Global emission of mercury to the atmosphere from anthropogenic sources in 2005 and projections to 2020. Atmos. Environ., 44: 2487-2499.
20. Pasieczna A. 2012. Rtęć w glebach obszarów zurbanizowanych Polski .Prz. Geol., 60: 46-58.
21. Polskie Towarzystwo Gleboznawcze. 2009. Klasyfikacja uziarnienia gleb i utworów mineralnych – PTG 2008. Rocz. Glebozn., 60(2): 5-16.
22. Program Ochrony Środowiska i Plan Gospodarki Odpadami dla powiatu zgorzeleckiego na lata 2005–2008 z perspektywą na lata 2009–2012. 2004. Charakterystyka Powiatu Zgorzeleckiego i stanu środowiska na terenie opracowania. www.powiat.zgorzelec.pl [Access 20.08.2013].
23. Raport 2011 KSE. http://www.pse.pl/index.php?dzid=171&did=1053
24. Smolińska B. 2010. Metody oczyszczania gleb zanieczyszczonych rtęcią. ZN Polit. Łódzkiej Biotechnology and Food Science, 1081: 121-136.
25. Symanowicz B., Kalembasa S. 2010. Wpływ odpadowych węgli brunatnych i osadów ściekowych oraz ich mieszanin na zawartość kobaltu, litu i glinu w glebie i roślinie. Acta Agroph., 15(1): 167-175.
26. Szczepocka A. 2005. Kryteria oceny zanieczyszczeń gleb metalami ciężkimi. Zesz. Nauk. SGSP, 32: 13-28.
27. Szopka K., Karczewska A., Kabała C., Jezierski P., Bogacz A. 2010. Zawartość rtęci w poziomach powierzchniowych gleb leśnych Karkonoskiego Parku Narodowego w rejonie Szklarskiej Poręby. Ochr. Śr. Zasobów Nat., 42: 167-175.
28. Tahri M., Benyyaïch F., Bounakhla M., Bilal E., Gruffat J.J., Moutte J., Garcia D. 2005. Multivariate analysis of heavy metal contents in soils, sediments and water in the region of Meknes (Central Morocco). Environ.Monit.Assess., 102: 405-417.
29. Wilcke W., Müller S., Kanchanakool N., Zech W. 1998. Urban soil contamination in Bangkok: heavy metal and aluminium partitioning in topsoils. Geoderma, 3-4 (86): 211-228.
30. http://www.worldcoal.org/resources/coal-statistics/ [Access 17.01.2014]
31. Zołotajkin M., Ciba J., Kluczka J., Skwira M., Smoliński A. 2011. Exchangeable and bioavailable aluminium in the mountains forest soil of Barania Góra range (Silesian Beskids, Poland). Water Air Soil Poll., 216: 571-580.

32

BACTERIAL POPULATION DYNAMICS IN WASTE OILY EMULSIONS FROM THE METAL-PROCESSING INDUSTRY

Paweł Kaszycki[1], Paulina Supel[1], Przemysław Petryszak[1]

[1] Department of Biochemistry, Institute of Plant Biology and Biotechnology, Faculty of Horticulture, University of Agriculture in Kraków, 29. Listopada Ave., 31-425 Kraków, Poland, e-mail: p.kaszycki@ogr.ur.krakow.pl; p.supel@ogr.ur.krakow.pl; p.petryszak@ogr.ur.krakow.pl

ABSTRACT

Oil-containing wastewaters are regarded as main industrial pollutants of soil and water environments. They can occur as free-floating oil, unstable or stable oil-in-water (O/W) emulsions, and in the case of extreme organic load, as water-in-oil (W/O) emulsions. In this study two types of oily effluents, a typical O/W emulsion marked as E1 and a W/O emulsion E2, both discharged by local metal processing plants were examined to test their toxicity to microbial communities and the ability to serve as nutrient sources for bacterial growth. The organic contaminant load of the samples was evaluated on the basis of chemical oxygen demand (COD) parameter values and was equal to 48 200 mg $O_2 \cdot dm^{-3}$ and >300 000 mg $O_2 \cdot dm^{-3}$ for E1 and E2, respectively. Both emulsions proved to be non toxic to bacterial communities and were shown to contain biodiverse autochthonous microflora consisting of several bacterial strains adapted to the presence of xenobiotics (the total of $1.36 \cdot 10^6$ CFU$\cdot$cm^{-3} and $1.72 \cdot 10^5$ CFU$\cdot$cm^{-3} was determined for E1 and E2, respectively). These indigenous bacteria as well as exogenously inoculated specialized allochthonous microorganisms were biostimulated so as to proliferate within the wastewater environment whose organic content served as the only source of carbon. The most favorable cultivation conditions were determined as fully aerobic growth at the temperature of 25 °C. In 9 to 18 day-tests, autochthonous as well as bioaugmented allochthonous bacterial population dynamics were monitored. For both emulsions tested there was a dramatic increase (up to three orders of magnitude) in bacterial frequency, as compared to the respective initial values. The resultant high biomass densities suggest that the effluents are susceptible to bioremediation. A preliminary xenobiotic biodegradation test confirmed that mixed auto- and allochthonous bacterial consortia obtained upon inoculation of the samples with microbiocenoses preselected for efficient hydrocarbon biodegradation led to a decrease in the organic pollution level.

Keywords: bioremediation, biostimulation, oily emulsions, petroleum-derived contaminants, autochthonous microorganisms, microbial consortia.

INTRODUCTION

Oily emulsions are common waste products of petrochemical, metal and food processing industries as well as transportation and domestic activities [Coca et al. 2011, Abdel-Raouf 2012]. These pollutants bear severe environmental risk due to their high organic load and chemical-physical characteristics that endanger human health and have strong negative impact on soil and aquatic life. Among these, oily effluents discharged by metal manufacturing facilities are particularly hazardous since they are chemically very complex [Effluent Limitations... 2003, Coca et al. 2011] and may be colonized with a variety of microorganisms (bacteria and fungi), including human pathogens [Saha, Donofrio 2012]. Four main types of fluids are used in metal processing technologies i.e. straight (neat) oils, soluble oils, semisynthetic fluids, and synthetic fluids

[Geier, Lessmann 2006, MacAdam et al. 2012]. Straight oils are usually very dense, water non-miscible preparations, composed primarily of mineral oil with numerous organic additives including surfactants [Geier, Lessmann 2006, Coca et al. 2011]. When discharged, they can become admixed with the aqueous phase, and finally form water-in-oil (W/O) emulsions in which oil acts as a dispersion medium. The latter three fluid types make up a dominating category of the so-called water-based fluids, applied as oil-in-water (O/W) emulsion suspensions in which water serves as a continuous phase, constituting up to 95% of the emulsion volume. Mineral oil-based O/W emulsions may contain a number (up to 60 of the total of 300 identified) of chemical additives [Rabenstein et al. 2009] that include emulsifiers, corrosion inhibitors, antifoam and anti-fog substances, heavy metal ions, pressure agents, stabilizers, buffers, lubricants, and, in particular, a variety of biocides [Geier, Lessmann 2006, Bakalova et al. 2008]. Biocides are added to reduce the risk of microbial contamination that often affects workers at their workplaces [Geier, Lessmann 2006] and causes metalworking fluid biodeterioration and emulsion breakdown [da Silva et al. 2001, Theaker, Thompson 2010].

Numerous microbial species were found to inhabit metalworking fluids, especially the water-based emulsions [van der Gast et al. 2003, Bakalova et al. 2007, Saha, Donofrio 2012]. Furthermore, many authors point to a very limited strain detectability employing standard microbiological culture-based techniques due to the presence (up to 95% of the total population) of the so-called VBNC-type strains (*viable but non-cultivable*). Thus, alternative, culture-independent identification methods are being proposed [van der Gast et al. 2003, Selvaraju et al. 2008, Gilbert et al. 2010, Saha, Donofrio 2012, Trafny 2013]. Many of the hitherto identified strains were found strongly pathogenic, *e.g.* those of the genera *Mycobacterium, Klebsiella, Aspergillus* and *Staphylococcus* [Bakalova et al. 2007, Wang et al. 2007, Saha, Donofrio 2012]. It should be stressed here, however, that the preventive usage of biocidal chemicals may dramatically enhance the resultant environmental toxicity of the emulsions and can make the biological treatment of these wastewaters impossible. On the other hand, the occurrence of non-pathogenic strains of such genera as *Pseudomonas, Bacillus, Agrobacterium, Acinetobacter, Alcaligenes, Stenotrophomonas, Methylobacterium, Rhodococcus* and many others [van der Gast et al. 2003, Saha, Donofrio 2012] might be beneficial in terms of the potential use of these autochthons for bioremediation purposes [Cheng et al. 2005, Bakalova et al. 2007, Saha, Donofrio 2012].

Together with the microbiological risk, the presence of aggressive chemical agents and oil-derived xenobiotics in water-miscible emulsions makes these fluids particularly hazardous. They can cause adverse health effects for humans, and when released to the environment they become strongly ecotoxic, negatively affecting natural ecosystems [Tillie-Leblond et al. 2011, Mirer 2010, Grijalbo et al. 2013, Lazarević et al. 2013, Gerulova et al. 2011].

Taking into account the above information regarding great chemical variability, broad range of contamination load, and microbiological diversity of the waste emulsions discharged by the metal industry facilities it is of high importance to perform appropriate toxicity and biodegradability testing of effluents originating from different sources. The results of such tests should bring valuable data necessary to elaborate optimal disposal/cleanup strategies.

The aim of this study was to test whether the mineral oil-based waste effluents of the metal-processing industry, generated upon technological processes as oily emulsions with high organic load, can be biologically treated using auto- and/or allochthonous bacterial communities. This aim was achieved by examining toxicity of the emulsions, evaluating their colonization with indigenous microorganisms, and testing both the conditions and growth potential of autochthonous and allochthonous bacteria.

MATERIALS AND METHODS

Oily emulsions

Two types of oily emulsions discharged by local metal processing plants were examined for their toxicity, occurrence of indigenous microflora and susceptibility to colonization with allochthonous microorganisms:

- **E1** - a spent metalworking fluid (MWF) consisting of mixed hydrocarbons and surfactants as well as other chemical additives, a typical oil-in-water (O/W) emulsion discharged from a metal cutting and forming process. It was a whitish, dense, homogeneous liquid, with no smell, characterized by a high organic contaminant load determined as the chemical oxygen demand parameter (COD) of 48 200 mg $O_2 \cdot dm^{-3}$.

- **E2** - an oily effluent generated upon metal cooling and machine tool lubrication, discharged as a water-in-oil emulsion (W/O) with a very high organic content (COD above 300 000 mg $O_2 \cdot dm^{-3}$). The sample was light grey with dark surface smudges and detectable smell of petroleum-like products, a dense liquid, heterogeneous and unstable, easily undergoing spontaneous phase separation.

The detailed characteristics of both emulsions are given in Table 1.

Microbiological monitoring

Emulsion samples were typically incubated at aerobic conditions in 300 cm^3 flasks rotary-shaken with laboratory thermostated shakers, each flask containing 50 cm^3 of a tested fluid. Specific experimental setups are given in the text.

Determination of bacterial frequency, bacterial survivability and cell population dynamics were done using standard microbiological techniques based on a Koch method of culture surface plating onto agar-solidified optimal media (2.5% enriched agar, Biocorp, Poland). To check for the presence of yeast and other fungi, a Sabouraud medium supplemented with an antibiotic cocktail (Biocorp, Poland) was used. CFU (colony-forming units) were counted after 3 days of incubation at 37 °C. Cell frequencies were expressed as $CFU \cdot cm^{-3}$ of aqueous samples. In order to obtain homogeneous cell suspensions, prior to dilutions, bacterial samples were sonicated under mild conditions for 10 min with a laboratory ultrasound washer (UN-2 Unitra/Unima, Poland).

Source of allochthonous microorganisms

In the tests involving allochthonous microorganisms the studied emulsions were inoculated with the microbial consortium ZB-01 developed in Biochemistry Department of University of Agriculture in Krakow (volume proportions given in the text). This aerobic bacterial biocenosis has been widely used in environmental bioremediation projects to trigger and/or stimulate biodegradation of petroleum-oil hydrocarbons in contaminated waters and soils [Kaszycki et al. 2001, Kołoczek, Kaszycki 2006, Kaszycki et al. 2010, Kaszycki et al. 2011]. It consists of a number of environmental strains, selected and isolated over years from sites heavily polluted with organic compounds. Typically, the initial cell density of the consortium used as *inoculum* exceeded $5 \cdot 10^8$ $CFU \cdot cm^{-3}$).

Toxicity tests of waste emulsions

Emulsion toxicity tests were done based on the survivability determination of the ZB-01 bacterial consortium (see the description above) inoculated to either E1 or E2. Two variant volume proportions (effluent: ZB-01) were examined, that is 1:1 and 9:1. The samples were then incubated for 7 and 9 days, respectively, and checked for the cell population density.

Chemical oxygen demand determination

The COD analyses were carried out employing a dichromate procedure according to the protocols of standardized automated ampoule-based mineralization-analytical system Hach-Lange DR5000 (Germany).

Extraction of organic substances and determination of contamination load

The organic content of emulsion samples was extracted using petroleum ether (fractions boiling at 40–60 °C, POCh, Poland). Briefly, a 10 cm^3 sample was weighed, thoroughly mixed, acidified with 1 cm^3 18% HCl and dried with approx.

Table 1. Characteristics of the waste oily emulsions

Sample No.	Emulsion type	Source (industrial process)	pH	COD [mg $O_2 \cdot dm^{-3}$]	Ether-extractable organic fraction [$mg \cdot dm^{-3}$]	Occurrence of bacteria [$CFU \cdot cm^{-3}$]	Bacterial colony morpho-types	Occurrence of fungi & yeast [$CFU \cdot cm^{-3}$]
E1	Oil-in-Water (O/W)	mineral-oil based metal-working fluid (metal cutting and forming)	6.67	48 200	10 100	$1.36 \cdot 10^6$	4	0
E2	Water-in-Oil (W/O)	mixture of oils from machine lubricating and cooling systems	7.20	>300 000*	290 895	$1.72 \cdot 10^5$	3	0

Explanation: * – a value beyond the measurable limit.

30 g of anhydrous $MgSO_4$. The sample was then placed in an extraction thimble made of nitrocellulose (Sigma-Aldrich) and extracted (for approximately 6 h) in a 65 cm^3 Soxhlet apparatus at 40–60 extraction cycles. All the extractions were carried out at least in duplicates. After evaporating the excess solvent at 85 °C, the flasks containing extracted material were dried at 105 °C for 1.5 h, then cooled in a vacuum desiccator and finally, the residual material (high-boiling organic fraction) was weighed.

All the chemicals were of analytical grade. In microbiological analyses fully sterile conditions were applied. The mean experimental error was established based on repetitive tests as 10%, 12% and 15% for the analyses of COD, organic substances content and for the determination of bacterial cell population, respectively.

RESULTS AND DISCUSSION

Occurrence of indigenous microflora

Both emulsions were shown to contain autochthonous microflora consisting of several bacterial strains adapted to the presence of xenobiotics. The total of $1.36 \cdot 10^6$ CFU·cm^{-3} and $1.72 \cdot 10^5$ CFU·cm^{-3} were determined for E1 and E2, respectively. Although no detailed strain identification was made, general biodiversity of the indigenous bacteria could be preliminary evaluated based on the number of distinct colony morphotypes. Four morphotypes found for E1, and three for E2 might suggest a relatively low biodiversity when compared to 8–15 colony types typically detected on plates with the multispecies microbial ZB-01 consortium (not shown). However, it should be pointed out here that many of the indigenous strains might belong to a group of non-cultivable (VBNC) microorganisms, which is a finding typical for metalworking fluids [van der Gast et al. 2003, Gilbert et al. 2010, Saha, Donofrio 2012, Trafny 2013]. In such a case the observed biodiversity would be limited and the total number of bacteria reduced due to detection of culturable strains only. Interestingly, the studied samples lacked in any yeast or fungi as could be excluded by the appropriate tests employing fungi-selective media (Table 1). Note that no fungi occurred throughout all the tests performed later.

As cited earlier, a number of studies on metalworking fluids collected from different industrial sources proved the occurrence of biodiverse indigenous microflora. According to other authors, the measured population densities varied depending on the emulsion source and ranged from 10^5 to 10^7 CFU·cm^{-3} [Veillette et al. 2004, Liu et al. 2010, Saha, Donofrio 2012] with one reported case of over 10^8 CFU·cm^{-3} [Mattsby-Baltzer et al. 1989]. Unlike our observations, fungi (mainly yeast and molds) were also detected in some samples [Liu et al. 2010, Saha, Donofrio 2012].

Toxicity tests

As might be expected from the presence of numerous autochthonous bacteria, none of the emulsions were found toxic to exogenously applied microbial consortia. Upon inoculation with the ZB-01 biocenosis, neither short-term (a 2-day toxicity test, not shown) nor long-term observations (7–9 days, see Table 2 and Figure 1) revealed any toxic effect that would lead to the loss of bacterial survivability.

Table 2. Bacterial cell population densities in emulsions E1 and E2 admixed (1:1 v/v) with the ZB-01 preparation, as determined immediately upon inoculation ("start") and after 7-day incubation at 25 °C ("end")

Sample	Start [CFU · cm^{-3}]	End (7 days) [CFU · cm^{-3}]
E1	$1.21 \cdot 10^8$	$1.48 \cdot 10^9$
E2	$1.22 \cdot 10^8$	$2.24 \cdot 10^9$

On the contrary, a strong tendency was observed for further proliferation of the allochthonous (together with autochthonous) bacteria over time for both of the examined inocula variants. The above effect is shown in Table 2 and in Figure 1 where the oil samples were inoculated at high and low initial densities (that is, ZB-01 admixed at volume proportions 1:1 and 1:9), respectively. The above results indicate that the lack of toxicity combined with microorganism biostimulation by aeration and shaking enabled bacteria to efficiently assimilate and/or biodegrade heterogeneous xenobiotics present in the treated samples. Apparently both of the studied waste effluents had not been supplemented with any biocidal agents, which might result in disabling possibility of biological treatment. Our findings are thus very important in terms of elaborating the most effective oily-waste management method since the results strongly suggest that the effluents might be susceptible to biological remediation.

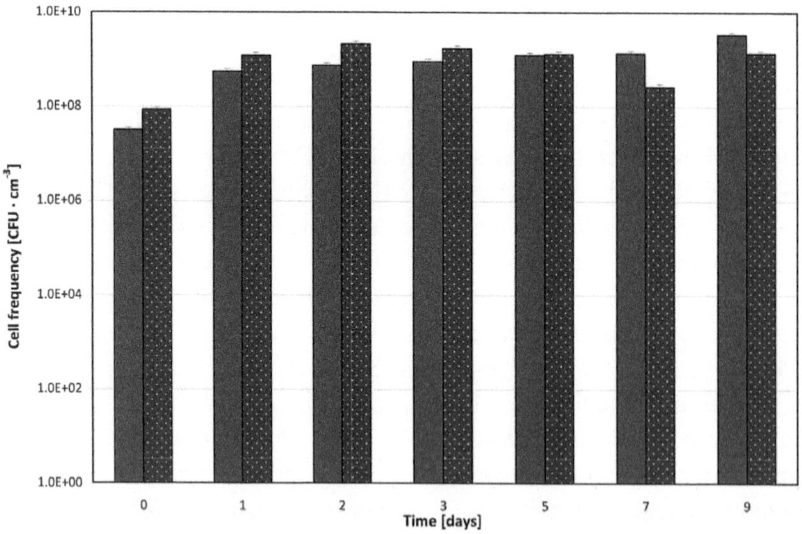

Figure 1. Emulsion toxicity test (E1 – solid bars and E2 – dotted bars): based on bacterial population observations in the samples inoculated with the ZB-01 consortium (diluted 10×)

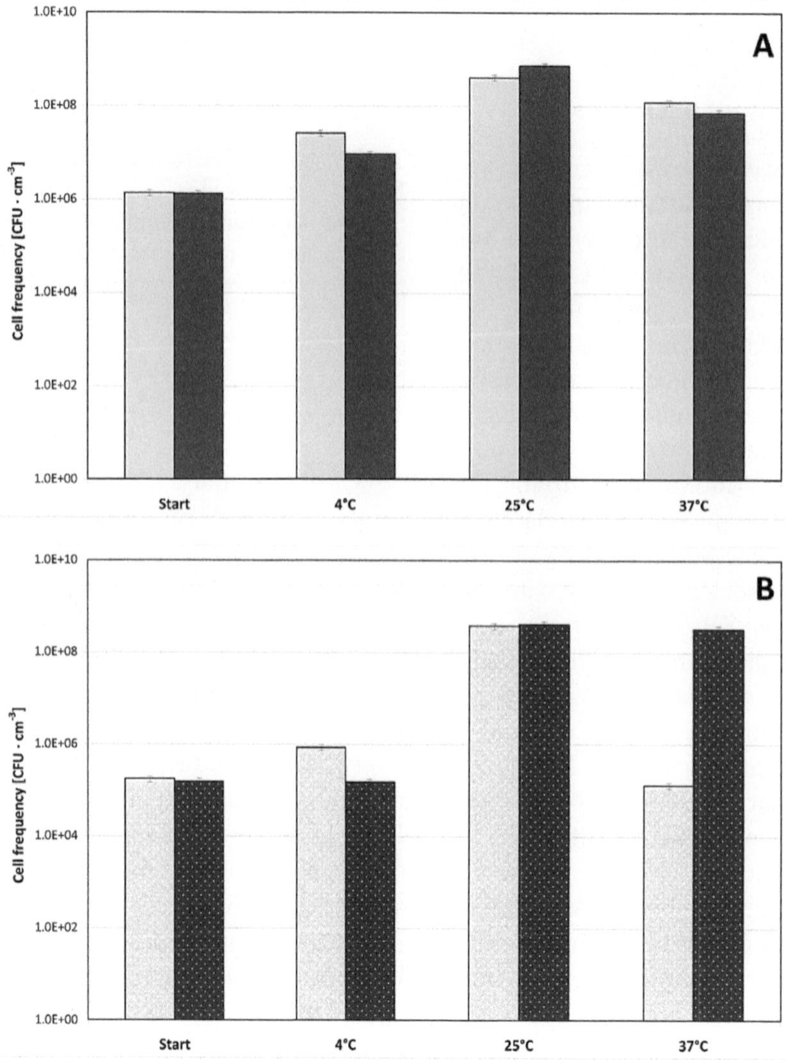

Figure 2. Temperature dependence of bacterial population determined in the E1 (A) and E2 (B) emulsions upon 6-day cultivation at 4 °C, 25 °C, and 37 °C; light and dark bars correspond to the frequency of autochthons (uninoculated samples), and autochthons + allochthons (samples inoculated with ZB-01 at 1000× dilution), respectively

Biostimulation of auto- and allochthonous microflora

In another set of experiments both the indigenous bacteria as well the introduced allochthonous microorganisms inoculated at 1000× dilution were biostimulated so as to proliferate within the emulsion wastewater environments. The cultivation conditions were described in Materials and Methods section. Since no trophic additives were supplemented it is inferred that the organic content of each emulsion served as the only source of carbon. Figure 2 presents the results of the test of temperature optima for bacteria in E1 and E2 (Figure 2 A and B, respectively). In a 6-day incubation at variant temperatures of 4 °C, 25 °C, and 37 °C a profound increase (three to four orders of magnitude) of both autochthonous and allochthonous bacterial frequency was observed, as compared to the initial population value (marked as "start"). For both effluents, the growth-stimulatory effect was pronounced the most at 25 °C and therefore any further testing was done at this temperature.

In a long-term (18 day) systematic test, bacterial cell population dynamics was monitored at 25 °C upon cultivation as described above. The results presented in Figure 3 confirm earlier data indicating bacterial growth within emulsions at favorable conditions. Furthermore, at least for the case of E1 it can be clearly seen that the introduction of allochthonous strains (dark bars) led to significantly larger microbial biomass density relative to uninoculated samples (light bars). The latter observation indicates that in attempts to elaborate a biological treatment method the bioaugmented process should be preferably considered.

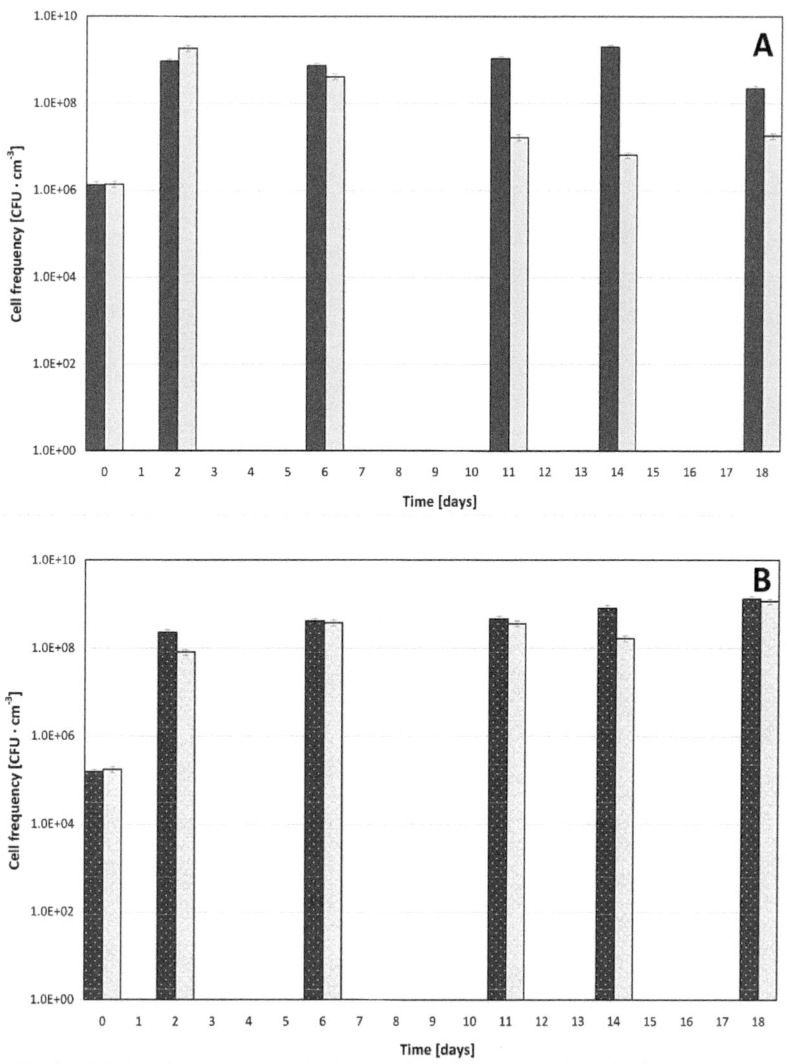

Figure 3. Bacterial cell population dynamics of the E1 (A) and E2 (B) emulsions as determined upon long-term cultivation at 25 °C; light and dark bars represent the uninoculated and inoculated samples, respectively, as in Figure 2

Preliminary analysis of E1 biodegradability

In order to check whether the observed dramatic increase of bacterial density correlated with the drop of the organic load due to xenobiotic carbon utilization, the E1 sample collected at the end of the test (day 18) was analyzed for the ether-extractable organic fraction content and COD (Figure 4 A and B, respectively) and compared with the initial value of untreated emulsion. Note that E2 was not examined due to its extreme load with organic compounds. In Figure 4 a decrease of both parameters can be seen, although the data obtained for COD (Figure 4 B) bear poor statistic reliability due to high experimental error and need to be further confirmed. At the same time the results presented in Figure 4 suggest that higher activity was revealed for the mixed bacterial consortium formed by both autho- and inoculated allochthons. Taken together these preliminary data, some optimization actions should be taken to improve the kinetics and the final yield of bioremediation process. Such actions are now planned to include the dilution of the waste effluent prior to biodegradation as well as inoculation with higher density of allochthonous bacteria. The results will be reported in a separate paper.

CONCLUSION

- The tested oily emulsions generated as waste products of metal industries are non toxic to bacteria and may thus be microbiologically treated.
- They are populated with biodiverse autochthonous bacterial species. These bacteria are well adapted to the emulsion environment and for that reason they appear as good candidate strains to form a biocenosis that would be capable of efficient bioremediation of the emulsions.

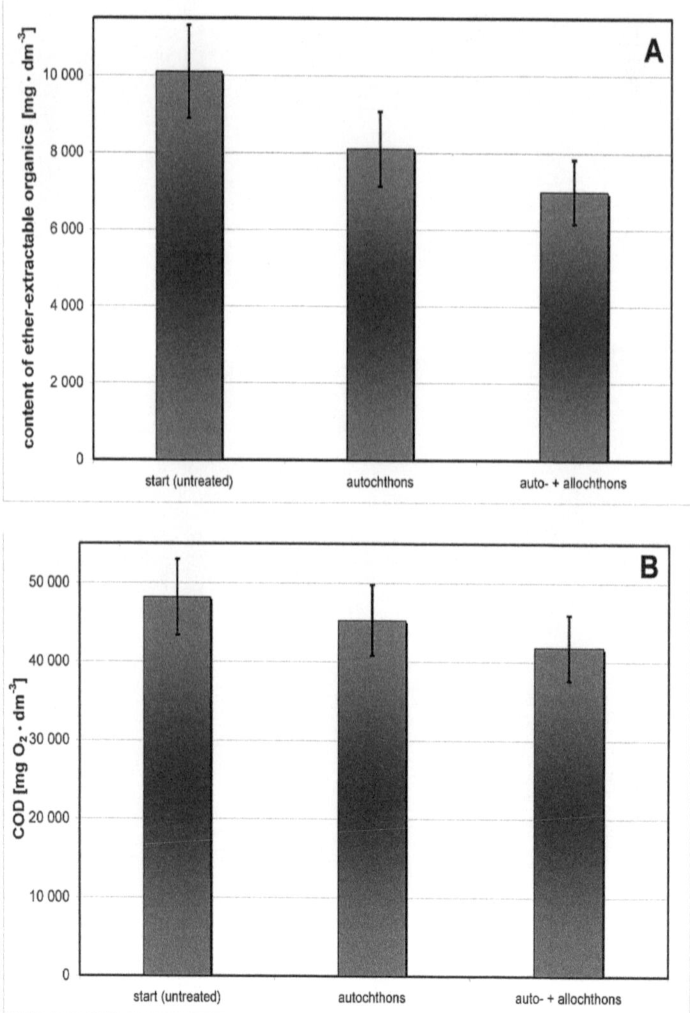

Figure 4. Results of 18-day bioremediation of undiluted E1 as observed with the method of organic compounds extraction (A) and COD determination (B)

- Microbial biomass grows upon oily xenobiotic utilization under aerobic conditions at the temperature optimum of 25 °C. The highest cell frequency is obtained when the emulsions become bioaugmented by inoculation with allochthonous bacteria preselected for efficient hydrocarbon biodegradation.
- Oily emulsion E1 can undergo biological treatment under conditions enabling bacterial proliferation, however the kinetics as well as the final yield of the bioprocess needs further optimization to make the microbiological method more economically competitive.

Acknowledgement

The work was financially supported by the grant for scientific research no. 3500, approved by the Polish Ministry of Science and Higher Education and by Mo-BRUK S.A., Niecew 68, Korzenna, Poland.

REFERENCES

1. Abdel-Raouf E.M. 2012. Crude Oil Emulsions - Composition Stability and Characterization. Croatia: InTech.
2. Bakalova S., Doycheva A., Ivanova I., Groudeva V., Dimkov R. 2007. Bacterial microflora of contaminated metalworking fluids. Biotechnol. & Biotechnol. Eq., 21(4), 437-441.
3. Bakalova S., Mincheva V., Doycheva A., Groudeva V., Dimkov R. 2008. Microbial toxicity of ethanoloamines. Biotechnol & Biotechnol Eq., 22(2), 716-720.
4. Cheng C., Phipps D., Alkhaddar R.M. 2005. Treatment of spent metalworking fluids. Water Res., 39, 4051-4063.
5. Coca J., Gutierrez G., Benito J.M. 2011. Treatment of oily wastewater. [In:] Water Purification and Management, J. Coca-Prados, G. Gutierrez-Cervello (Eds.), NATO Science for Peace and Security Series - C: Environmental Security, Dordrecht: Springer.
6. da Silva E.J., Bianchi E.C., de Aguiar P.R. 2001. A Review of grinding fluids – performances and management. Revista de Ciencia & Technologia, 8(18), 67-77.
7. Effluent Limitations Guidelines and New Source Performance Standards for the Metal Products and Machinery Point Source Category; Final Rule. Part II. 2003. U.S. Environmental Protection Agency, Rules and Regulations, Federal Register, 68(92), 25685-25745.
8. Geier J., Lessmann H. 2006. Metalworking fluids. [In:] Contact dermaitis. 3rd edition, P.J. Frosch, T. Menne, J-P. Lepoittevin (Eds.), Berlin Heidelberg: Springer.
9. Gerulova K., Mihalkova A., Sergovicova M., Guoth A., Nadasska Z. 2011. Exotoxicity and biodegradability assessment of metalworking fluids by activated sludge bacteria. Research Papers: Faculty of Materials Science and Technology in Trnava, 19(31), 45-56.
10. Gilbert Y., Veillette M., Duchaine C. 2010. Metalworking fluids biodiversity characterization. J. Appl. Microbiol., 108(2), 437-449.
11. Grijalbo L., Fernandez-Pascual M., Garcia-Seco D., Gutierrez-Manero F.J., Lucas J.A. 2013. Spent metal working fluids produced alterations on photosynthetic parameters and cell-ultrastructure of leaves and roots of maize plants. J. Hazard. Mater., 260, 220-230.
12. Kaszycki P., Pawlik M., Petryszak P., Kołoczek H. 2010. Aerobic process for in situ bioremediation of petroleum-derived contamination of soil: a field study based on laboratory microcosm tests. Ecol. Chem. Eng. A, 17(4-5), 405-414.
13. Kaszycki P., Petryszak P., Pawlik M., Kołoczek H. 2011. *Ex situ* bioremediation of soil polluted with oily waste: use of specialized microbial consortia for process bioaugmentation. Ecol. Chem. Eng. S, 18(1), 83-92.
14. Kaszycki P., Szumilas P., Kołoczek H. 2001. Biopreparat przeznaczony do likwidacji środowiskowych skażeń węglowodorami i ich pochodnymi. Inż. Ekol., 4, 15–22.
15. Kołoczek H., Kaszycki P. 2006. Bioremediacja zanieczyszczeń rafineryjnych w środowisku gruntowo-wodnym. [In:] Metody usuwania zanieczyszczeń węglowodorowych ze środowiska gruntowo--wodnego (Methods of hydrocarbon contaminants removal from ground-water environment), S. Rychlicki (Ed.), Uczelniane Wyd. Nauk.-Dydakt., AGH, Kraków.
16. Lazarević F.B., Krstić I.M., Lazić M.L., Savić D.S., Skala D.U., Vejlković V.B. 2013. Scaling up the chemical treatment of spent oil-in-water emulsions from non-ferrous metal-processing plant. Hem. Ind., 67(1), 59-68.
17. Liu H-M., Lin Y-H., Tsai M-Y., Lin W-H. 2010. Occurrence and characterization of culturable bacteria and fungi in metalworking environments. Aerobiologia, 26, 339-350.
18. MacAdam J., Ozgencil H., Autin O., Pidou M., Temple C., Parsons S., Jefferson B. 2012. Incorporating biodegradation and advanced oxidation processes in the treatment of spent metalworking fluids. Environ. Technol., 33(24), 2741-2750.

19. Mattsby-Baltzer I., Sandin M., Ahlström B., Allenmark S., Edebo M., Falsen E. 1989. Microbial growth and accumulation in industrial metal-working fluids. Appl. Environ. Microbiol., 55, 2681-2689.

20. Mirer F.E. 2010. New Evidence on the Health Hazards and Control of Metalworking Fluids Since Completion of the OSHA Advisory Committee Report. Am. J. Ind. Med., 53(8), 792-801.

21. Rabenstein A., Koch T., Remesch M., Brinksmeier E., Kuever J. 2009. Microbial degradation of water miscible metal working fluids. Int. Biodeter. Biodegr., 63(8), 1023-1029.

22. Saha R., Donofrio R.S. 2012. Microbiology of metalworking fluids. Appl. Microbiol. Biotechnol., 92, 1119-1130.

23. Selvaraju S.B., Khan I.U.H., Yadav J.S. 2008. Specific detection and quantification of culturable and non-culturable mycobacteria in metalworking fluids by fluorescence-based methods. Lett. Appl. Microbiol., 47(5), 451-456.

24. Theaker D., Thompson I. 2010. The industrial consequences of microbial deterioration of metal working fluid. [In:] Handbook of hydrocarbon and lipid microbiology, K.N. Timmis (Ed.), Berlin Heidelberg: Springer-Verlag.

25. Tillie-Leblond I., Grennouillet F., Reboux G., Roussel S., Chouraki B., Lothois C., Dalphin J-C., Wallaert B., Millon L. 2011. Hypersensitivity pneumonitis and metalworking fluids contaminated by mycobacteria. Eur. Respir. J., 37, 640-647.

26. Trafny E.A. 2013. Microorganisms in metalworking fluids: current issues in research and management. Int. J. Occup. Med. Environ. Health., 26(1), 4-15.

27. van der Gast C.J., Whiteley A.S., Lilley A.K., Knowles C.J., Thompson I.P. 2003. Bacterial community structure and function in a metal-working fluid. Environ. Microbiol., 5(6), 453-461.

28. Veillette M., Thorne P.S., Gordon T., Duchaine C. 2004. Six month tracking of microbial growth in a metalworking fluid after system cleaning and recharging. Ann. Occup. Hyg., 48, 541-546.

29. Wang H., Reponen T., Lee S.A., White E., Grinshpun S.A. 2007. Size distribution of airborne mist and endotoxin-containing particles in metalworking fluid environments. J. Occup. Environ. Hyg., 4(3), 157-165.

33

ZINC AND LEAD IN BOTTOM SEDIMENTS AND AQUATIC PLANTS IN RIVER NAREW

Elżbieta Skorbiłowicz[1]

[1] Bialystok Technical University, Wiejska 45A, 15-351 Bialystok, Poland, e-mail: e.skorbilowicz@pb.edu.pl

ABSTRACT

Aquatic ecosystems are a valuable part of natural environment. The increasing level of pollution in waters transforming biocoenoses and other adverse effects of the impact of toxic substances have contributed to the development of biological monitoring. The aim of the study was to determine the changes in contents of zinc and lead in bottom sediments and roots of aquatic plants: *Phragmites australis* and *Acorus calamus* in the river Narew. There were 14 points on the river, from where samples of bottom sediments and plant material were collected. The contents of lead and zinc were determined by means of flame atomic absorption spectrophotometry using Varian device. It was proven that bottom sediments were characterized by low contents of zinc and lead except from two sampling points: in Bondary and Narew. Achieved results of analyzes of plant material showed a slight exceeding in the case of lead. Spatial distribution of zinc and lead contents in examined roots of plants coincided with their contents in bottom sediments, which was also confirmed by statistical analysis. It was proven that aquatic plants had greater tendency for accumulation of metals than bottom sediments.

Keywords: river, bottom sediment, aquatic plants, zinc, lead.

INTRODUCTION

Aquatic ecosystems are a valuable part of natural environment. They influence climate, water retention, maintaining an adequate level of groundwater, create ecological corridors, and furthermore play many socio-economic functions [Sadowska 2012]. The development of civilization, industry, agriculture, and transport, contributes to increased environmental pollution, including aquatic ecosystems [Samecka-Cymerman Kempers 2007, Huin et al. 2010]. Among the toxic components reaching the natural environment, heavy metals are a potential threat to the functioning of ecosystems [Helios-Rybicka 1991, Saygideger Dogan 2005, Skorbiłowicz 2012]. Increasing level of pollution in waters transforming the biocoenoses and other adverse effects of the impact of toxic substances contributed to the development of biological monitoring [Kolada Ciecierska 2008, Zimny 2006]. Monitoring assesses the extent of changes and possible negative effects on the basis of observations made on living organisms. Organisms are the fundamental indicator of all processes running within the aquatic environment [Jankowski, 1994]. For years, macrophytes have been used in bio-indication of aquatic ecosystems. Their structure is adapted for a continuous contact with water. They have an extensive system of absorption of various microelements from the environment, both necessary for life, and harmful ones [Rabajczyk, Jóźwiak 2008]

The aim of the study is to determine changes in the contents of zinc and lead in bottom sediments and roots of aquatic plants: *Phragmites australis* and *Acorus calamus* in the river Narew.

MATERIAL AND METHODS

Area of the study

The study object – river Narew flows through the north-eastern part of Poland, it has a length of 484 km, out of which 455 km flows in Poland, while the rest flows through the territory of Belarus. Narew river is a right tributary of the Vistula river. It is a typical lowland river, whose catchment covers an area of 75 175.2 km^2. Hydrological conditions occurring in the Narew river catchment are typical for lowland rivers. The areas, through which the Narew river flows, from natural and ecological point of view, are one of the most valuable in Poland. The most important of them include: Bialowieza National Park, Narew National Park, and Biebrza National Park.

Factories localized in the Narew river catchment area rely mainly on raw materials of agricultural origin: sugar mills, slaughterhouses, dairies, as well as fruit and vegetable processing plants. Among other branches of industry in Podlasie, they are light and wood industry facilities, but agro-food industry is of the dominant importance.

Industrial and municipal wastewater treatment plants functioning in the catchment, in most cases meet the requirements specified in the water-legal permits, however, when assessing the overall quality of the water, the impact of wastewaters from larger towns located within the catchment, is visible [WIOŚ 2012].

Analyzing the area crossed by the Narew river, it may be noted that municipalities are essentially agricultural. Consequence of this is the surface runoff from fields, which include fertilizers and plant protection means (that contain a group of active compounds of arsenic, copper, mercury, zinc, or lead).

Table 1. List of measurement points localized on the Narew river

Point	Nearest town	Kilometer of the river course
1	Bondary	431.7
2	Narew	410.0
3	Suraz	355.3
4	Zoltki	302.2
5	Łomza	203.5
6	Nowogrod	181.1
7	Ostroleka	147.0
8	Rozan	117.0
9	Gnojno	78.0
10	Pultusk	63.5
11	Lubienica	58.0
12	Wierzbica	41.0
13	Dede	20.0
14	Nowy Dwor Mazowiecki	3.0

Field and laboratory works

Fourteen points were assigned on the river Narew, where samples of bottom sediments and plant material were collected. They were located in such a way to allow for analysis of changes in water environment quality in the river, taking into account the tributaries and wastewater discharge points. Table 1 shows the location of the measurement points. Environmental sampling took place in summer 2012. Bottom sediments were collected in the coastal zone, where the deposition of suspended material occurs. Several individual surface samples of bottom sediments (from a depth of 10 cm) from underneath the water were collected in every selected point. The samples were then dried in air till the "air-dried" and stored until determinations [Lis and Pasieczna 1995]. Prior to chemical analysis, bottom sediment sample was dried at 40 °C and sieved through a nylon 0.2 mm mesh. Bottom sediments were mineralized using nitric acid with addition of hydrogen peroxide [Salminen et al. 2005] in a closed microwave system CEM Mars-5. An analysis of the granulometric composition of sediments by means of sieve method was also performed, involving the separation of raw air-dry sediment samples for granulometric fractions in an appropriate set of sieves (hole size of: 2.0 mm, 1.0 mm, 0.2 mm, 0.1 mm, 0.063 mm, 0.02 mm) connected to a mechanical shaker. Tests of organic matter were carried out by annealing at a temperature of 550 °C, while the reaction of sediments - potentiometrically in water. Samples of aquatic plants (*Phragmites australis* and *Acorus calamus*) were collected from the same area as bottom sediments. Each plant test sample was a combination of at least 5 individual samples. After arrival in the laboratory, harvested plant roots were thoroughly washed with tap water to remove sediments and periphyton and rinsed with distilled water. Then they were dried till the constant weight at 80 °C [Mazej and Germ 2009]. After drying, samples were homogenized and mineralized in nitric acid with hydrogen peroxide using a microwave system CEM Mars-5. According to Du Laing et al. [2003], the above method is most suitable for determination of heavy metals in *Phragmites*. The contents of lead and zinc were determined by flame atomic absorption spectrophotometry on a Varian device.

Results of the content of tested metals were given in relation to air-dry sediments and compared with literature data as well as arithmetic means of these metals contents in bottom sediments for Poland (fraction <0.2 mm) [Lis and Pasieczna 1995] and Europe (fraction <0.15 mm) (Salminen et al. 2005). To assess the extent of sediment contamination by heavy metals, proposed classification of aquatic sediments in Poland based on geochemical criteria (Bojakowska and Sokołowska 1998) and on the basis of threshold values taking into account the harmful effects of pollutants accumulated in sediments to aquatic organisms, was applied [Bojakowska 2001]. The results of zinc and lead contents in common reed (*Phragmites australis*), and ordinary calamus (*Acorus calamus*) were based on the dry weight of plants and compared with literature data. In contrast, the physiological standard of metals contents for plants was given according to data presented by Kabata-Pendias, Pendias [1999]. Levels of zinc and lead bio-concentration indicator as a ratio of these metals in examined roots of aquatic plants to their contents in bottom sediments, were calculated.

A licensed version of Statistica 10 software was used for statistical analysis of test results. Arithmetic means, medians, standard deviations, and Spearman correlation coefficients were calculated. The results were also subject to Ward's agglomeration cluster analysis based on the concept of a distance of objects or variables in a multidimensional space.

RESULTS AND DISCUSSION

Particle size analysis of bottom sediments from the Narew river showed characteristics of loose sand. Only at the measuring point Narew it was found that there was a poor loamy sand. Two fractions showed the largest percentage: 1.0-0.2 mm (mean 55.75 ± 22.28%) and 0.2-0.1 mm (mean 25.88 ± 16.97%). Analyzed fractions 0.063-0.02 mm and <0.02 mm were absent at points in Bondary, Suraz, Debe, and Nowy Dwor Mazowiecki (Table 2). The content of organic matter in samples of bottom sediments ranged from 0.05% to 1.59%. The highest content of organic matter was recorded in samples collected in the towns of Gnojno (1.59%) and Novgorod (1.17%). The lowest content not exceeding 0.3% was found at the measuring points: Suraz, Lomza, Rozan, Pultusk, Lubienica, Wierzbica, Debe, and Nowy Dwor Mazowiecki (Table 3). The pH of bottom sediment from the river Narew in most cases was neutral or slightly alkaline in the range of pH 7.27 to 8.10.

Zinc content in dry mass of bottom sediments changed over a wide range. The lowest value was recorded at point Debe (25.9 mg · kg^{-1}), while the largest in points Bondary (175.8 mg · kg^{-1}) and Narew (127.1 mg · kg^{-1}). At points from Ostroleka to the Nowy Dwor Mazowiecki, there were no specific trends related to zinc content in bottom sediments and its contents were at the level of geochemical background. The amount of lead in bottom sediments varied from 5.9 mg · kg^{-1} up to 40.5 mg · kg^{-1}. Usually higher contents of this element were observed at two measurement points: Bondary (39.8 mg · kg^{-1}) and Narew (40.5 mg · kg^{-1}) and four times reduced ones starting from the measurement point in Suraz. The increased content was recorded only in point Wierzbica (19.5 mg · kg^{-1}).

Analyzing the arithmetic mean values of zinc content in bottom sediments in Poland amounting to 247 mg · kg^{-1} and Europe 120 mg · kg^{-1}, it is clear that significantly lower contents of this element were achieved in the present study (average 55.0 mg · kg^{-1}). Bojakowska and Sokołowska [1998] reported that lead content in uncontaminated sediments generally did not exceed 30 mg · kg^{-1}. The pattern of lead variability in bottom sediments substantially coincides with the geochemical background of this element in soils, and the Geochemical Atlas of Poland indicates that the arithmetic mean for lead is 68 mg · kg^{-1} (Lis and Pasieczna 1995), while for Europe 38.6 mg · kg^{-1} (Salmin et al. 2005). Considerably lower contents were obtained in this study (average 13.6 mg · kg^{-1}).

Comparing the contents of analyzed elements in bottom sediments of the Narew river, similar or lower metal concentrations in the sediments from European and Asian rivers, were recorded, which ranged (mg · kg^{-1}): 15-160 (Pb), and 37-303 (Zn) [Demirak et al. 2006, Lin et al. 2008, Skorbiłowicz 2012, 2014].

Within the same measurement points, plant material consisting of reed roots and calamus was collected, in which contents of zinc and lead were also determined. Zinc content at common reed growing in the river Narew ranged from 56.5 mg · kg^{-1} to 205.2 mg · kg^{-1}. The highest levels were recorded in Bondary (205.2 mg · kg^{-1}) and Narew

Table 2. Contents of granulometric fractions in bottom sediments of the river Narew

River	Measurement point	Content of granulometric fraction [% of weights]						
	Bondary	>2 mm	2.0–1.0 mm	1.0–0.2 mm	0.2–0.1 mm	0.1–0.063 mm	0.063–0.02 mm	<0.02 mm
	Narew	0.14	6.96	74.76	14.52	3.6	n.w.	n.w.
	Suraz	1.60	7.80	28.62	23.42	18.28	20.24	n.w.
	Zoltki	n.w.	2.68	59.04	32.48	5.76	n.w.	n.w.
	Łomza	n.w.	2.70	52.90	35.20	5.90	3.18	0.08
	Nowogrod	1.72	5.60	56.42	26.78	7.64	1.84	n.w.
	Ostroleka	3.68	8.90	30.62	19.84	34.10	1.84	1.00
Narew	Rozan	n.w.	1.02	19.04	64.94	11.64	2.86	0.44
	Gnojno	0.18	1.32	21.00	54.84	22.48	0.14	n.w
	Pultusk	1.66	3.02	60.56	27.82	3.82	3.12	n.w.
	Lubienica	1.70	4.36	75.98	17.00	0.74	0.16	n.w.
	Wierzbica	4.38	10.38	67.10	15.54	1.90	0.70	n.w.
	Dede	4.20	1.19	84.56	8.32	0.44	0.08	n.w.
	Nowy Dwor Mazowiecki	6.38	11.80	78.52	2.76	0.48	n.w.	n.w
Fundamental statistics								
Minimum		0.14	1.02	19.04	2.76	0.44	0.08	–
Maximum		6.38	11.80	84.56	64.94	34.10	20.24	–
Arithmetic mean		2.59	5.12	55.75	25.88	8.53	3.41	–
Median		1.72	4.19	58.80	21.63	4.79	1.84	–
Standard deviation		1.91	3.53	22.28	16.97	9.93	6.04	–

n.w. – not detected by available method

Table 3. Results of tested indicators in bottom sediments and roots of *Phragmites australis* and *Acorus calamus*

River	Measurement point	Bottom sediments				*Phragmites Australis*				*Acorus calamus*			
		pH in water	Org. matter	Zn	Pb	Zn	Pb	BCF		Zn	Pb	BCF	
			%	mg·kg^{-1}DM		mg·kg^{-1}DM		Zn	Pb	mg·kg^{-1}DM		Zn	Pb
	Bondary	7.47	0.38	175.8	39.8	205.2	59.5	1.16	1.49	232.3	51.6	1.32	1.29
	Narew	7.51	0.59	127.1	40.5	151.3	34.0	1.19	0.84	147.1	45.5	1.15	1.12
	Suraz	7.67	0.21	42.3	10.3	68.9	48.0	1.62	4.66	94.5	52.3	2.23	5.01
	Zoltki	7.99	0.36	58.9	11.2	74.7	29.3	1.23	2.59	102.1	32.5	1.74	2.90
	Łomza	8.11	0.19	41.2	9.5	70.3	27.1	1.76	2.85	91.3	29.3	2.21	3.08
	Nowogrod	7.39	1.17	62.3	8.3	113.6	23.0	1.82	2.78	121.3	32.1	1.94	3.86
	Ostroleka	7.52	0.57	43.1	7.5	65.7	40.2	1.52	5.34	84.7	16.5	1.96	2.2
Narew	Rozan	7.27	0.29	29.9	6.1	56.5	56.5	1.89	9.26	67.3	36.1	2.25	5.91
	Gnojno	7.53	1.59	35.8	5.9	91.2	21.0	2.54	3.55	79.1	19.3	2.21	3.27
	Pultusk	8.09	0.28	41.4	11.5	96.4	54.0	2.32	4.69	102.3	30.5	2.47	2.65
	Lubienica	7.81	0.14	26.7	6.5	84.3	27.1	3.15	4.16	86.9	18.5	3.25	2.85
	Wierzbica	7.91	0.06	27.3	19.5	65.2	46.0	2.39	2.35	76.3	54.5	2.79	2.79
	Dede	7.99	0.05	25.9	5.9	79.1	19.5	1.12	3.31	62.1	12.0	2.39	2.03
	Nowy Dwor Mazowiecki	7.68	0.11	33.1	7.8	91.3	18.0	2.76	2.31	67.6	15.0	2.04	1.92
Fundamental statistics													
Minimum		7.27	0.05	25.9	5.9	56.5	18.0	1.12	0.84	62.1	12.0	1.15	1.12
Maximum		8.11	1.59	175.8	40.5	205.2	59.5	3.15	9.26	232.3	54.5	3.25	5.91
Arithmetic mean		–	0.43	55.0	13.6	93.8	35.9	1.89	3.58	101.2	31.9	2.14	2.92
Median		–	0.29	41.3	8.9	81.7	31.6	1.79	3.08	89.1	31.6	2.21	2.82
Standard deviation		–	0.44	43.35	11.78	40.5	14.44	0.64	2.06	44.5	14.56	0.53	1.31

(151.3 mg · kg^{-1}), while content of this element in bottom sediments decreased to a value below 100 mg · kg^{-1} (natural, optimal value according to Kabata-Pendias, Pendias [1999], the point in Nowogrod was the exception (113.6 mg · kg^{-1}). Similar trends were found in calamus. The highest content was recorded in Bondary (232.2 mg · kg^{-1}) and Narew (147.1 mg · kg^{-1}), and a pronounced decrease in the zinc content at sweet flag was visible from the measurement point Ostroleka, which ranged from 67.3 mg · kg^{-1} (Rozan) to 102.3 mg · kg^{-1} (Pultusk). Lower concentrations of zinc in reed occurred in lakes of Suwalki Landscape Park and they ranged from 5.719 to content 27.689 mg Zn · kg^{-1} DM [Dobicki 2004].

The lead content in roots of common reed varied from 18.0 mg · kg^{-1} to 59.5 mg · kg^{-1} , while in calamus 12.0 mg · kg^{-1} to 54.5 mg · kg^{-1} with the greatest amount of this element in the case of common reed in the following towns: Bondary (59.5 mg · kg^{-1}), Rozan (56.5 mg · kg^{-1}), and Pultusk (54.0 mg · kg^{-1}). Sweet flag accumulated the largest quantities of lead at the following measurement points: Wierzbica (54.5 mg · kg^{-1}), Suraz (52.3 mg · kg^{-1}), and Bondary (51.6 mg · kg^{-1}). Kabata-Pendias and Pendias [1999] indicated the excessive amounts at levels ranging 30–300 mg Pb · kg^{-1} DM. None of the analyzed samples showed the exceeding the maximum value of 300 mg · kg^{-1}, whereas the value 30 mg · kg^{-1} was exceeded by about 50% of samples tested. Significantly lower values of Pb were recorded in the upper Narew river and its tributaries in 2002–2007. Average contents of this element in roots of reeds and calamus did not exceed 15 mg · kg^{-1} DM [Skorbiłowicz 2012].

To estimate the size of the impurities in the trophic chains, the bio-concentration factor (BCF) is defined, which uses information about the concentrations of substances in an organism and the outside environment. A high bio-concentration factor can prove the increased bioavailability of a given substance for a particular organism. Aquatic organisms living in contaminated environment easily absorb toxic substances, even in small concentrations [Sadowska 2012]. The highest bioaccumulation appears to be for heavy metals due to their slow metabolism in an organism [Lamdrum et al. 1996]. Bio-concentration factor for zinc in common reed reached the mean value of 1.89 ± 0.64 and in 80% it ranged from 1.1 to 2.4. Its highest values were reported at measurement points: Gnojno (2.54), Nowy Dwor Mazowiecki (2.76), and Lubienica (3.15). A similar range of BCF values occurred in the calamus: also 80% of the results ranged within 1.1–2.4 at slightly higher mean value of 2.14 ± 0.53. Also, the largest accumulation of zinc in calamus occurred in Lubienica (3.25). The test plants showed higher bio-concentration factor for lead at a large diversity of results with an average of: reed - 3.58 ± 2.06, sweet flag – 2.92 ± 1.31. It was shown that lead was the most accumulated in the common reed in Rozan (9.26), while calamus at the point in Suraz (5.01).

The results allow to conclude that at the measurement points Bondary and Narew, the accumulation conditions for elevated zinc and lead amounts occurred in bottom sediments and roots of plants. The metal contents at these points were two-three times higher than at neighboring points. This may be an effect of supply of unspecified substances from the area of Belarus to the river Narew waters. Siemianowka reservoir may also have large impact on the quality of river at the point Bondary. Algal blooms are frequent in the reservoir. The bottom of a shallow reservoir with a long water exchange time consists of marshland peat soils. However, in the Narew village, cleaned and treated household and industrial wastewater, as well as storm water, including agricultural-type pollution, are discharged into the river. Considering the zinc and lead concentration in bottom sediments and plants along the entire length of the river, a downward trend is visible (Figures 1–6).

The test material, in the form of numerical contents, was subject to statistical analysis. The analyzes showed a significant statistical correlation of zinc content in bottom sediments vs. its quantity in the roots of sweet flag ($r = 0.86$), organic matter ($r = 0.72$), and granulometric factions 0.1–0.063 mm ($r = 0.56$) and 0.063–0.02 mm ($r = 0.70$). It was also statistically confirmed that zinc content in the roots of *Acorus calamus* depends on the amount of organic matter in the sediment ($r = 0.54$), and zinc content in *Phragmites australis* ($r = 0.64$). Also significant correlations between the content of Pb in sediments vs. its content in the roots of *Acorus calamus* ($r = 0.71$) were found. There were positive correlations between the content of organic matter in bottom sediment and the percentage of two granulometric fractions 0.1–0.063 mm ($r = 0.69$) and 0.063–0.020 ($r = 0.70$) (Table 4).

During the analysis of the data obtained, Ward cluster analysis (CA) was carried out in relation to the variation of zinc and lead contents in bottom

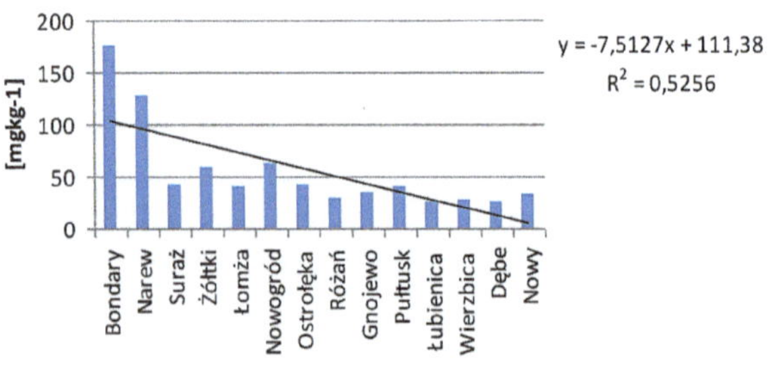

Figure 1. Zinc content in bottom sediments from the river Narew

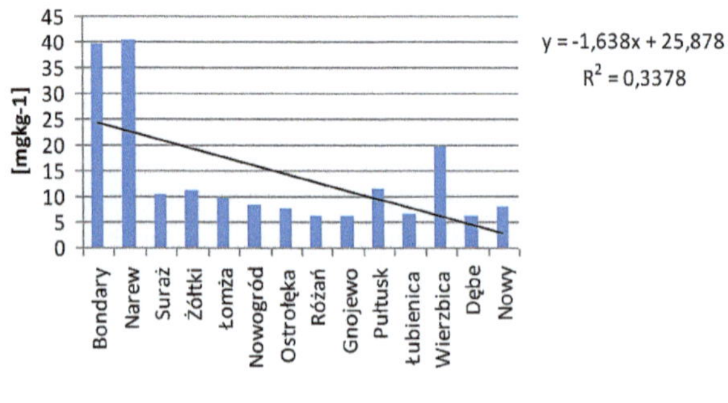

Figure 4. Lead content in bottom sediments from the river Narew

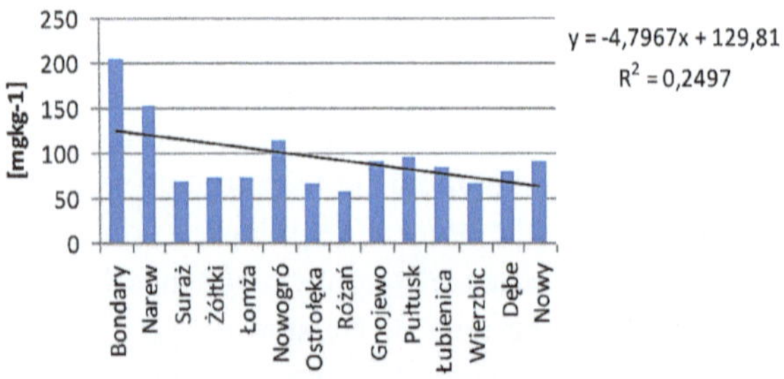

Figure 2. Zinc content in calamus from the river Narew

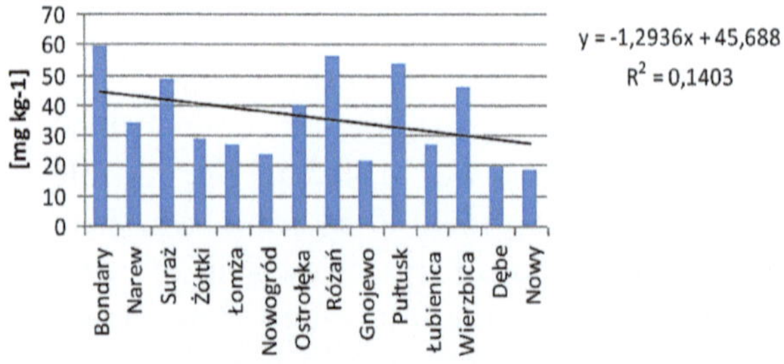

Figure 5. Lead content in calamus from the river Narew

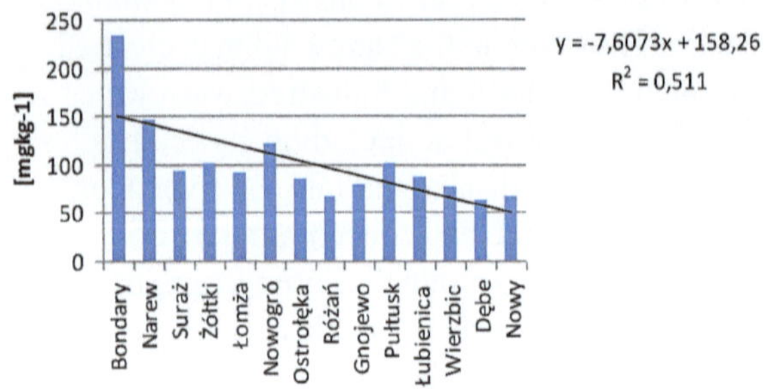

Figure 3. Zinc content in reed from the river Narew

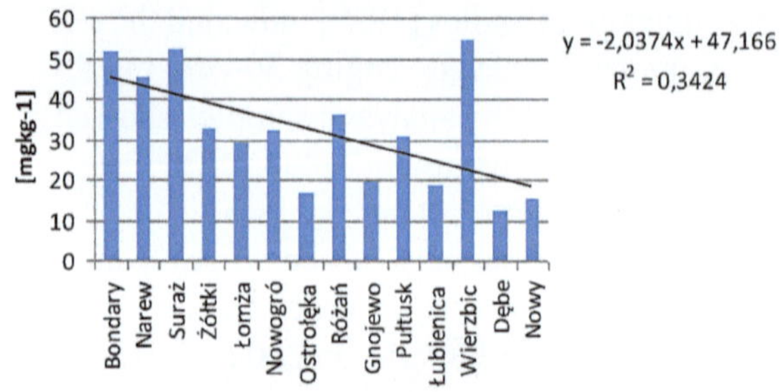

Figure 6. Lead content in reed from the river Narew

Table 4. Dependencies between zinc and lead contents in bottom sediments and plant roots vs. factors determining their concentrations ($p<0.05$)

Dependencies	Spearman coefficient
Zn^{o} – Zn^{Tat}	0.86
Pb^{o} – Pb^{Tat}	0.71
Zn^{o} – sub. org.	0.72
Zn^{o} – $F^{0,1-0,063}$	0.56
Zn^{o} – $F^{0,063-0,02}$	0.70
Zn^{Tat} – Zn^{Trz}	0.63
Zn^{Tat} – sub. org	0.54
Zn^{Trz} – $F^{2,0-1,0}$	0.71
Sub. org – $F^{0,1-0,063}$	0.69
Sub. org – $F^{0,063-0,02}$	0.70

Comments: ^{o} – bottom sediments; ^{Tat} – calamus; ^{Trz} – reed; F – granulometric fractions.
Sub. org organic matter

sediments and roots of reed and calamus, as well as the tested grain size of sediments, pH of sediments and organic matter. The Ward model consists of two main groups and four subgroups. The analysis of the first main group showed that lead content in reed and calamus roots was associated with the fraction of 1.0–0.2 mm, the amount of which was the highest in studied sediment – about 56% of weights. Besides the first major group, the diagram also illustrates the relationship of zinc content in reeds and calamus vs. its amount in bottom sediments. The second cluster indicates that the accumulation of lead in sediments is mainly connected with the fraction of 0.1–0.063 mm and 0.2–0.1 mm (together about 65% of weights), as well as with organic matter, pH, and fractions 2.0–1.0 mm and 0.063–0.02 mm (Figure 7).

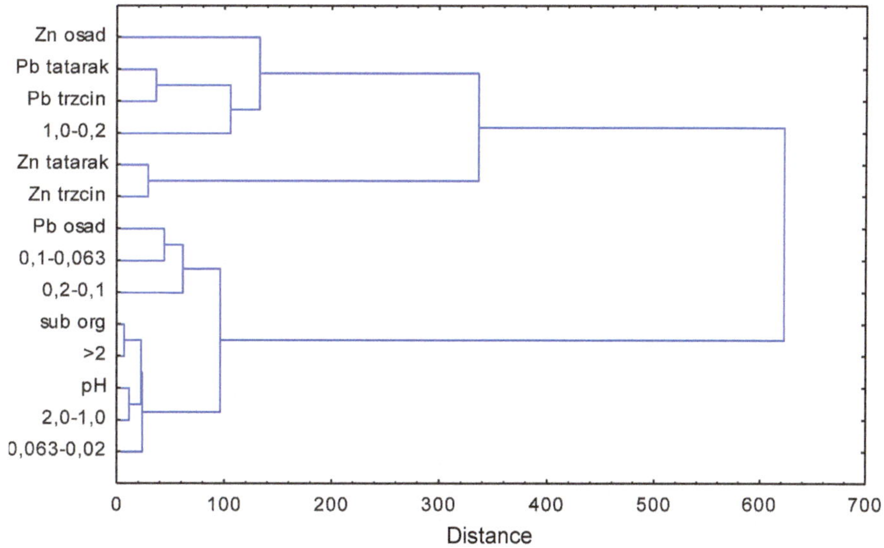

Figure 7. Ward cluster analysis (CA) results from river Narew

CONCLUSIONS

1. It was proved that zinc and lead in bottom sediments of the river Narew occurred in small amounts except for two measurements points: in Bondary (175.8 mgZn·kg^{-1}, 39.8 mgPb·kg^{-1}) and Narew (127.1 mgZn·kg^{-1}, 40.5 mgPb·kg^{-1}).
2. The achieved analysis results of plant material indicate a slight exceeding in the case of lead.
3. Zinc and lead concentrations in bottom sediments and plant roots considerably depended on the measurement point. Spatial distribution of these metals contents in the tested plant roots correlates with their contents in sediments. It was proved that aquatic plants are characterized by a greater tendency to accumulate zinc and lead rather than bottom sediments.
4. Sediments particle size and organic matter are factors that affect the zinc and lead contents in bottom sediments also in test plants, which was statistically confirmed.

REFERENCES

1. Bojakowska I., 2001. Kryteria oceny zanieczyszczenia osadów wodnych. Przegląd Geologiczny 49 (3), 213–218.
2. Bojakowska I., Sokołowska G., 1998. Geochemiczne klasy czystości osadów wodnych. Przegląd Geologiczny 46 (1), 49–54.
3. Demirezen D., Aksoy A., 2004. Accumulation of heavy metals in Typha angustifolia (L.) and Potamogeton pectinatus (L.) living in Sultan Marsh (Kayseri, Turkey). Chemosphere 56, 685–696.
4. Dobicki W., 2004. Biodostępność metali ciężkich w środowisku jezior Suwalskiego Parku Krajobrazowego. Wydawnictwo Akademii Rolniczej we Wrocławiu, pp. 99.
5. Du Laing G, Tack fmg, Verloo Mg., 2003. Performance of selected destruction methods for the determination of heavy metals in reed plants (Phragmites australis). Anal Chim Acta, 497, 191–198.
6. Helios-Rybicka E., 1991. Akumulacja i mobilizacja metali ciężkich w osadach środowiska wodnego: osady datowane jako wskaźnik chronologiczny. Konferencja: Geologiczne aspekty ochrony środowiska, Kraków, 18–23.
7. Hu J., Zheng A., Pei D., Shi G., 2010. Bioaccumulation and chemical forms of cadmium, copper and lead in aquatic plants. Brazilian Archives of Biology and Technology Aninternational Journal, 53 (1), 235–240.
8. Jankowski W., 1994. Zastosowanie bioindykacji w praktyce monitoring środowiska na przykładzie północno-wschodniej Polski, Państwowa Inspekcja Ochrony Środowiska, Warszawa.
9. Kabata-Pendias, Pendias, 1999. Biogeochemia pierwiastków śladowych, Wydawnictwo Naukowe PWN, Warszawa.
10. Kolada A., Ciecierska H., 2008. Terenowe metody badania makrofitów w jeziorach w świetle monitoringu biologicznego wód zgodnego z Ramową Dyrektywą Wodną, Ochrona Środowiska i Zasobow Naturalnych, 37, 9–23.
11. Landrum P., Harkey G., Kukkonen J., 1996. Evaluation of organic contaminant exposure in aquatic organisms: The significance of bioconcentration adn bioaccumulation. [In:] Newman M.C., Jagoe C.H.

(eds.) "Ecotoxicology. A hierarchical treatment", CRC/ Lewis Publishers, Nowy Jork, 85–131.

12. Lin CH., He M., Zhou Y., Guo W., Yang Z., 2008. Distribution and contamination assessment of heavy metals in sediment of the Second Songhua River, China, Environ Monit. Assess., 137, 329–342

13. Lis J., Pasieczna A., 1995. Atlas geochemiczny Polski w skali 1: 2 500 000, Państw. Inst. Geol., Warszawa, pp. 72.

14. Mazej Z., Germ, M., 2009. Trace element accumulation and distribution in four aquatic macrophytes. Chemosphere 74, 642–647.

15. Rabajczyk A., Jóźwiak M.A., 2008. Możliwości wykorzystania makrofitów jako bioindykatorów metali ciężkich zdeponowanych w osadach dennych (The possibilities of using macrophytes as bioindicators of heavy metals occuring in sediments). Kieleckie Towarzystwo Naukowe, Monitoring Środowiska Przyrodniczego 9, 19–26.

16. Sadowska U. 2012. Ranga bioindykacji w ekotoksykologii wód, Studia Ecologiae et Bioethicae. UKSW, 10(2), 33–52.

17. Salminen R. (Ed.), 2005. Geochemical atlas of Europe. Part 1 - Background Information, Methodology and Maps.

18. Samecka-Cymerman A., Kempers A.J., 2007. Heavy metals in aquatic macrophytes from two small rivers polluted by urban, agricultural and textile industry sewages SW Poland. Arch. Environ. Contam. Toxicol. 53, 198–206.

19. Saygideger S., Dogan M., 2005. Variation of Lead, Cadmium, Copper, and Zinc in Aquatic Macrophytes from the Seyhan River, Adana, Turkey. Bull. Environ. Contam. Toxicol. 74, 545–551.

20. Skorbiłowicz E. 2014. Assessment of heavy metals contients in bottom sediments of Bug River. Journal of Ecological Engineering 15 (3), 82–89.

21. Skorbiłowicz E. 2012. Studia nad rozmieszczeniem niektórych metali w środowisku wodnym zlewni górnej Narwi.

22. WIOŚ Białystok. Ocena stanu ekologicznego, potencjału ekologicznego i stanu chemicznego rzek województwa podlaskiego w roku 2011, 2012.

23. Zimny H., 2006. Ekologiczna ocena stanu środowiska. Bioindykacja i biomonitoring. Agencja Reklamowo-Wydawnicza Arkadiusz Grzegorczyk, Warszawa.

Permissions

All chapters in this book were first published in JEE, by Polish Society of Ecological Engineering (PTIE); hereby published with permission under the Creative Commons Attribution License or equivalent. Every chapter published in this book has been scrutinized by our experts. Their significance has been extensively debated. The topics covered herein carry significant findings which will fuel the growth of the discipline. They may even be implemented as practical applications or may be referred to as a beginning point for another development.

The contributors of this book come from diverse backgrounds, making this book a truly international effort. This book will bring forth new frontiers with its revolutionizing research information and detailed analysis of the nascent developments around the world.

We would like to thank all the contributing authors for lending their expertise to make the book truly unique. They have played a crucial role in the development of this book. Without their invaluable contributions this book wouldn't have been possible. They have made vital efforts to compile up to date information on the varied aspects of this subject to make this book a valuable addition to the collection of many professionals and students.

This book was conceptualized with the vision of imparting up-to-date information and advanced data in this field. To ensure the same, a matchless editorial board was set up. Every individual on the board went through rigorous rounds of assessment to prove their worth. After which they invested a large part of their time researching and compiling the most relevant data for our readers.

The editorial board has been involved in producing this book since its inception. They have spent rigorous hours researching and exploring the diverse topics which have resulted in the successful publishing of this book. They have passed on their knowledge of decades through this book. To expedite this challenging task, the publisher supported the team at every step. A small team of assistant editors was also appointed to further simplify the editing procedure and attain best results for the readers.

Apart from the editorial board, the designing team has also invested a significant amount of their time in understanding the subject and creating the most relevant covers. They scrutinized every image to scout for the most suitable representation of the subject and create an appropriate cover for the book.

The publishing team has been an ardent support to the editorial, designing and production team. Their endless efforts to recruit the best for this project, has resulted in the accomplishment of this book. They are a veteran in the field of academics and their pool of knowledge is as vast as their experience in printing. Their expertise and guidance has proved useful at every step. Their uncompromising quality standards have made this book an exceptional effort. Their encouragement from time to time has been an inspiration for everyone.

The publisher and the editorial board hope that this book will prove to be a valuable piece of knowledge for researchers, students, practitioners and scholars across the globe.

List of Contributors

Joannna Szczykowska
Department of Technology in Environmental Engineering and Protection, Faculty of Civil and Environ-mental Engineering, Bialystok University of Technology, ul. Wiejska 45 b, 15-351 Białystok, Poland

Anna Siemieniuk
Department of Technology in Environmental Engineering and Protection, Faculty of Civil and Environ-mental Engineering, Bialystok University of Technology, ul. Wiejska 45 b, 15-351 Białystok, Poland

Józefa Wiater
Department of Technology in Environmental Engineering and Protection, Faculty of Civil and Environ-mental Engineering, Bialystok University of Technology, ul. Wiejska 45 b, 15-351 Białystok, Poland

Krzysztof Pulikowski
Institute of Environmental Engineering, Wrocław University of Environmental and Life Sciences, pl. Grunwaldzki 24, 50-363 Wrocław, Poland

Katarzyna Pawęska
Institute of Environmental Engineering, Wrocław University of Environmental and Life Sciences, pl. Grunwaldzki 24, 50-363 Wrocław, Poland

Aleksandra Bawiec
Institute of Environmental Engineering, Wrocław University of Environmental and Life Sciences, pl. Grunwaldzki 24, 50-363 Wrocław, Poland

Mariusz Kucharski
Institute of Soil Science and Plant Cultivation - State Research Institute, Department of Weed Science and Tillage Systems, Orzechowa 61, 50-540 Wrocław, Poland

Olga Kalitowska
Institute of Soil Science and Plant Cultivation - State Research Institute, Department of Weed Science and Tillage Systems, Orzechowa 61, 50-540 Wrocław, Poland

Magdalena Daria Vaverková
Faculty of Agronomy, Mendel University in Brno, Zemědělská 1665/1, 613 00 Brno, Czech Republic

Dana Adamcová
Faculty of Agronomy, Mendel University in Brno, Zemědělská 1665/1, 613 00 Brno, Czech Republic

Kalina Orłowska
Department of Agricultural and Environmental Chemistry, Faculty of Agriculture and Economics, University of Agriculture in Krakow, 21 A. Mickiewicza Ave., 30-120 Kraków, Poland

Monika Mierzwa-Hersztek
Department of Agricultural and Environmental Chemistry, Faculty of Agriculture and Economics, University of Agriculture in Krakow, 21 A. Mickiewicza Ave., 30-120 Kraków, Poland

Michał Kopeć
Department of Agricultural and Environmental Chemistry, Faculty of Agriculture and Economics, University of Agriculture in Krakow, 21 A. Mickiewicza Ave., 30-120 Kraków, Poland

Katarzyna Ignatowicz
Department of Technology in Engineering and Environmental Protection, Białystok University of Technology, Wiejska 45a, 15-351 Białystok, Poland

Jacek Piekarski
Department of Water-Sludge Technology and Waste Utilization, Koszalin University of Technology, Śniadeckich 2, 75-543 Koszalin, Poland

Bożena Dębska
Department of Environmental Chemistry, University of Technology and Life Sciences in Bydgoszcz, 6 Bernardyńska Str., 85-029 Bydgoszcz, Poland

Magdalena Banach-Szott
Department of Environmental Chemistry, University of Technology and Life Sciences in Bydgoszcz, 6 Bernardyńska Str., 85-029 Bydgoszcz, Poland

Ewa Rosa
Department of Environmental Chemistry, University of Technology and Life Sciences in Bydgoszcz, 6 Bernardyńska Str., 85-029 Bydgoszcz, Poland

Mirosław Kobierski
Department of Soil Science and Soil Protection, University of Technology and Life Sciences in Bydgoszcz, Bernardyńska 6/8, 85-029 Bydgoszcz, Poland

Ewa J. Lipińska
Subcarpathian Voivodeship Inspector for Environmental Protection Voivodeship Inspectorate for Environmen-tal Protection in Rzeszów, Gen. M. Langiewicza 26 Str., 35-101 Rzeszów, Poland

Jacek Kubiak
Department of Hydrochemistry and Aquatic Biological Resources, West Pomeranian University of Techno- logy in Szczecin, Kazimierza Królewicza 4, 71-550 Szczecin, Poland

Sylwia Machula
Department of Hydrochemistry and Aquatic Biological Resources, West Pomeranian University of Techno- logy in Szczecin, Kazimierza Królewicza 4, 71-550 Szczecin, Poland

Agnieszka Kisło
Department of Technology in Engineering and Environmental Protection, Faculty of Civil and Environmental Engineering, University of Technology in Bialystok, 45A Wiejska Str., 15-351 Bialystok, Poland

Iwona Skoczko
Department of Technology in Engineering and Environmental Protection, Faculty of Civil and Environmental Engineering, University of Technology in Bialystok, 45A Wiejska Str., 15-351 Bialystok, Poland

Batool Ahansazan
Departmen of Chemical Industries, Valiasr Technical College of Tehran, Sepah Str., Tehran, Iran

Baharan Moazenipour
Technical Faculty, University of Tehran, Enghelab Ave., Tehran, Iran

Elżbieta Skorbiłowicz
Department of Technology in Engineering and Environmental Protection, Technical University in Bialystok, Wiejska 45A, 15-351 Bialystok, Poland

Anna Samborska
Department of Technology in Engineering and Environmental Protection, Technical University in Bialystok, Wiejska 45A, 15-351 Bialystok, Poland

Janina Piekutin
Bialystok University of Technology, 45A Wiejska Str., 15-351 Bialystok, Poland

Mirosław Wyszkowski
Department of Environmental Chemistry, University of Warmia and Mazury in Olsztyn, 4 Łódzki Sq., 10-727 Olsztyn, Poland

Arkadiusz Chełstowski
Department of Environmental Chemistry, University of Warmia and Mazury in Olsztyn, 4 Łódzki Sq., 10-727 Olsztyn, Poland

Zdzisław Ciećko
Department of Environmental Chemistry, University of Warmia and Mazury in Olsztyn, 4 Łódzki Sq., 10-727 Olsztyn, Poland

Radosław Szostek
Department of Environmental Chemistry, University of Warmia and Mazury in Olsztyn, 4 Łódzki Sq., 10-727 Olsztyn, Poland

Mirosław Skorbiłowicz
Białystok University of Technology, Wiejska 45A, 15-351 Białystok, Poland

Piotr Ofman
Białystok University of Technology, Wiejska 45A, 15-351 Białystok, Poland

Dorota Porowska
Institute of Hydrogeology and Engineering Geology, Faculty of Geology, University of Warsaw, Żwirki i Wigury 93, 02-089 Warsaw, Poland

Agnieszka Józefowska
Department of Soil Science and Soil Protection, University of Agriculture in Krakow, 21 Mickiewicza Ave., 31-120 Kraków, Poland

Anna Miechówka
Department of Soil Science and Soil Protection, University of Agriculture in Krakow, 21 Mickiewicza Ave., 31-120 Kraków, Poland

Michał Gąsiorek
Department of Soil Science and Soil Protection, University of Agriculture in Krakow, 21 Mickiewicza Ave., 31-120 Kraków, Poland

Paweł Zadrożny
Department of Soil Science and Soil Protection, University of Agriculture in Krakow, 21 Mickiewicza Ave., 31-120 Kraków, Poland

Barbara Filipek-Mazur
Department of Agricultural and Environmental Chemistry, University of Agriculture in Kraków, A. Mickiewicza 21, 31-120 Kraków, Poland

Monika Tabak
Department of Agricultural and Environmental Chemistry, University of Agriculture in Kraków, A. Mickiewicza 21, 31-120 Kraków, Poland

Olga Gorczyca
Department of Agricultural and Environmental Chemistry, University of Agriculture in Kraków, A. Mickiewicza 21, 31-120 Kraków, Poland

Katarzyna Malinowska
Department of Plant Physiology and Biochemistry, Faculty of Environmental Management and Agriculture, West Pomeranian University of Technology in Szczecin, 17 Słowackiego Str., 71-434 Szczecin, Poland

Ryszard Malinowski
Department of Soil Science, Faculty of Environmental Management and Agriculture, West Pomeranian University of Technology in Szczecin, 17 Słowackiego Str., 71-434 Szczecin, Poland

Andrzej Stera
Department of Plant Physiology and Biochemistry, Faculty of Environmental Management and Agriculture, West Pomeranian University of Technology in Szczecin, 17 Słowackiego Str., 71-434 Szczecin, Poland

Jacek Antonkiewicz
Department of Agricultural and Environmental Chemistry, University of Agriculture in Krakow, 21 Mickiewicza Ave., 31-120 Krakow, Poland

Mariusz Głowacki
Department of Land Protection, Opole University, Oleska 22, 45-052 Opole, Poland

Tomasz Ciesielczuk
Department of Land Protection, Opole University, Oleska 22, 45-052 Opole, Poland

Veranika Sivitskaya
Department of Environmental Chemistry, University of Warmia and Mazury in Olsztyn, 4 Łódzki Sq., 10-727 Olsztyn, Poland

Justyna Rybak
Department of Environmental Protection, Wroclaw University of Technology, Wybrzeże Wyspiańskiego 27, 50-370 Wrocław, Poland

Jan Kalembkiewicz
Department of Inorganic and Analytical Chemistry, Faculty of Chemistry, Rzeszow University of Technology, 6 Powstancow Warszawy Ave., 35-959 Rzeszow, Poland

Elżbieta Sitarz-Palczak
Department of Inorganic and Analytical Chemistry, Faculty of Chemistry, Rzeszow University of Technology, 6 Powstancow Warszawy Ave., 35-959 Rzeszow, Poland

Monika Arasimowicz
Department of Agricultural and Environmental Chemistry, University of Agriculture, 21 A. Mickiewicza Ave., 31-120 Kraków, Poland

Barbara Wiśniowska-Kielian
Department of Agricultural and Environmental Chemistry, University of Agriculture, 21 A. Mickiewicza Ave., 31-120 Kraków, Poland

Marcin Niemiec
Department of Agricultural and Environmental Chemistry, University of Agriculture, 21 A. Mickiewicza Ave., 31-120 Kraków, Poland

Elżbieta Skorbiłowicz
Department of Technology in Engineering and Environmental Protection, Faculty of Civil Engineering and Environmental Engineering, Bialystok University of Technology, 45A Wiejska Str., 15-351 Bialystok, Poland

Teresa Noga
Department of Biological Foundations of Agriculture and Environmental Education, Faculty of Biology and Agriculture, University of Rzeszów, Ćwiklińskiej 2, 35-601 Rzeszów, Poland

Jadwiga Stanek-Tarkowska
Department of Soil Studies, Environmental Chemistry and Hydrology, Faculty of Biology and Agriculture, University of Rzeszów, Ćwiklińskiej 2, 35-601 Rzeszów, Poland

Anita Pajączek
Department of Soil Studies, Environmental Chemistry and Hydrology, Faculty of Biology and Agriculture, University of Rzeszów, Ćwiklińskiej 2, 35-601 Rzeszów, Poland

Natalia Kochman
International Doctoral Studies in Natural Sciences, Faculty of Biology and Agriculture, University of Rzeszów, Ćwiklińskiej 2, 35-601 Rzeszów, Poland

Łukasz Peszek
International Doctoral Studies in Natural Sciences, Faculty of Biology and Agriculture, University of Rzeszów, Ćwiklińskiej 2, 35-601 Rzeszów, Poland

Magdalena Vaverková
Department of Applied and Landscape Ecology, Faculty of Agronomy, Mendel University in Brno, Zemědělská 1, 613 00 Brno, Czech Republic

Dana Adamcová
Department of Applied and Landscape Ecology, Faculty of Agronomy, Mendel University in Brno, Zemědělská 1, 613 00 Brno, Czech Republic

Agnieszka Parzych
Environmental Chemistry Research Unit, Institute of Biology and Environmental Protection, Pomeranian University in Slupsk, Arciszewskiego St. 22b, 76-200 Słupsk, Poland

Jerzy Jonczak
Department of Geoecology and Geoinformation, Pomeranian University in Słupsk, Partyzantów St. 27, 76- 200 Słupsk, Poland

Ewa Kucharczak-Moryl
Department of Biochemistry, Pharmacology and Toxicology, Wroclaw University of Environmental and Life Sciences, 31 Norwida Str., 50-375 Wroclaw, Poland

Andrzej Moryl
Institute of Environmental Engineering, Wrocław University of Environmental and Life Sciences, 24 Grunwaldzki Sq., 50-363 Wrocław, Poland

Jolanta Dąbrowska
Institute of Environmental Engineering, Wrocław University of Environmental and Life Sciences, 24 Grunwaldzki Sq., 50-363 Wrocław, Poland

Romuald Żmuda
Institute of Environmental Protection and Development, Wrocław University of Environmental and Life Sciences, 24 Grunwaldzki Sq., 50-363 Wrocwaw, Poland

Elżbieta Skorbiłowicz
Bialystok Technical University, Wiejska 45A, 15-351 Bialystok, Poland

Paweł Kaszycki
Department of Biochemistry, Institute of Plant Biology and Biotechnology, Faculty of Horticulture, University of Agriculture in Kraków, 29. Listopada Ave., 31-425 Kraków, Poland

Paulina Supel
Department of Biochemistry, Institute of Plant Biology and Biotechnology, Faculty of Horticulture, University of Agriculture in Kraków, 29. Listopada Ave., 31-425 Kraków, Poland

Przemysław Petryszak
Department of Biochemistry, Institute of Plant Biology and Biotechnology, Faculty of Horticulture, University of Agriculture in Kraków, 29. Listopada Ave., 31-425 Kraków, Poland